PHYSICS OF
THE HOT PLASMA
IN THE
MAGNETOSPHERE

NOBEL SYMPOSIUM COMMITTEE

STIG RAMEL, *Chairman* • Executive Director, Nobel Foundation

LAMEK HULTHÉN • Chairman, Nobel Committee for Physics

ARNE FREDGA • Chairman, Nobel Committee for Chemistry

BENGT GUSTAFSSON • Secretary, Nobel Committee for Medicine

LARS GYLLENSTEN • Member, Swedish Academy (Literature)

TIM GREVE • Director, Norwegian Nobel Institute (Peace)

ERIK LUNDBERG • Chairman, Prize Committee for Economic Sciences

NILS-ERIC SVENSSON • Executive Director, Bank of Sweden Tercentenary Foundation

PHYSICS OF THE HOT PLASMA IN THE MAGNETOSPHERE

Edited by
Bengt Hultqvist

Kiruna Geophysical Institute
Kiruna, Sweden

and

Lennart Stenflo

Department of Plasma Physics
Umeå University
Umeå, Sweden

PLENUM PRESS □ NEW YORK AND LONDON

Library of Congress Cataloging in Publication Data

Nobel Symposium, 30th, Kiruna Geophysical Institute, 1975.
 Physics of the hot plasma in the magnetosphere.

 Includes bibliographies and index.
 1. Magnetosphere—Congresses. 2. High temperature plasmas—Con-
gresses. I. Hultqvist, Bengt. II. Stenflo, Lennart. III. Title.
QC809.M35N6 1975   538'.766   75-30725
ISBN-13:978-1-4613-4439-1   e-ISBN-13:978-1-4613-4437-7
DOI: 10.1007/978-1-4613-4437-7

Proceedings of the thirtieth Nobel Symposium held April 2-4, 1975
at Kiruna Geophysical Institute, Kiruna, Sweden

© 1975 Plenum Press, New York
Softcover reprint of the hardcover 1st edition 1975

A Division of Plenum Publishing Corporation
227 West 17th Street, New York, N.Y. 10011

United Kingdom edition published by Plenum Press, London
A Division of Plenum Publishing Company, Ltd.
Davis House (4th Floor), 8 Scrubs Lane, Harlesden, London, NW10 6SE, England

EARLIER NOBEL SYMPOSIA

1 ● Muscular Afferents and Motor Control—*Edited by Ragnar Granit*

2 ● Prostaglandins—*Edited by Sune Bergström and Bengt Samuelsson*

3 ● Gamma globulins—*Edited by Johan Killander*

4 ● Current Problems of Lower Vertebrate Phylogeny—*Edited by Tor Ørvig*

5 ● Fast Reactions and Primary Processes in Chemical Kinetics—*Edited by Stig Claesson*

6 ● Problems of International Literary Understanding—*Edited by Karl Ragnar Gierow*

7 ● International Protection of Human Rights—*Edited by Asbjörn Eide and August Schou*

8 ● Elementary Particle Theory—*Edited by Nils Svartholm*

9 ● Mass Motions in Solar Flares and Related Phenomena—*Edited by Yngve Öhman*

10 ● Disorders of the Skull Base Region—*Edited by Carl-Axel Hamberger and Jan Wersäll*

11 ● Symmetry and Function of Biological Systems at the Macromolecular Level—*Edited by Arne Engström and Bror Strandberg*

12 ● Radiocarbon Variations and Absolute Chronology—*Edited by Ingrid U Olsson*

13 ● Pathogenesis of Diabetes Mellitus—*Edited by Erol Cerasi and Rolf Luft*

14 ● The Place of Value in a World of Facts—*Edited by Arne Tiselius and Sam Nilsson*

15 ● Control of Human Fertility—*Edited by Egon Diczfalusy and Ulf Borell*

16 ● Frontiers in Gastrointestinal Hormone Research—*Edited by Sven Andersson*

17 ● Small States in International Relations—*Edited by August Schou and Arne Olav Brundtland*

18 ● Cancelled

19 ● Cancelled

20 ● The Changing Chemistry of the Oceans—*Edited by David Dyrssen and Daniel Jagner*

21 ● From Plasma to Planet—*Edited by Aina Elvius*

22 ● ESR Applications to Polymer Research—*Edited by Per-Olof Kinell and Bengt Rånby*

23 ● Chromosome Identification-Technique and Applications in Biology and Medicine—*Edited by Torbjörn Caspersson and Lore Zech*

24 ● Collective Properties of Physical Systems—*Edited by Bengt Lundqvist and Stig Lundqvist*

25 ● Chemistry in Botanical Classification—*Edited by Gerd Bendz and Johan Santesson*

26 ● Coordination in the Field of Science and Technology—*Edited by August Schou and Finn Sollie*

27 ● Super-Heavy Elements—*Edited by Sven Gösta Nilsson and Nils Robert Nilsson*

28 ● Somatomedins and Some Other Growth Factors—*To be published*

29 ● Man, Environment, and Resources—*Edited by Torgny Segerstedt and Sam Nilsson*

Symposia 1-17 and 20-22 were published by Almqvist & Wiksell, Stockholm and John Wiley & Sons, New York; Symposia 23-25 by Nobel Foundation, Stockholm and Academic Press, New York; Symposium 26 by the Norwegian Nobel Institute, Universitetsforlaget, Oslo; Symposium 27 by Nobel Foundation, Stockholm and Almqvist & Wiksell International, Stockholm; Symposium 28 to be published by Academic Press, New York, and Symposium 29 by Nobel Foundation, Stockholm and Trycksaksservice AB, Stockholm.

PREFACE

Nobel Symposium No. 30 on the Physics of the Hot Plasma in the Magnetosphere was held at Kiruna Geophysical Institute, Kiruna, Sweden from April 2-4, 1975. Some 40 leading experts from America, USSR, and Western Europe attended the Symposium.

The purpose of the meeting was to review and discuss the physics of the hot plasma in the magnetosphere with special emphasis on unsolved problems on which attention needs to be focused during the International Magnetospheric Study 1976-1978. The field is very extensive and complete coverage of all aspects was of course not possible. The radiation belts proper were, for instance, not covered. There were no formal contributed papers, but much time was devoted to discussion. These proceedings contain all review papers except the one by R.Z. Sagdeev. They are ordered by subject, starting, after the introductory lecture, with the problem of how the plasma enters the magnetosphere and ending with the question of the interaction with the ionosphere.

The Organizing Committee for the symposium was composed of the following Swedish scientists: E.-Å. Brunberg, C.G. Fälthammar, I. Hulthén, B. Hultqvist (chairman), L. Stenflo, and H. Wilhelmsson.

The Symposium was financed by the Nobel Foundation through grants from the Tercentenary Foundation of the Bank of Sweden, by the Swedish Board for Space Activities, and the Royal Swedish Academy of Sciences, which is gratefully acknowledged. Appreciated contributions "in natura" were also received from the town of Kiruna and the LKAB Company.

A special effort for rapid publication has been made by contributors and publisher. This effort is very much appreciated.

We owe many thanks to a large number of staff members of Kiruna Geophysical Institute, who made it possible to carry the meeting through in a successful way. In particular Christina Jurén, Eivor Söderquist, and Georg Gastafsson should be mentioned as they carried large fractions of the workload.

Bengt Hultqvist

Lennart Stenflo

CONTENTS

Introductory Lecture

Electric Current Structure of the Magnetosphere   1
  H. Alfvén

Entry of the Hot Plasma into the Magnetosphere

Entry of Solar Wind Plasma into the Magnetosphere   23
  G. Haerendel and G. Paschmann

Composition of the Hot Plasmas in the Magnetosphere   45
  R.G. Johnson, R.D. Sharp, and E.G. Shelley

Distribution within the Magnetosphere
and Large Scale Dynamical Processes

Magnetospheric Plasma Regions and Boundaries   69
  W.J. Heikkila

Auroral Electron Beams near the Magnetic Equator   91
  C.E. McIlwain

A Study of Auroral Displays Photographed from
    the DMSP-2 and ISIS-2 Satellites 113
  S.-I. Akasofu

Recent Observations Relating to the Dynamics and
    Origin of the Magnetotail Plasma Sheet 137
  E.W. Hones, Jr.

Hot Plasma Dynamics within Geostationary Altitudes 159
  D.J. Williams

Acceleration/Heating and Instabilities/Turbulence

Acceleration Processes in the Plasma Sheet 187
    J.W. Dungey

VLF Electrostatic Waves in the Magnetosphere 201
    M. Ashour-Abdalla and C.F. Kennel

Double Layers   229
    L.P. Block

Plasma Turbulence in the Magnetosphere with Special
        Regard to Plasma Heating 251
    A.A. Galeev

Characteristics of Instabilities in the Magnetosphere
        Deduced from Wave Observations 271
    F.L. Scarf

Some Experimentally Determined Characteristics of the
        Turbulence in the Magnetosphere   291
    B. Hultqvist

Interaction with the Ionosphere

Evidence for the Low Altitude Acceleration of
        Auroral Particles   319
    D.S. Evans

Mechanisms for Driving Birkeland Currents 341
    R. Boström

List of Participants 363

Index   367

ELECTRIC CURRENT STRUCTURE OF THE MAGNETOSPHERE

Hannes Alfvén

Royal Institute of Technology

S-100 44 Stockholm 70, Sweden

INTRODUCTION

The first approach to magnetospheric theory was based on a mathematically elegant formalism which, however, was highly idealized and derived without contact with experiment. It led to the Chapman-Ferraro and later to the Axford-Hines theories. The laboratory confrontation between this formalism and the real plasma took place when an attempt was made to construct thermonuclear reactors based on this formalism. This led to the thermonuclear crisis, which demonstrated the inadequacy of the simplifying assumptions on which the theoretical formalism was based. It became obvious that the formalism described the properties of a fictitious medium, a "pseudo-plasma", which in basic respects is vastly different from the real plasma.

The most important task in cosmical electrodynamics of today is a "second approach" aiming at a replacement of pseudo-plasma theories by an empirically based description.

Because in the pseudo-plasma theories the electric current is traditionally eliminated so that it does not appear explicitly, the formalism is not suited for describing phenomena associated with currents. Such phenomena are anomalous resistivity, formation of electrostatic double layers, and current disruption. Most serious is that the transfer of energy cannot be described in an adequate way. (The "field line reconnection" concept is misleading.)

1

Magnetospheric activity is basically due to the immersion of the magnetosphere in the solar wind electric field. Energy from the solar wind is transferred to the magnetosphere by two current systems:

a. The magnetopause current which may transfer up to 1 TW.

b. The magnetotail current which may transfer a few GW.

The electric field produces a sunward convection in the magnetosphere. This transfers energy to the auroral zones through:

c. The auroral current systems, consisting of Birkeland currents and auroral electrojets. Total power: several GW.

d. There is further a ring current, partly connected with the auroral system.

The energy release by a, b, and c is partly in the form of accelerated particles (~10 keV). Examples of acceleration mechanisms are the electric double layers, either produced by Birkeland currents, or associated with the disruption ("flare") of the magnetotail current at magnetic substorms. A model of the magnetospheric currents is given.

A similar approach is used in analyzing the cosmogonic problem of angular momentum transfer, the current system in interplanetary space and in the Jovian magnetosphere, and finally to a model of double radio sources.

THE BIRKELAND APPROACH TO THE MAGNETOSPHERIC PHENOMENA

The exploration of what is now called the magnetosphere was started by Birkeland (1908, 1913). He approached the problem along two parallel lines:

1. He observed the aurora and the magnetic perturbations, and tried to construct the current system during magnetic disturbances. He realized that the current system could not be confined to the upper atmosphere but must include what Dessler has called "Birkeland currents" (Dessler, 1967).

As we know today Birkeland's approach was essentially correct, although of course the conclusions he could draw were limited.

2. He made (what today is called) plasma experiments in order to study the basic physical phenomena which produced the aurora. His "terrella" experiment demonstrated that if a magnetized sphere is immersed in a plasma, the plasma has a tendency to penetrate to "auroral regions" encircling the poles and he concluded that this was the basic phenomenon underlying the auroral zone phenomena.

As we know today, his basic conclusions were correct although plasma physics at his time was not enough developed to allow any more detailed mechanisms to be worked out.

Birkeland never tried to work out a mathematical theory of the phenomena he observed in nature and in his laboratory. He had a very good reason for not doing so: it was impossible to do it at that time. Before it could be done a development of plasma physics was necessary which did not take place until the 1960's. Also crucial observational data were not available until space research supplied them at about the same time. Although much work of permanent value could be done in the meantime the theoretical and observational prerequisites for an understanding of the basic phenomena in the magnetosphere have not been present until today. It is the task of the 1970's to build up the theory of the magnetosphere--essentially from scratch.

THE PSEUDO-PLASMA APPROACH

In the meantime another development had taken place. Mathematically minded scientists, who did not understand Birkeland's approach, tried to construct theories of the magnetic storms and of the structure of the magnetosphere. As we know now, this was a brave attempt to do the impossible. In fact, it was comparable to the effort to understand the energy source of stars before the rise of nuclear physics.

It was thought that a straight-forward mathematical development of the kinetic theory of gases could lead to a complete theory of an ionized gas, --with modern terminology--a plasma. It was also thought that the dynamics of a magnetized plasma could be derived by a straightforward combination of ordinary hydrodynamics

with Maxwell's equations. By such theoretical work several important properties of a plasma **were discovered,**
which later were confirmed by experiments. Most important was that this theoretical work triggered off the
thermonuclear research, which has given us a much better understanding of the properties of plasmas. But to
the extent it was not guided by experiments the theoretical work went wrong. In fact the sophisticated mathematical theories did not describe a real plasma but a
fictitious medium which we shall call the "pseudo-plasma".

The application of such concepts to the magnetosphere led to the development of theories which were
"generally accepted" as long as they could not be checked by space observations. In one of the theories it
was claimed that magnetic storms were caused by the arrival of a non-magnetized plasma--we know now that the
magnetization of all space plasma is essential and that
the electric field in interplanetary space is a decisive factor for geomagnetic activity. In another pseudo-
plasma theory it was claimed that the parallel electric field in the magnetosphere was necessarily zero.
We know now that this is not the case, and that the
formation of electric double layers are likely to be
of decisive importance.

It is quite natural that speculative theories of
this kind are proposed--speculation is an essential
part of all scientific activity. However, some of these theories soon became "generally accepted" to an extent that they were considered sacrosanct and constitu-
ted the credo of powerful schools. The regrettable result of this was that criticism of them was suppressed,
and that the pseudo-plasma concept still is presented
in text books and review articles as if it were correct. In this way the theories constitute an obstacle
to further progress.

Hence, what is most urgently needed in order to
clarify the structure of the magnetosphere is to draw
a clear line between the work which is based on the
pseudo-plasma concept and the work based on the properties of a real plasma--as we know it from experiments
(Alfvén, 1968, 1971; Alfvén and Arrhenius, 1974, 1975).

TABLE 1

<u>COSMIC ELECTRODYNAMICS</u>

First Approach Pseudo-plasma	Second Approach Real plasma
Homogeneous models	Space plasmas often have a complicated inhomogeneous structure
Conductivity $\sigma = \infty$	σ depends on current and often suddenly vanishes
Electric field $E_{\parallel}$ along magnetic field = 0	$E_{\parallel}$ often $\neq 0$
Magnetic field lines are "frozen-in" and "move" with the plasma	Frozen-in picture is often completely misleading
Electrostatic double layers are neglected	Electrostatic double layers are of decisive importance in low-density plasma
Instabilities are neglected	Many plasma configurations are unrealistic because they are unstable
Electromagnetic conditions are illustrated by magnetic field line pictures	It is equally important to draw the current lines and discuss the electric circuit
Energy release calculated from "magnetic field line reconnection"	Energy release $P = Ei$. Field line reconnection is a misleading concept
Filamentary structures and current sheets are neglected or treated inadequately	Currents produce filaments or flow in thin sheets
Maxwellian velocity distribution	Non-Maxwellian effects are often decisive
Theories are mathematically elegant and very "well-developed"	Theories are not very well developed and are partly phenomenological

   As shown by Table 1 the pseudo-plasma has a number
of properties which the real plasma has not. A number
of basic phenomena which a real plasma has, is not accounted for by the pseudo-plasma theories. Hence, any
resemblance between observed phenomena and phenomena
described by pseudo-plasma theories is coincidental.

   To criticize is often simpler than to construct,
but in the present situation it is also more important
as a first phase. Not until we have got rid of the bal-
last of pseudo-plasma theories is it possible to concentrate the efforts on constructive work.

LABORATORY EXPERIMENTS WITH REFERENCE TO MAGNETOSPHERIC
THEORY

   The first fix point in an attempt to clarify the
structure of the magnetosphere must necessarily be the
properties of a laboratory plasma. Modern thermonuclear
research has developed a technique of diagnostics which
is much superior to what we can hope for in the investigation of space plasmas for a long time. The laboratory
research has revealed that a plasma has a number of properties which no doubt must be of importance also in cosmical plasmas. Examples of such properties are the formation of electric double layers (see Block's lecture)
a drastic increase in resistivity when the electron velocity approaches the ion sound velocity, the existence
of a critical velocity between a magnetized plasma and
a non-ionized gas, and of course a large number of instabilities, which often are different from the pseudo-
-plasma instabilities.

   From the side of the pseudo-plasma scientists it
has been pointed out that laboratory conditions differ
so much from cosmic conditions that a transfer of information from the laboratory to cosmos cannot be made.
This is partially true. Although already Birkeland's
experiment revealed some properties of a laboratory
plasma which are applicable to cosmic conditions a
more detailed transfer of knowledge from laboratory
to cosmos is a very difficult procedure. It requires a
careful study of scaling laws, and a study of which
phenomena can be simulated and which cannot. Further,
a number of experiments must be done in the laboratory
which are especially designed to clarify cosmic phenomena. It is usually not possible to make "scale model
experiments" which give a true picture of cosmic phenomena scaled down to laboratory size.

Much work has now been done about the relation between laboratory and cosmic phenomena, and although much remains to be done we can already apply to cosmos much of the knowledge from laboratory work. A recent survey is given by Fälthammar (1974).

SPACE EXPERIMENTS

The next important task is to make a number of experiments in space especially designed to bridge the gap. In a number of cases one wants to study the same basic phenomenon in the laboratory and in space. Only in this way is it possible to check the scaling laws experimentally.

THE IMPORTANCE OF ELECTRIC CURRENTS

One of the unfortunate features of pseudo-plasma theories is that the electric current is often eliminated in the formalism. The hydromagnetic equations contain a large number of variables and, seen from a purely mathematical point of view, it is essential to eliminate as many of these as possible in order to make the mathematical problem tractable. One of the variables which for this reason is eliminated is the electric current, and often the theories do not contain the currents explicitly. However, a current has a number of properties associated with the particle aspect and the circuit aspect and by eliminating the current we easily lose sight of these. Examples are the increase in resistivity when the electron velocity approaches the ion sound velocity, the formation of double layers (Block, 1972; Carlqvist, 1969, 1972, 1973), the kink instability, and current disruption phenomena. Equally important is that the circuit of the current, including the electromotive force and the transfer of energy, are lost. This has disastrous consequences for this approach.

The main purpose of this paper is to demonstrate that a better understanding of magnetospheric activity is possible if we derive the current system. As currents are seldom measured directly, the current picture is essentially found from magnetic measurements with the use of Maxwell's first equation

$$\text{curl } H = 4\pi i$$

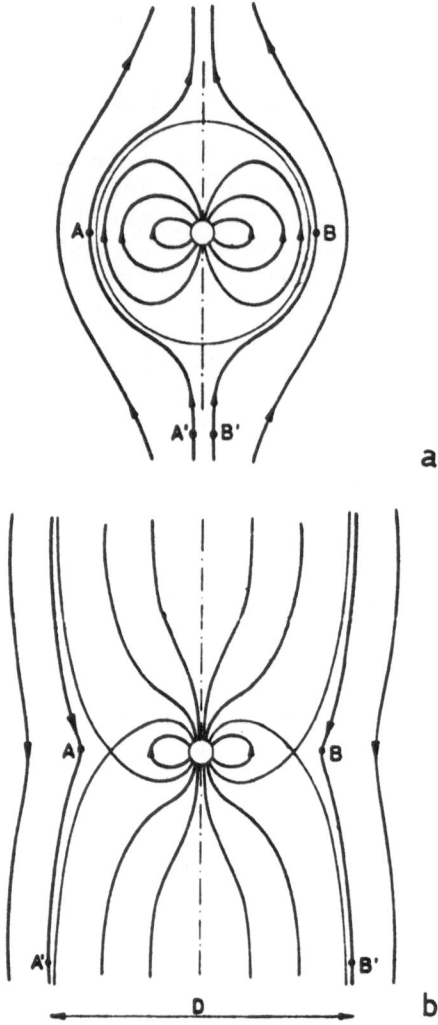

Figure 1. The combined magnetic field of a dipole field
and a parallel (1 a) and antiparallel (1 b) homogeneous
field. In the first case, corresponding to a northbound
interplanetary field, the electrostatic coupling between
the solar wind field and the magnetosphere is small; in
the second case it is strong.

The electric current system which we are presenting
here, is essentially a synthesis of current systems derived by Boström (1964), Rostoker and Boström (1974),
Zmuda and Armstrong (1974), Alfvén-Fälthammar (1971)
and Heikkila (1974).

THE MAGNETOSPHERIC ACTIVITY AS REGULATED BY THE SOLAR WIND ELECTRIC FIELD

If the solar wind has a magnetic field $\vec{B}$ and a velocity $\vec{v}$ in relation to the earth, its electric field--
seen from the coordinate system of the earth--is

$$\vec{E} = -\frac{1}{c}\,\vec{v}\times\vec{B}$$

Two idealized models of the combined interplanetary
and terrestrial magnetic fields are shown in Fig.1.

If the interplanetary magnetic field is parallel
to the earth's dipole (southward), there is a strong
coupling between the solar wind and the magnetosphere.
If the direction is the reverse (northward), the coupling is small. We shall confine our discussion to the
first alternative. We have a dawn-dusk directed electric
field, and as such a field produces a sunward convection
in the magnetosphere, this flow is dominant.

TRANSFER OF ENERGY FROM THE SOLAR WIND TO THE MAGNETO-SPHERE

The magnetospheric activity is due to a transfer
of solar wind kinetic energy to the magnetosphere. This
transfer is difficult to understand by the pseudo-plasma
treatment which calculates the energy release by means
of "magnetic field line reconnection" which often is
grossly misleading. Forgetting the "frozen in" picture
we shall here discuss a model which gives the basic
mechanism for energy transfer.

Suppose that there is a homogeneous magnetic
field B_0 in the $-z$ direction, and a plasma with density ρ flowing with the velocity v_0 in the y direction in
a volume limited by the plane surfaces

$x = \pm\frac{1}{2}\,x_0$ and $z = \pm\frac{1}{2}\,z_0$ (see Fig.2a). This means that

there is an electric field in the x direction
$E_0 = \frac{v_0}{c}\,B_0$. Along the x-axis we place an insulated

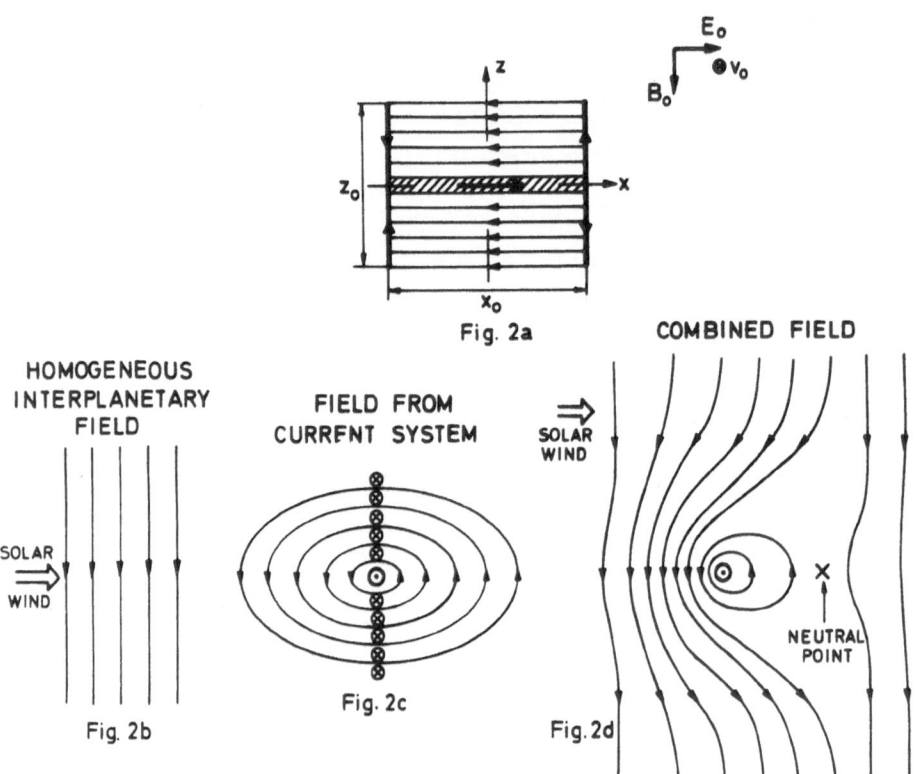

Figure 2.   Model of energy transfer from a moving plasma
to a fix conductor.   (2a) A fix conductor (shaded) in a
plasma which is magnetized in the -x direction and moves
in the y direction draws an electric current which closes
through the plasma.   The magnetic field from this current
system (2c) combines with the homogeneous magnetic field
(2b) to the field configuration (2d).

conductor with resistance R connecting two electrodes
at $x = \frac{1}{2} x_o$ and $x = -\frac{1}{2} x_o$ with the extension in the
y-direction between $y = +\frac{1}{2} y_o$ and $y = -\frac{1}{2} y_o$. This conductor connects two regions in the plasma which in ab-
sence of the disturbance caused by the conductor would
have the voltage difference $V_o = E_o x_o$. This causes a
current which flows through the conductor producing a
voltage drop $V_1 = R I$ which is $< V_o$.

Neglecting the voltage drop at the current passage
between electrodes and plasma and also the plasma resistance parallel to the magnetic field, the surfaces
$x = +\frac{1}{2} x_o$ and $x = -\frac{1}{2} x_o$ have the voltage difference V_1.
This means that at the passage from $y = -\frac{1}{2} y_o$ to
$y = +\frac{1}{2} y_o$ the plasma velocity is decreased from v_o to

$$v_1 = \frac{cE_1}{B_o} = \frac{cV_1}{x_o B_o} \quad .$$

Assuming y_o and the relative difference $\frac{\Delta v}{v_o} = \frac{v_o - v_1}{v_o}$ to
be small, the kinetic energy of a plasma sheath with
thickness y_o is decreased by

$$W_k = \rho x_o y_o z_o v_o \Delta v$$

which means that it delivers the power

$$P_k = x_o z_o \rho v_o^2 \Delta v \quad .$$

The retardation of the plasma causes an inertia
drift v_i in the x-direction

$$v_i = \frac{c}{eB_o} m \frac{dv}{dt}$$

where m is the mass of an ion or electron.
As $\frac{dv}{dt} = \Delta v \frac{v_o}{y_o}$

the total current drawn from the plasma is

$$I = y_o z_o \, nev_i = z_o \frac{c\rho}{B_o} v_o \, \Delta v$$

where n is the number density, and

$$\rho = n \, (m_e + m_i) \quad .$$

The power P delivered to the conductor is

$$P = V_1 I \approx V_o I = x_o z_o \rho v_o^2 \, \Delta v$$

which of course equals the rate of decrease of kinetic energy in the plasma. This will be further clarified by Boström at this symposium.

FORMATION OF A NEUTRAL SURFACE IN THE WAKE

The current in the conductor and in the plasma produces a secondary magnetic field as depicted in Fig.2b which is superimposed on the primary homogeneous field. The resulting field has the shape shown in Fig.2c. It has a neutral line behind the conductor (at approximately $y = 2 \, I/B_o$). The electric field $E_1 = \frac{V_1}{x_o}$ in this region may produce a secondary discharge along this neutral line. Further, the magnetic field from this current I_W may change the neutral line into a neutral surface in the x-y plane. The power delivered at this neutral surface is

$$P_W = I_W \cdot V_I = I_W \, x_o \, E_1 \quad .$$

It should be stressed again that in our treatment the magnetic field lines have the Maxwellian meaning: they are fictitious lines indicating only the direction of the magnetic field. The magnetic field we have derived is <u>static</u>, caused by the <u>time independent</u> combined current systems, viz. a Helmholtz coil in infinity giving a homogeneous magnetic field and the secondary system of time constant currents which we have derived. The picture of "frozen-in" magnetic field lines is not useful.

COMPARISON WITH THE FROZEN-IN FIELD LINE PICTURE

Comparing our results with the customary "frozen-in" picture we find that it is similar in certain respects:

Upstreams from the conductor the primary magnetic field is increased by the field from the induced current system. In the frozen-in picture this corresponds to the piling up of magnetic field lines in front of an obstacle. The field lines in our analysis get a shape similar to the frozen-in picture.

We have no use for the concept of "energy release by magnetic field line reconnection". Indeed, this monstrous concept is a product of the frozen-in picture in absurdum, and is often grossly misleading.

APPLICATION TO THE MAGNETOSPHERE

We may apply our model to the magnetosphere in case the interplanetary magnetic field is southward so that we have a circle with B = 0 in the combined field (cf Fig. 1). The upstream part of this circle allows a current to be drawn from the electric field of the solar wind, and this takes the place of the current in the conductor of our model. The downstream part of the circle permits also a current to flow in analogy to the wake current in our model.

Both currents spread from currents at neutral lines to surface currents (see Fig.3). For obvious geometrical reasons the upstream current spreads to a surface essentially in the x-z plane and can be identified with the magnetopause sheath current; the wake current spreads to a current in the x-y plane, which we identify with the neutral sheath current in the magnetotail.

The magnetopause current system may transfer up to 1 TW of solar wind energy to the magnetosphere, whereas the tail current delivers several GW. See Alfvén-Fälthammar (1971).

MAGNETOSPHERIC CONVECTION AND AURORAL CURRENT SYSTEMS

The interplanetary electric field seems to penetrate also to the region of closed field lines in the magnetosphere, where it causes a sunward plasma drift ("convection"). It is not quite clear whether the "constant of dielectrics" of the magnetosphere is larger or smaller than in the interplanetary space (which means whether the magnetospheric electric field is smaller or larger than the solar wind field

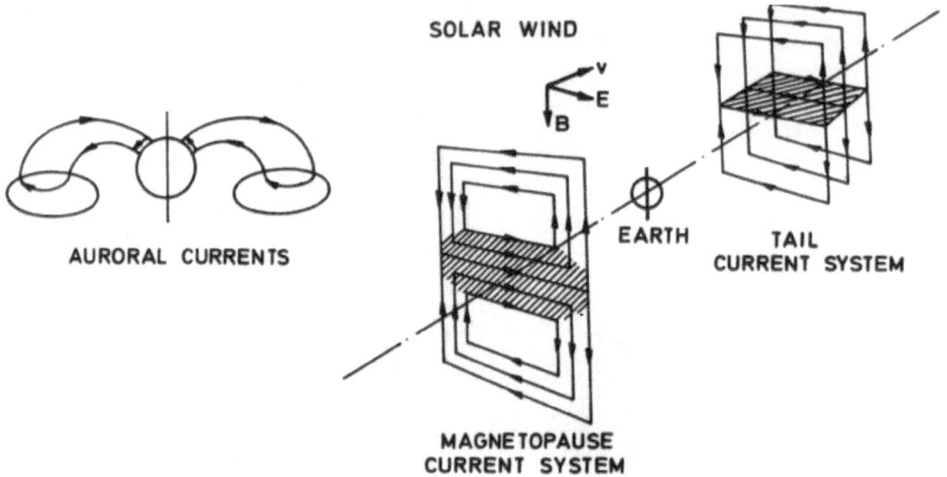

Figure 3. The current system in the magnetosphere consists
of the magnetopause system, the tail system, and the auroral
current system.

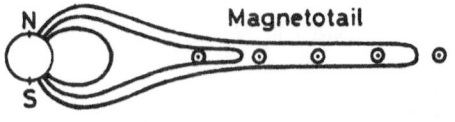

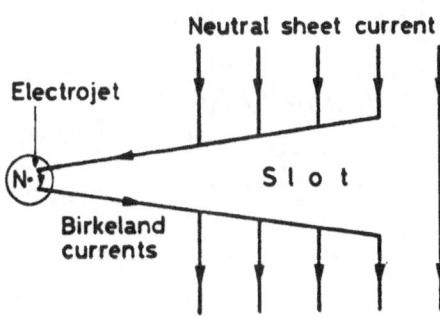

Figure 4. Magnetic field in the tail and sheet current
producing it. The disruption of the sheet current causes
a magnetic substorm with Birkeland currents to the auroral
zones (Boström, 1974).

because of electric charge at the border between the
magnetosphere and the solar wind). There are some
reasons for believing that it is not very far from
unity, which we will assume here. It is much disturbed
close to the earth, partly by the electric field due to
the earth's rotation.

The interaction between the magnetospheric plasma
and the ionosphere of the earth produces a current
system, which begins to be fairly well explored by space
measurements (see Zmuda and Armstrong, 1974). It consists of a sheath current flowing towards the earth in
the morning and away from the earth in the evening.
Further there are similar currents somewhat more equatorward, but their directions are reversed (Fig. 3).

There is no doubt a coupling between the solar
wind-magnetosphere current system, the tail current,
and the auroral current system but this is not yet very
well understood.

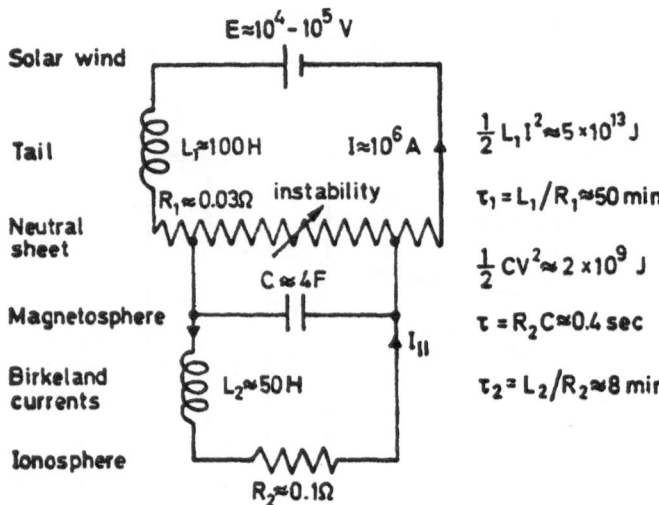

Figure 5. Electric circuit of a magnetic substorm (Boström, 1974).

When the high energy particles in the van Allen belts drift in the inhomogeneous magnetic field, they produce a ring current. This may be partially connected to the auroral current system.

EQUIVALENT CIRCUITS OF MAGNETOSPHERE DYNAMICS

By basing our treatment of the magnetosphere on the current system we get a different approach to many basic problems. Fig. 4 shows the theory of magnetic substorms derived from the model of current disruption, and Fig. 5 shows the equivalent circuit of a substorm. (Boström, 1974).

APPLICATION TO THE COSMOGONIC PROBLEM

It is obvious that the observed auroral current system transfers angular momentum from the earth to the surrounding plasma--indeed, the difference in angular velocity causes an electro-motive force. The general situation is similar to the model of transfer of angular

momentum from a spinning central body to a surrounding
plasma, which is essential to the understanding of how
the secondary bodies obtained their orbital momenta
(Alfvén and Arrhenius 1974, 1975). In all discussions
of cosmogonic models it is essential to reduce the speculative element as far as possible, and this can be
done by considering the cosmogonic transfer of angular
momentum as an extrapolation of observed magnetospheric
phenomena.

CURRENT SYSTEM OF THE HELIOSPHERE

The currents of the auroral current system in the
magnetosphere are so small that their magnetic field
is only a rather small perturbation of the dipole field.
We may ask what happens if the currents produced by the
interaction of the central body and the surrounding
plasma become so large that their magnetic field is comparable to or larger than the dipole field.

Fig. 6a shows the combined current system of the
northern and southern hemispheres of the terrestrial
magnetosphere. A current loop has a tendency to expand.
If we try to visualize what will be the result of ex-
pansions of both the northern and the southern auroral
loop systems we find that the innermost shell which in
Fig. 6a carries the currents towards the central body
will move closer toward the equatorial plane. At the
same time the outermost shell with currents away from
the central body will expand and move closer towards
the pole. In an extreme case we will have a radial
sheath of inwards currents in the equatorial plane and
outward currents along the axis both from the north pole
and the south pole. (Fig. 6b) The axial currents may
have a tendency to pinch. These two parts of the system
must close through currents to the equatorial plane
from the axial regions. Hence the resultant current system is likely to look like in Fig. 6b.

The current system in interplanetary space and the
heliosphere seems to be of the general type of Fig. 6b.
The magnetic field lines in interplanetary space are
known to be spirals with the magnetic field directed
away from the sun to the south and towards the sun to
the north of a "neutral sheet"' (Rosenberg and Coleman
1969) which in reality means a current sheet. The surface density of the current can be calculated from
Maxwell's equations and consists of a radial part i_r and
an azimuthal part i_ϕ:

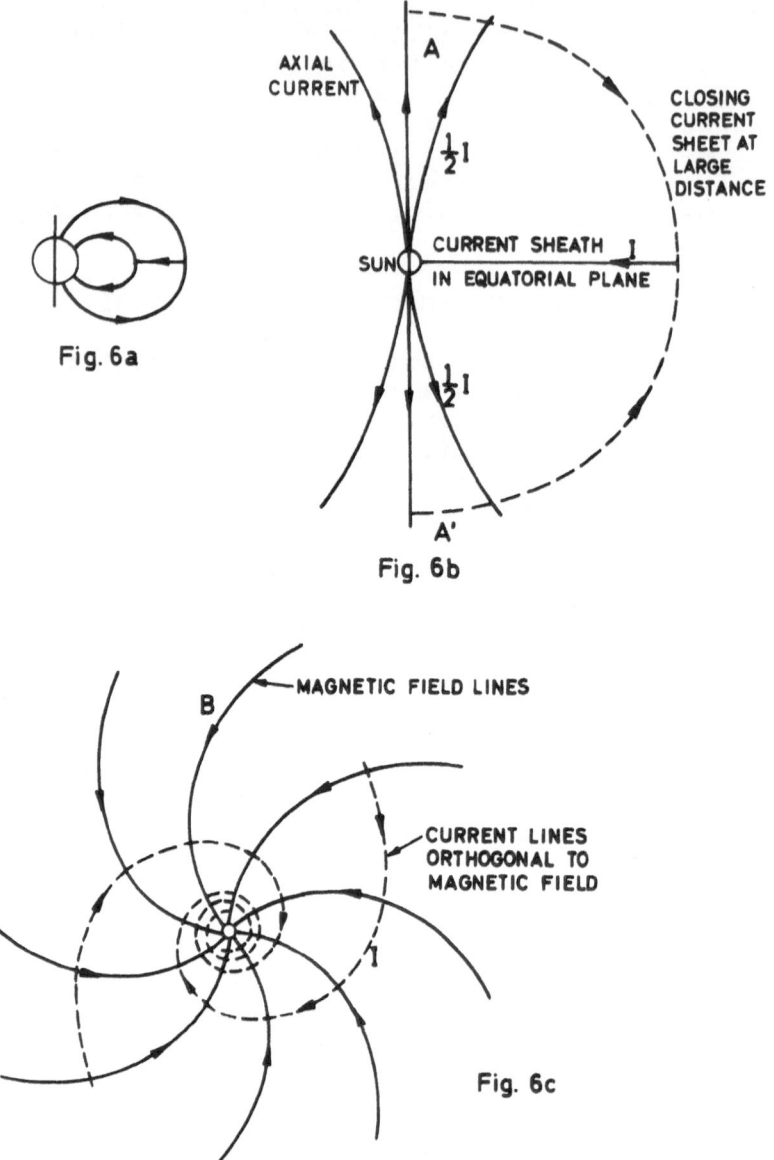

Figure 6. Current system in the heliosphere (interplanetary space).
Poleward displacement of the outermost current sheet in the auroral
current system (6a) may transform it into the system (6b), which
characterizes the current system in interplanetary space. In the
equatorial plane there is a radial sheet current which close to the
sun becomes increasingly spiralized (6c). This current must close
through axial currents and a current sheet at large distance
(not yet discovered).

$$i_r = \frac{H_\phi}{2\pi} \; : \quad i_\phi = \frac{H_r}{2\pi}$$

where H_ϕ and H_r are the azimuthal and radial components of the field. The values are supposed to be the same to the north and to the south of the current sheath.

The current flows in a tight spiral close to the sun but with increasing solar distance it widens, at the earth's distance it makes about 45° with the vector radius and becomes almost radial at larger distances.

The total radial current is

$$I_r = 2\pi \, R \, i_r = R H_\phi$$

With $H_\phi = 2\cdot10^{-5}$ gauss at $R = 1.5\cdot10^{13}$ cm we find $I_r = 3\cdot10^8$ emu or

$$I_r = 3\cdot10^9 \text{ A} = 3 \text{ GA}.$$

Implying that if there is a north-south symmetry the axial currents are each

$$I = 1.5 \text{ GA}.$$

It is not known whether the axial currents are pinched or flow over extended cones, nor is it known how far out the current system is closed by shell currents. The electro-motive force of the current system is due to the rotation of the magnetized sun, acting as a unipolar inductor.

THE JOVIAN MAGNETOSPHERE

The Jovian magnetosphere seems to be similar to the heliosphere. The magnetic field near the equatorial plane is a dipole field close to the planet but at larger distances it becomes more radial with an increasing tangential component at very large distances. At $R = 80 \, R_J = 6\cdot10^{11}$ cm the field makes 30° angle with the vector radius and the total field is about $5\,\gamma$ (E. Smith, 1975). From this we can deduce a radial current

$$I = 6 \cdot 10^{11} \cdot 5 \cdot 10^{-5} \sin 30^{\circ} \text{ emu}$$

or

$$I = 1.5 \cdot 10^{8} \text{ A}$$

The space probes never reached such high latitudes that any trace of the axial currents were detected.

The current shell joining the radial sheath with the axial current must have been traversed both at the inward and outward paths, but has probably been mistaken for one of the several "magnetopauses" the spacecraft passed.

APPLICATION TO THE THEORY OF DOUBLE RADIO SOURCES

Finally it should be pointed out that the current system of Fig. 6b may transfer energy from the rotational energy of the central body to any part of the circuit. For example, under certain conditions most of the energy may be released in the region A and A', situated at large distances and symmetrically in relation to the central body. This means that if the central body is a galaxy it may transfer a large part of its rotational energy to two regions situated symmetrically on both sides of the galaxy. An emission of this energy as synchrotron radiation may explain the double radio sources.

REFERENCES

Alfvén, H., The second approach to cosmical electro-
    dynamics, Annales de Géophysique, 24, 1, 1968

Alfvén, H., Relations between cosmic and laboratory
    plasma physics, Cosmic Plasma Physics, ed. Karl
    Schindler, Plenum Press, New York, 1971

Alfvén, H. and G. Arrhenius, Structure and evolutionary
    history of the solar system, III, Astr.Space Sci.,
    21, 117, 1973

Alfvén, H. and G. Arrhenius, NASA Publ., 1975, in press

Alfvén, H. and C.-G. Fälthammar, A new approach to the
    theory of the magnetosphere, Cosmic Electrodynamics,
    2, 78, 1971

Birkeland, C., The Norwegian auroral polaris expedition,
    1902-1903, Broggers Printing Office, Christiania,
    1908, 1913

Block, L., Potential double layers in the ionosphere,
    Cosmic Electrodynamics, 3, 349-376, 1972

Boström, R., A model of the auroral electrojet, J. Geophys.
    Res., 69, 23, 1964

Boström, R., Ionosphere-Magnetosphere Coupling, Magneto-
    spheric Physics, ed. B.M.McCormac, p. 45-59, D.
    Reidel Publ. Co., Dordrecht, Holland, 1974

Carlqvist, P., Current limitation and solar flares, Solar
    Physics, 7, 377, 1969

Carlqvist, P., On the formation of double layers in
    plasmas, Cosmic Electrodynamics, 3, 377-388, 1972

Carlqvist, P., Double layers and two-stream instability
    in solar flares, TRITA-EPP-73-05, Royal Institute of
    Technology, Stockholm, 1973

Dessler, A.J., Field-aligned currents in the magnetosphere,
    J. Geophys.Res., 72, 3, 1967

Fälthammar, C.-G., Laboratory experiments of magnetospheric
    interest, Space Sci. Rev., 15, 803-825, 1974

Heikkila, W., See paper presented at this symposium

Rosenberg, R.L. and P.J. Coleman, Jr., Heliographic
    latitude dependence of the dominant polarity of the
    interplanetary magnetic field, J. Geophys.Res., $\underline{74}$,
    24, 5611-5622, 1969

Rostoker, G. and R. Boström, A mechanism for driving the
    gross Birkeland current configuration in the auroral
    oval, TRITA-EPP-74-25, Royal Institute of Technology,
    Stockholm, 1974

Smith, E., Science, 1975, in print

Zmuda, A.J. and J.C. Armstrong, The diurnal flow pattern
    of field-aligned currents, J. Geophys. Res., $\underline{79}$,
    4611, 1974

ENTRY OF SOLAR WIND PLASMA INTO THE MAGNETOSPHERE

Gerhard Haerendel and Götz Paschmann

Max-Planck-Institut für Physik und Astrophysik
Institut für extraterrestrische Physik
8046 Garching b. München, Germany

1.   INTRODUCTION

It is not our intention to review data and theoretical ideas bearing on the general subject of entry of solar wind plasma into the magnetosphere. We want to restrict ourselves to the entry through the polar cusp or cleft, on which recently available HEOS2 data contribute some new information. It has long been established by a number of measurements with low altitude as well as eccentric orbiting satellites (Heikkila and Winningham, 1971; Frank and Ackerson, 1971; Frank, 1971; Paschmann et al., 1974) that plasma closely resembling that found in the magnetosheath and, therefore, thought to originate there has easy access to the inner magnetosphere. At low altitudes, the cusp region extends between 08 and 16 local time and has a latitudinal width of a few degrees centered in the high 70° depending on the magnetic activity (Burch, 1972). Although investigated at medium (Frank, 1971) and high altitudes (Scarf et al., 1974) its connection to the dayside magnetopause has not been elucidated so far. On the night side, however, it appears to be connected to the plasma mantle (Rosenbauer et al., 1975) a layer of outward flowing plasma, mostly several R_E thick, that covers the inside of the tail magnetopause. The low latitude flanks of this layer were already discovered by Akasofu et al. (1973).

The central questions remain still unanswered: Where and by which process does the solar wind plasma

enter the magnetosphere? Does the magnetic field topology in the distant cusp resemble that of a closed or
an open magnetosphere, as drawn by Dungey (1961). And
if it is open, does the plasma enter simply by flowing
parallel to open magnetic field lines? In an attempt to
attack these questions by inspection of data obtained
in the relevant regions of space one finds oneself con-
fronted with the problem that there does not exist a
self-consistent model of an open magnetosphere, so that
its observational signatures remain uncertain (Vasyliunas, 1974).

In mid-1973 and mid-1974, the HEOS2 spacecraft
frequently traversed the northern distant cusp region
during midday at a range of latitudes tailward and
equatorward of the theoretical neutral point, thus providing observations in one of the remaining observational gaps in the magnetosphere. Data from the MPI plasma
experiment and the Imperial College magnetometer are
being combined in this study, of which a fuller account
is to be given elsewhere (Paschmann et al., 1975). The
main finding is that plasma mantle and mid-altitude
cusp appear to be connected to a plasma layer inside the
dayside magnetopause whose characteristics suggest that
it is the "port of entry" of solar wind on the front
side of the magnetosphere. We name it the "entry layer".
At the time of the Nobel Colloquium and the time of
writing not all the relevant data have been inspected.
We are presenting a few case studies here in order to
demonstrate the typical features of the entry layer and
draw some tentative conclusions from these findings.

2.   HEOS2 MEASUREMENTS IN THE DISTANT POLAR CUSP

(a)   Instrumentation

The MPI plasma detector which has been described
by Rosenbauer et al. (1975) consists of two hemispheri-
cal electrostatic analyzers (for protons and electrons)
covering the energy range from 100 eV to 40 keV. The
polar acceptance angle is 28° FWHM centered at the
equatorial plane of the satellite. During each spin
period (6 seconds), the intensity is sampled at a fixed
energy in 8 sectors for protons and 4 for electrons. A
complete spectrum is taken in 86 seconds and repeated
every 256 seconds. It should be remembered during the
forthcoming discussion that the measured velocity
distribution represents essentially a two-dimensional
cut through the real one. The relatively long time

needed for scanning through a full spectrum causes time
aliasing at times of high variability of the plasma,
e.g. close to the magnetopause. During the period from
which we have selected the data to be discussed below,
the spin axis of HEOS2 was oriented perpendicular to
the sun-earth line in the ecliptic plane. We will confine our discussion to the protons, because frequently
the bulk of the electrons falls outside the energy
range covered by the instrument.

The magnetic field was measured with the Imperial
College three-axis flux-gate magnetometer (Hedgecock,
1974). The measurement of each component was $\pm$ 144γ
with a digital resolution of $\pm$ 0.125γ for components
less than 16γ, and of $\pm$ 1.0γ for larger components. Regular field samplings took place every 32 seconds. By
use of core memory faster sampling rates were achieved
on a limited duty cycle.

(b)   A Plasma Layer inside the Dayside Magnetopause

When HEOS2, on an outbound orbit, approaches the
magnetopause at midday hours and latitudes below the
expected position of the neutral point it is often
observed that more or less coincident with a rather
fast disappearance (within a few 1000 km) of the ring
current protons and electrons, there appears a plasma
component with much the same density and energy spectrum as found later in the magnetosheath. The average
direction of the magnetic field is still consistent
with dayside closed field lines, although its magnitude
does no longer fall off like a vacuum dipole field
confined by the magnetopause. Since plasma and magnetic
pressures are about equal ($\beta \approx 1$), the magnetic field is
inflated and the magnetopause shifted locally outward
with respect to the position of a vacuum magnetopause.
Fluctuations of the magnitude of B of the same amplitude
as in the magnetosheath are present. The field direction
is more steady and exhibits a systematic tilt with
respect to the local meridian as if caused by a tangential drag exerted by the solar wind. The most
distinguishing feature of the low energy plasma component is its irregular flow behavior. The flow speed is
in general substantially lower than in the magnetosheath and sometimes even becomes undetectable. The
direction can change several times while the spacecraft
traverses the layer and is consistent with dominant
flow either parallel or anti-parallel to the magnetic
field.

At the magnetopause which we identify with the
first sharp rotation of the magnetic field into a con-
sistently new direction, there is normally a sharp
transition to the typical magnetosheath flow with a few
exceptions where a stagnating or rather turbulent
external plasma seems to exist or where we have so far
been unable to identify the magnetopause by its magnetic signature.

We will now inspect more closely a few of the
measurements from which we deduce the existence of this
dayside plasma boundary layer. Figure 1 shows the
plasma and field data on an outbound pass on day 169 of
1973 between 05 and 12 UT. From top to bottom we plot
the proton density, N_p^*, its temperature, T_p^*, speed,
V_p^*, and direction angle, ϕ_v^*, of the bulk flow in the
equatorial plane of the satellite which is almost perpendicular to the ecliptic and contains the sun-earth
line. (The star at the various quantities is meant as a
reminder of the limited range of velocity space from
which the parameters were derived.) The definition of
flow direction is meteorological, i.e. it refers to the
direction of arrival of protons averaged over the part
of velocity space that is covered by the instrument. If
V_p^* falls below 60 km/sec (dashed line), the measurements are no longer meaningful, and we refrain from
showing the direction, ϕ_v^*. In calculating the density
we have extrapolated the measurements by assuming inde-
pendence of the polar angle. A more thorough discussion
of these procedures was given by Rosenbauer et al.
(1975). In spite of some uncertainties we regard density, temperature, and flow velocity in most cases as
accurate within a factor of $2:V_p^*$ is, of course,
essentially only a projection into the X-Z plane in
solar ecliptic coordinates.

10-minute averages of the magnetic field azimuth,
ϕ_B, elevation angle, Λ_B, in the satellite system and
its magnitude, B, are shown by solid lines in the lower three panels. The letters S, N, E, M refer to south,
north, evening, morning. The lowest panel contains, in
addition, the energy density of the protons, including
the kinetic energy of the bulk flow, if we refer to the
ordinate on the r.h.s. The scales on the left (in
Gammas) and on the right (from $4 \cdot 10^{-12}$ to $2 \cdot 10^{-7}$ ergs
cm^{-3}) have been chosen such as to allow direct comparison of the magnetic and plasma pressures. If the
measured flow component is parallel (anti-parallel) to
the component of $\underline{B}$ in the equatorial plane of the
spacecraft, dots and solid line are separated by 180°

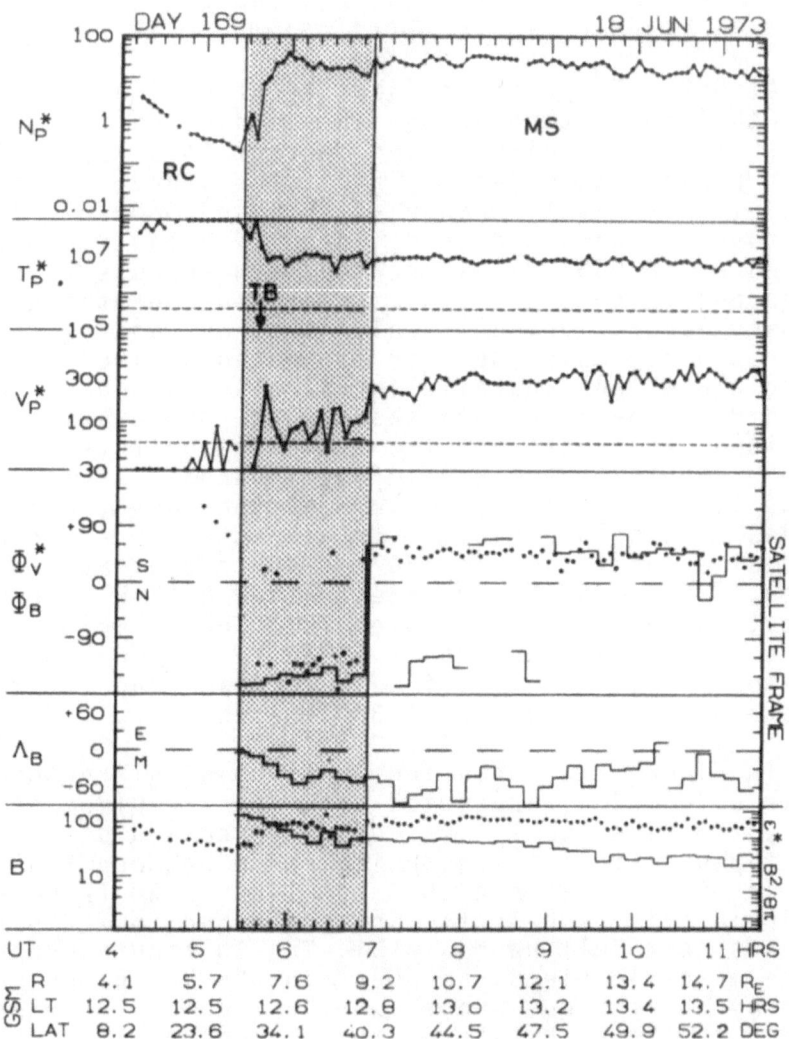

FIGURE 1: Proton parameters (dots) and magnetic field data (solid lines) as function of time on day 169. From top to bottom: density, N_p^*, temperature, T_p^*, speed, V_p^*, azimuthal direction angles ϕ_v^* and ϕ_B and elevation, Λ_B, in satellite frame, magnetic field strength, B, in Gammas and energy density, ϵ^*, in 10^{-12} ergs cm^{-3}.

(coincide). Below the UT abscissa we add the geocentric
distance, R , local time, LT , and latitude, LAT , in
solar-magnetospheric coordinates (GSM), in which the
X-direction points to the sun and the X-Z plane contains the earth's dipole axis.

At 0520 UT the proton density starts increasing
while the temperature drops. This indicates the transition from the hot dilute ring current to a layer of low
energy plasma whose density and temperature, in this
example, show little change after about 0600 UT. V_p^*
is strongly fluctuating and generally below the level
reached after 0657 UT, where the magnetopause is encountered. This is the typical signature of the dayside
boundary layer. Typical is also that ϕ_V^* and ϕ_B are
either closely coincident or separated by 180° , indicating almost anti-parallel or parallel orientation of
$\underline{V}_p$ and $\underline{B}$ in the plane of measurement. On several
occasions, the vector of the proton velocity switches
by 180°. At the inner edge of the boundary layer, Λ_B
starts to turn strongly toward morning hours, and
reaches a much larger tilt than expected in a vacuum
field at a local time of 1247 in GSM coordinates or,
better, 1306 in SM coordinates. After 0657 UT the plasma
parameters become more stable. The flow velocity points
on the average 45° upward in GSM as expected beyond the
magnetopause at high latitudes. The magnitude of the
magnetosheath flow is about 300 km/sec.

In Figure 2 we show the location of six outbound
orbits of HEOS2 as projected into the X-Z plane of the
GSM system. The local times of intersection of the
magnetopause along these orbits range between 1100 and
1330 in GSM and between 1000 and 1430 in SM (Z parallel
to dipole axis, sun-earth line contained in X-Z plane).
The dots mark full hours in UT. The 10-minute averages
of the magnetic field component in the X-Z plane are
shown by short lines whose lengths are logarithmically
related to the magnitude of this component. The relation
to SM latitudes can be easily estimated by noting the
indicated tilt of the dipole axis at the time of magnetopause traversal. For better orientation of the reader
we have sketched by dashed lines part of a magnetopause
contour as expected in a closed magnetosphere and the
expected demarcation line between dayside and tail
field lines in the noon-midnight meridian plane. Only
close to the particular point of intersection through
the magnetopause have we chosen the slope of the bound-
ary in accordance with the orientation of $\underline{B}$ and V_p^* both

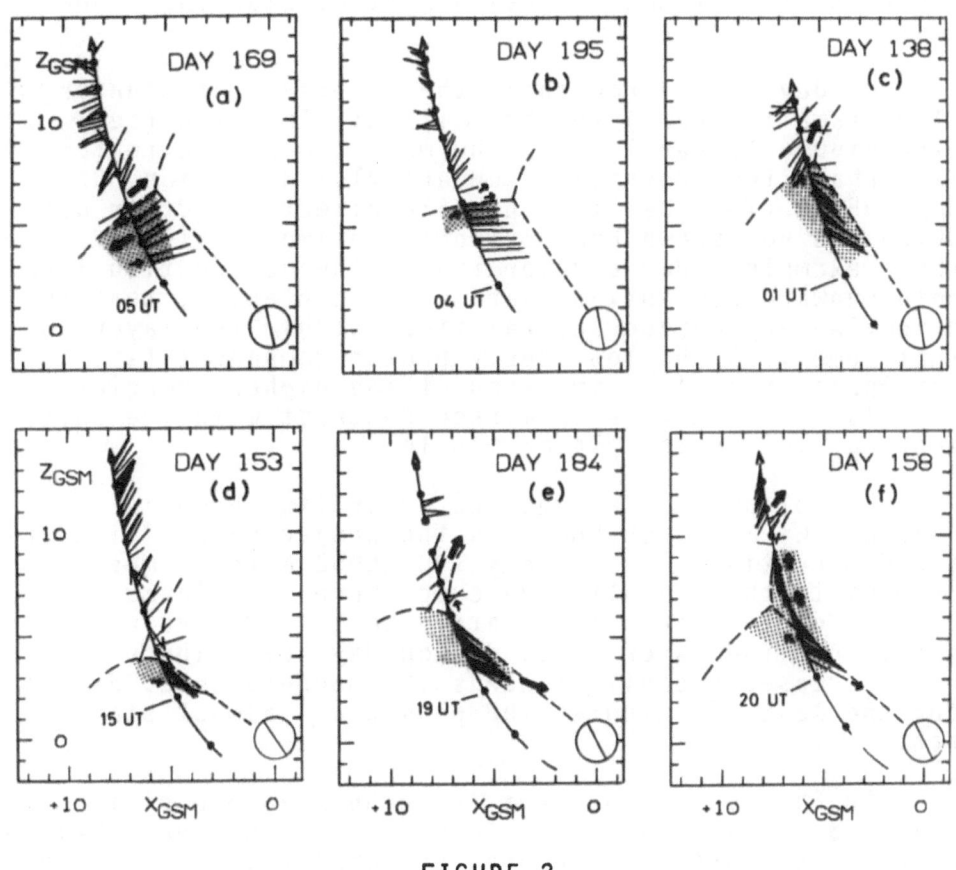

FIGURE 2

Location of 6 outbound passes of HEOS2 projected into
the X-Z plane of the GSM coordinate system. The inclination of the earth's dipole axis at the time of magnetopause traversal is indicated. Dots along the orbit
mark full hours. 10-minute averages of the X-Z component
of B are shown by short lines in logarithmic scaling.
Conjectured contours of magnetopause and demarcation
line are shown by dashed lines. The width of the cusp
plasma layer is indicated by grey shading.

inside and outside the magnetopause. By grey shading we
indicate the width of the boundary layer of low energy
plasma in the vicinity of the orbit. The arrows give
the dominant direction of flow, if there is any, in the
layer and in the magnetosheath just outside the magnetopause.

On day 169 we recognize the existence of counter-
flows (small arrows) to the dominant direction (see
also Figure 1, panel ϕ_v^*). The most remarkable feature
is perhaps the prevailing sunward flow direction. On
day 195 we find just the opposite direction. Other data
which are not shown here support the implication of
these examples that equatorward of the demarcation line
both sunward and anti-sunward flows can exist, that the
direction may change several times within the layer and
that sometimes the low energy proton component ($\leqslant 1$ keV)
can oppose the flow direction of the higher energies.
This lack of order is in marked contrast with the anti-
sunward flow in the plasma mantle.

By following the sequence of orbits shown in
Figure 2 we approach and pass the projected position of
the demarcation line. On day 138 HEOS2 moved almost
tangent to the mid-altitude cusp. Here the plasma
density did not jump immediately to its full value after
entry into the layer as it did on day 169 (Figure 1),
but increased steadily towards the magnetopause. Except
for the last few minutes the plasma was almost stag-
nant.

In Figure 3 we present one other example of the
plasma and field data as a function of time (day 184) in
order to demonstrate the greater complexity of data in
the vicinity of the demarcation line. Here we have one
of the few cases in which the magnetopause as indicated
by the first sharp rotation of the magnetic field at
2118 UT (see Figure 5) does not coincide with a clear
change of the flow characteristics. In moving outward
we see for another hour a somewhat turbulent flow of
generally lower magnitude which is mostly neither
parallel nor anti-parallel to the projection of $\underline{B}$. Only
after 2210 UT it develops smoothly into a steady magnetosheath flow. We interprete the observations between
2118 and 2210 UT as the expression of an essentially
stagnating turbulent plasma related to the existence of
an indentation of the magnetopause somewhat reminiscent
of the cleft as drawn by Heikkila (1972). A similarly
turbulent flow was discovered on day 153 just outside

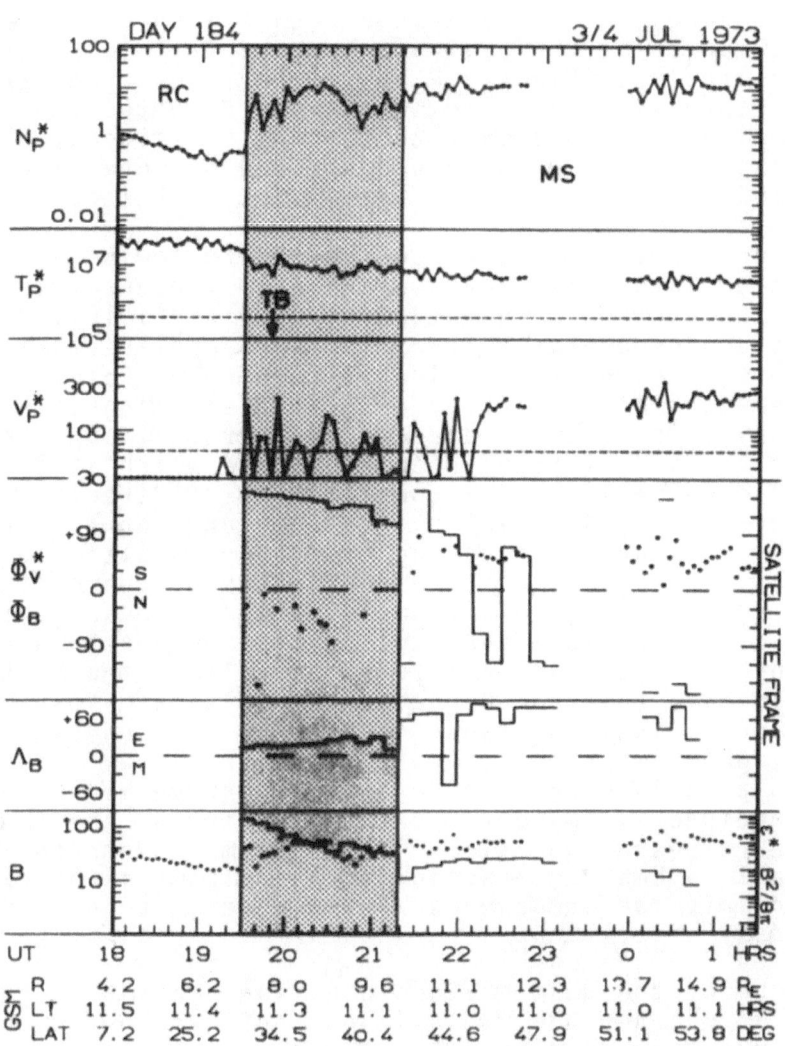

FIGURE 3:   Same as Figure 1 for day 184.

the magnetopause, again a case suggesting an indentation or cleft-like shape of the magnetopause. Below we
will discuss the geometry of the boundary more thorough-
ly.

On the last pass shown in Figure 2 (day 158) HEOS2
moved through the mid-altitude cusp and entered the
plasma mantle at the beginning of the tail lobe. As
soon as the magnetic field has clearly the orientation
of the tail field we find an outward directed flow and
a velocity distribution with the typical signature of
the plasma mantle which was described by Rosenbauer et
al. (1975).

In general our data confirm the following picture:
On field lines which have, apart from the described
tilt, the average orientation of dayside field lines we
find a confused plasma flow. The same applies to the
equatorward edge of the mid-altitude cusp. Poleward of
the demarcation line, i.e. on tail field lines, the
boundary layer flow is directed outward. It apparently
drains the mid-altitude cusp region to which it is
connected. The lack of a consistent flow direction in
the dayside boundary layer, its apparent connection to
the low latitude part of the mid- and low-altitude cusp
regions, the existence of a dominantly poleward (tailward) convection (known from low altitude measurements)
leads us to conclude that the dayside boundary layer
and the cusp at lower altitudes as well as the mantle
are causally related: Magnetosheath plasma makes its
first entry into the magnetosphere in this dayside
layer, which we name accordingly "entry layer"; much of
this plasma is mirroring somewhere between magnetopause
and ionosphere, thereby eventually entering the mantle
as the corresponding field lines become stretched into
the tail; a smaller fraction of the plasma is precipitating into the atmosphere giving rise to a host of
secondary effects.

One of the most obvious questions to ask at this
point is whether the field lines confined in the entry
layer are open or closed. So far our statement has been
that the orientation of the field as well as the
position of the layer suggest that we are dealing with
dayside field lines. A substantial fraction of the time
there is no way to tell whether the field lines are
somewhere connected to the interplanetary field or not.
However, sometimes and mostly in the inner part of the
layer, we encounter pitch-angle distributions of the
electrons in the higher energy range ($\gtrsim 5$ keV) identify-

ing them as trapped particles and suggesting that the field lines are closed. This is quite in agreement with the findings of Burrows and McDiarmid (1972) for somewhat more energetic electrons at the equatorward side of the low altitude polar cusp region. Hence we conclude that, although part of the field lines of the entry layer may be open, a transfer of low energy plasma onto closed field lines does take place inside the entry layer.

When proceeding near noon to lower latitudes we find less evidence of the entry layer. Sometimes there are abrupt transitions to the magnetosheath or only few frames of plasma measurements (4 minutes each) carrying the signature of the entry layer. At the time of writing these data are still under investigation.

(c)   Magnetic Field and Plasma Flow at the Magnetopause

In an attempt to learn something about the entry process of the solar wind plasma we have inspected the magnetic field changes at the magnetopause as well as the detailed velocity distributions. So far no consistent picture could be established. One of the main shortcomings of our data is their low sampling rate. Even the temporal resolution of $\bar{B}$ of 32 sec is insufficient to establish the type of transition of B at the magneto-pause the way it has been done, for instance, by Sonnerup and Cahill (1967). However, a few conclusions can be drawn.

On day 169 (see Figures 1 and 2a) the orientation of the internal and external magnetic field was quite favorable for reconnection. If the field had actually been interconnected in the vicinity of the HEOS2 orbit the way it is normally conjectured in theories of merging, we would not expect a dominant plasma velocity in the sunward direction. Only in the last frame (4 minutes) before the encounter of the magnetopause did the plasma instrument measure an anti-sunward flow; for most of the layer the region of dominant entry was apparently located further tailward. In other cases, however, e.g. days 195 and 138, we see the plasma in the entry layer streaming away from the sun.

It has been frequently said that in the polar cusp magnetosheath plasma has "free access" to the magneto-sphere, without defining the precise meaning of this

term. Although the presence of the various plasma
layers with magnetosheath-type plasma strongly suggests
an "easy access", the plasma seems to encounter some
kind of barrier at the magnetopause leading to abrupt
changes of the phase space density. Figure 4 presents a
good example. It shows contour plots of the phase space
density for four frames around the encounter of the
magnetopause at 1553 UT on day 153. The maximum density
in units of 10^{-25} cm^{-6} sec^3 is written in the right hand
corners. There are three contours per decade. It should
be noted that the contours have been derived by inter-
polation from averages over only 8 sectors. The projection of the magnetic vector into the plane of the
distribution (equatorial plane of satellite, i.e.
roughly the solar-ecliptic X-Z plane) is indicated by
an arrow. Although the position is as close as conceivable to the region for which free access along open
field lines has been claimed, we find a clear signature
of the magnetopause as well as drastic and abrupt
changes of the phase space density. One of the shortcomings of single satellite measurements is, of course,
the difficulty to decide whether one sees a quasi-
stationary boundary or a travelling discontinuity.

Another interesting crossing of the magnetopause
occurred on day 184. Figure 5 contains the magnetic
field data for this transit with the highest time
resolution available (6 sec, storage mode). The external field in the region of turbulent flow after 2118 UT
is extremely variable and exhibits several sharp turns
and on one occasion (2121 UT) reaches a minimum value
of 1γ . The angle between the magnetic vector immediately before and after the first discontinuity at 2118 UT
was about 150^0. If these strongly anti-parallel fields
had the tendency to merge as, for instance, proposed by
Dungey (1961) we should see matter ejected along the
boundary. However, in this very region the velocity
distribution was almost perfectly isotropic.

In this figure we also see that the fluctuations of
the magnitude of $\underline{B}$ inside the entry layer are not much
weaker than in the magnetosheath.

In summary we can say that we are not able to
establish the existence or absence of an appreciable
normal component of B. There are, however, many cases
where we do not find an obvious confirmation of the
plasma flow as expected in a simple minded application
of the theory of merging (for a review see Vasyliunas
(1975)).

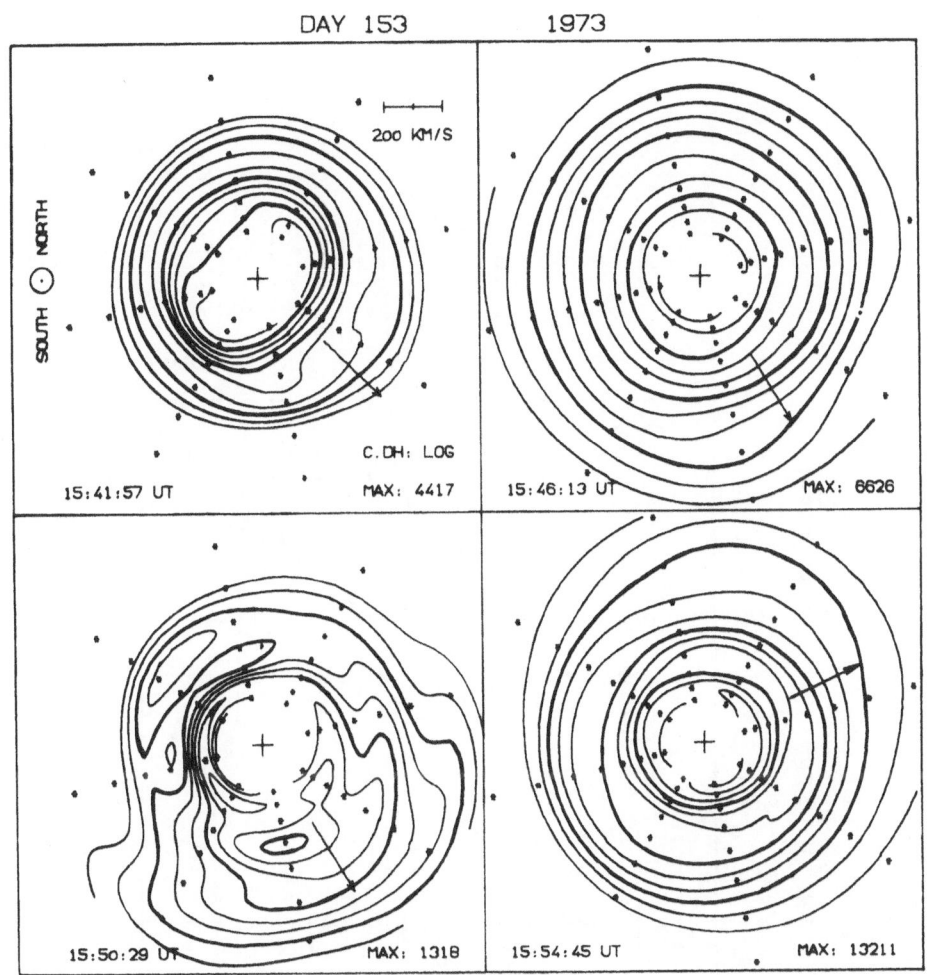

FIGURE 4

Contours of phase space density in units of 10^{-25} cm^{-6} sec^3. Maximum densities are given in right hand corners. There are 3 contours per decade, full decades being marked by thicker contours.

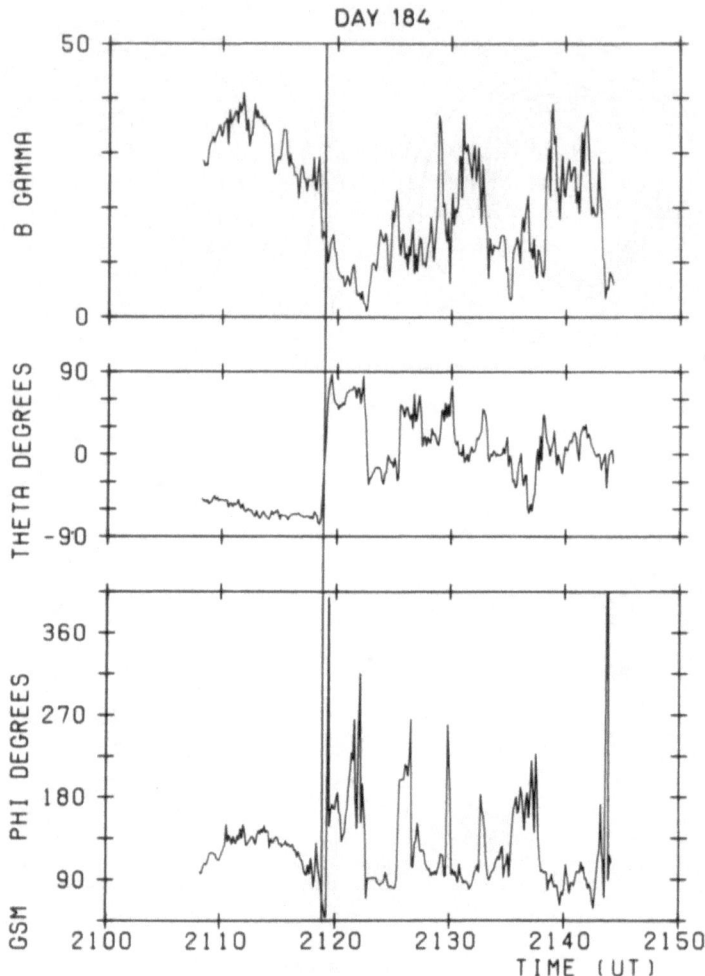

FIGURE 5: Magnetic field data of day 184 in GSM coordinate. Vertical line marks magnetopause.

(d)   Shape of the Magnetopause

Several attempts have been made to determine the
normal component of B at the magnetopause (see the review by Willis (1971) and Sonnerup and Ledley (1974)).
In the majority of cases it was found to be indistin-
guishably small. Therefore, we should be reasonably well
justified if we attempt to determine the normal vector
of the magnetopause by taking the cross-product between
the field vectors just inside and outside the magnetopause. In order to avoid complications by small scale
fluctuations of B we use averages of a few minutes. One
of the most interesting aims of such a study is to look
for the typical signature of a cleft. So far and not
having inspected all our data, we found only one
example (day 153) showing clearly a normal vector pointing towards the night side (in GSM coordinates) as one
should have on the sunward flank of the cleft. On days
184 and 158 we see a steeply rising magnetopause almost
exactly facing the sun just tailward of the demarcation
line. The observed orientations are also consistent
with the flow of the magnetosheath plasma found at the
magnetopause. Since the data of days 153, 184 and 158
span about 2.5 hours of LT in GSM and 4.5 hours of LT
in SM coordinates we have a good indication of an
elongated indentation reminiscent of a cleft. On the
other hand, we do not see the cleft penetrating very
deeply into the magnetosphere as suggested by the model
of Heikkila (1972).

Another interesting feature is that the normal
vector on day 138 is strongly tilted (by 35°) towards
morning in spite of a LT of 1330 in GSM and 1345 in SM
coordinates. A similar behavior, but less well expressed
was found on day 169. We conclude that even at latitudes
substantially below the demarcation line the magnetopause is not quite convex towards the solar wind, but
tends to have a groove in north-south direction.

From the limited data set underlying this discussion we are not able to establish the average shape of
the high latitude dayside magnetosphere. More data are
available and under investigation. We should note, however, that we have also encountered a few cases where
we still have difficulties to locate a magnetopause in
a way consistent with both magnetic and plasma measurements.

3. FORMATION OF THE ENTRY LAYER

From the data that we inspected so far we are unable to deduce a simple dependence of the existence or
thickness of the entry layer on the solar wind parameters. We see such a layer in almost all passes on the
dayside above 45° SM latitude. At lower latitudes it can
be recognized as well, but tends to be much thinner; in
several passes through this region it was completely
absent. The persistence of this layer at higher latitudes is in agreement with the continuous existence of
the low altitude polar cusp phenomena. The plasma
mantle which is apparently located on open field lines,
however, has been shown to become thin or be even absent if the interplanetary magnetic field points strongly northward (Sckopke et al., 1975). The frequent presence of more energetic electrons with a pitch-angle
distribution indicative of trapping, in the entry layer
as well as in thelow altitude polar cusp Burrows and Mc
Diarmid, 1972), presents evidence that in this layer
plasma is transferred to closed field lines. Such a
process may well be only weakly dependent on the interplanetary magnetic field.

A restricted region of entry of plasma could not be
established. The variability of the flow inside the entry
layer seems to indicate that entry can occur, although
not necessarily homogeneously distributed, over a wide
area of the dayside magnetopause. A simple influx of
magnetosheath plasma along open field lines is not
evident in our data. The magnetopause presents at least
some kind of barrier as expressed by abrupt changes in
the phase space density. It is, on the other hand, very
unlikely that the whole entry layer is filled by some
unknown diffusion process transverse to $\bar{B}$, because in
such a case one would hardly see the sharp rise of the
density to its full value at the inner edge of the
layer and an essentially constant value out to the
magnetopause, as for instance on day 169 (Figure 1).

One possibility to account for these observations
would be to postulate the existence of a transport by
some kind of eddy convection in the layer. The eddies
could be generated by the Kelvin-Helmholtz instability
at the magnetopause (Southwood, 1968; Boller and Stolov,
1973). The initial filling of the flux-tubes with magnetosheath plasma may be achieved by anomalous diffusion
in thin boundary layers (smaller or equal to the ion
gyro-radius), but the essential transport should be
convective. The role of field line merging in this

context is not clear, since it has the tendency to drive
plasma from both sides into the boundary layer and is,
therefore, opposing the entry of the dominant plasma
components, at least close to the merging region. The
concept of entry by diffusion and eddy convection, as
proposed by Haerendel (1974), seems to be supported by
the very irregular convection found by Heppner (1973)
and Mikkelsen and Stockflet Jørgensen (1975) in the low
altitude polar cusp. The decrease of ring current plasma
at the inner edge of the entry layer as well as the
existence of high energy electrons up to 1 MeV close to,
but also outside the magnetopause (Meng and Anderson,
1970; Domingo et al., 1974) may be due to the same
transport process.

4.   CURRENTS AND CONVECTION IN THE POLAR CUSP REGION

One of the striking features of the entry layer is
the bending of the magnetic lines of force away from the
noon meridian in the sense of a tangential force applied
to the magnetosphere by the solar wind. Whatever the
microscopic process may be, there must be an electric
current normal to the magnetopause related to the
magnetic stress that balances the mechanical force.
These normal currents can be estimated to be somewhat
less than 10^6 A. On the morning side they are directed
inward, on the evening side outward. At the inner edge
of the entry layer they have a non-vanishing divergence
and connect to a sheet of field-aligned currents. The
signature of these currents has been seen before by
Fairfield and Ness (1972) and Fredericks et al. (1973).
At least part of the field-aligned currents should close
through the ionosphere and, since this region acts as a
resistor, should set up an electric field which is
essentially pointing from dawn to dusk. The ionospheric
closure currents are consistent with the observed overall convective transport to the polar cap during midday
hours (Gurnett, 1972; Heppner, 1973; Mozer et al.,
1974). On the outside the current loop should close by
a westward directed current in the magnetopause, i.e.
in the sense as to reduce the Chapman-Ferraro currents.
The dynamo driving the current system is, according to
this picture, located in the entry layer to which
momentum connected with the magnetosheath flow towards
the flanks of the magnetosphere is transferred and
where it becomes eventually dissipated by coupling to
the lower magnetosphere. This is only a very rough
description of the net effect of a complex process

involving possibly the set up of convection cells,
anomalous diffusion, anomalous resistivity etc. We will
pursue this subject in a later paper. The generator
which we propose to be in the entry layer would have the
same properties and consequences as that postulated by
Heikkila (1974) except that he placed it outside the
magnetopause.

In the context of currents and convection in the
polar cusp region many interesting problems have yet to
be investigated. One of them relates to the double
current sheets observed by Zmuda and Armstrong (1974)
at low altitudes all along the auroral oval and extending into the polar cusp region. These currents are not
to be confused with the current system described above
which corresponds to a single sheet with inward directed current on the morning side and an outward directed
current on the evening side. The double current sheets
may well be causally related to the cusp plasma, but
could be driven by a dynamo at medium latitudes. It is
of particular interest that their ionospheric closure
currents are consistent with a north-south component of
the transverse electric field corresponding to an
observationally verified convection towards noon inside
the current sheet which has a thickness of several
degrees at low altitudes. Apparently the magnetic flux
which is transported to the polar cap on the dayside is
replenished by this convection. A new aspect of our
understanding of this region is added by recent rocket
measurements of our group in Greenland showing the precipitation of large fluxes of electrons and protons below 1 keV in the presence of convection towards noon in
the low latitude part of the polar cusp. This could mean
that the convection at the inner edge of the entry
layer is directed essentially sunward, i.e. opposite to
the mean flow after entry into the layer.

In closing we would like to express a warning. We
have all the time assumed that the plasma observed in
the entry layer has its origin in the magnetosheath.
Indeed a natural conclusion, in view of the similarity
of density and mean energy of both plasmas. Furthermore, the bending of the magnetic field lines indicating
a large scale tangential stress is supporting this
assumption. It is, however, not excluded that a substantial amount of ionospheric plasma is being added
and accelerated to similar energies. An investigation of
the ionic composition of the cusp plasma should be very
rewarding.

This report reflects a relatively early state of a still ongoing investigation. More data have yet to be digested. Thus some of our preliminary conclusions on the entry of solar wind plasma into the magnetosphere may subsequently become modified.

ACKNOWLEDGEMENTS

We are indebted to P.C. Hedgecock (Imperial College London) and H. Rosenbauer and N. Sckopke for permitting this early presentation of material to appear in a joint publication.

The MPI plasma experiment was supported by the German Bundesministerium für Forschung und Technologie under grant WRK 210.

REFERENCES

Akasofu, S.-I., E.W. Hones, Jr., S.J. Bame, J.R. Asbridge, and A.T.Y. Lui, Magnetotail and boundary layer plasmas at a geocentric distance of ~18 R_E: Vela 5 and 6 observations, J. Geophys. Res., 78, 7257, 1973.

Boller, B.R. and H.L. Stolov, Explorer 18 study of the stability of the magnetopause using a Kelvin-Helmholtz instability criterion: J. Geophys. Res., 78, 8078, 1973.

Burch, J.L., Precipitation of low-energy electrons at high latitudes: effects of interplanetary magnetic field and dipole tilt angle: J. Geophys. Res., 77, 6696, 1972.

Burrows, J.R., and I.B. McDiarmid, Trapped particle boundary regions: in Critical Problems of Magnetospheric Physics, Proc. COSPAR/IAGA/URSI Symposium, Madrid 1972, p. 83.

Domingo, V., D.E. Page, and K.-P. Wenzel, Energetic electrons at the magnetopause: in Correlated Interplanetary and Magnetospheric Processes, ed. by D.E. Page, Reidel Publ. Comp., Dordrecht-Holland, 159, 1974.

Dungey, J.W., Interplanetary magnetic field and the auroral zones: Phys. Rev. Letters, 6, 47, 1961.

Fairfield, D.H., and N.F. Ness, IMP 5 magnetic field measurements in the high-latitude outer magnetosphere near the noon meridian: J. Geophys. Res., 77, 611, 1972.

Frank, L.A., Plasma in the Earth's polar magnetosphere:
    J. Geophys. Res., 76, 5202, 1971.
Frank, L.A., and K.L. Ackerson, Observations of charged
    particle precipitation into the auroral zone:
    J. Geophys. Res., 76, 3612, 1971.
Fredricks, R.W., F.L. Scarf, and C.T. Russell, Field-
    aligned currents, plasma waves, and anomalous
    resistivity in the disturbed polar cusp: J. Geo-
    phys. Res., 78, 2133, 1973.
Gurnett, D.A., Electric field and plasma observations
    in the magnetosphere: in Critical Problems of
    Magnetopsheric Physics, Proc. COSPAR/IAGA/URSI
    Symposium, Madrid 1972, p. 123.
Haerendel, G., Die Spur der Magnetopause in der Magne-
    tosphäre, Mitteilungen der Astronomischen Gesell-
    schaft 35, 165; 1974.
Hedgecock, P.C., Magnetometer experiments in the
    European Space Research Organization's HEOS
    satellites, Space Sci. Instrum., 1, 53, 1975.
Heikkila, W.J., The morphology of auroral particle
    precipitation, Space Res. XII, 1343, 1972.
Heikkila, W.J., Outline of a magnetospheric theory,
    J. Geophys. Res., 79, 2496, 1974.
Heikkila, W.J., and J.D. Winningham, Penetration of
    magnetosheath plasma to low altitudes through the
    dayside magnetospheric cusps, J. Geophys. Res., 76,
    883, 1971.
Mikkelsen, I.S., and T. Stockflet Jørgensen, Plasma
    motion in the cusp region during quiet magnetic
    conditions: Danish Meteorological Institute,
    Copenhagen, preprint, 1975.
Meng, C.I., and K.A. Anderson, A Layer of Energetic
    Electrons (> 40 keV) near the Magnetopause: J. Geo-
    phys. Res., 75, 1827, 1970.
Mozer, F.S., W.D. Gonzales, F. Bogott, M.C. Kelley, and
    S. Schutz, High-latitude electric fields and the
    three-dimensional interaction between the inter-
    planetary and terrestrial fields, J. Geophys. Res.,
    79, 56, 1974.
Paschmann, G., H. Grünwaldt, M.D. Montgomery, H. Rosen-
    bauer, and N. Sckopke, Plasma observations in the
    high-latitude magnetosphere, in: Correlated Inter-
    planetary and Magnetospheric Observations, (D.E.
    Page, ed.), 249, Dordrecht, 1974.
Paschmann, G., G. Haerendel, N. Sckopke, H. Rosenbauer,
    and P.C. Hedgecock, Topology and plasma characte-
    ristics of the distant polar cusp region, Trans.
    Amer. Geophys. Union, 56, June 1975.

Rosenbauer, H., H. Grünwaldt, M.D. Montgomery,
    G. Paschmann, and N. Sckopke, HEOS2 plasma
    observations in the distant polar magnetosphere -
    the plasma mantle, J. Geophys. Res., 80, 1975
    (in press).
Scarf, F.L., R.W. Fredricks, I.M. Green, and C.T.
    Russell, Plasma waves in the dayside polar cusp,
    1. magnetospheric observations, J. Geophys. Res.,
    77, 2274, 1972.
Sckopke, N., G. Paschmann, H. Rosenbauer, and D.H. Fair-
    field, Relation between properties of the plasma
    mantle and the orientation of the interplanetary
    magnetic field, Max-Planck-Institut für extraterr.
    Physik, 8046 Garching, preprint, 1975.
Sonnerup, B.U.Ö., and Cahill, L.J., Jr., Magnetopause
    structure and attitude from Explorer 12 observa-
    tions, J. Geophys. Res., 72, 171, 1967.
Sonnerup, B.U.Ö., and Ledley, B.G., Magnetopause
    rotational forms, J. Geophys. Res., 79, 4309, 1974.
Southwood, D.J., The hydromagnetic stability of the
    magnetospheric boundary, Planet. Space Sci., 16,
    587, 1968.
Vasyliunas, V.M., Magnetospheric cleft symposium,
    Trans. Amer. Geophys. Union, 55, 60, 1974.
Vasyliunas, V.M., Theoretical models of magnetic field
    line merging, 1: Rev. Geophys. Space Phys., 13,
    303, 1975.
Willis, D.M., Structure of the magnetopause, Rev.
    Geophys. Space Phys., 9, 953, 1971.
Zmuda, A.J., and J.C. Armstrong, The diurnal flow
    pattern of field-aligned currents, J. Geophys.
    Res., 79, 4611, 1974.

COMPOSITION OF THE HOT PLASMAS IN THE MAGNETOSPHERE

R. G. Johnson, R. D. Sharp, and E. G. Shelley

Lockheed Palo Alto Research Laboratory (52-12/205)

3251 Hanover Street, Palo Alto, California 94304

INTRODUCTION

The investigation of the mass and charge composition of the energetic (keV) plasmas in the earth's magnetosphere represents one of the most important approaches to establishing the origin of the particles in the plasmas and to understanding the complex electro-dynamic processes occurring within or at the boundaries of the magnetosphere. The processes responsible for the injection, energization, transport, and loss of the plasma components are still largely unidentified and some of the processes are likely to be dependent on the mass and/or charge of the components. Thus, measurements of the differences in energy spectra, spatial distributions, and temporal behavior of the various ionic components may provide the key to identifying and characterizing the important processes. In this paper we shall limit our discussion of the composition of the energetic particles in the magnetosphere primarily to particle energies less than 50 keV. The composition measurements at higher energies and their importance in understanding the magnetospheric processes have recently been reviewed by West (1975) and Krimigis (1973).

The most plausible candidates for the source of the ionic components of the hot magnetospheric plasma are the solar wind (Meinel, 1951; Axford, 1970) and the ionosphere (Van Allen, 1962; Axford, 1970). Axford (1969, 1970) has pointed out that the measurements of the He^3/He^4 abundance ratio and the charge state of the energetic helium ions within the magnetosphere could be used to determine the origin of the ions. In the ionosphere the He^3/He^4 ratio is about 10^{-5} (Axford, 1969), whereas in the solar wind the ratio is in the range 10^{-3} to 10^{-4} (Bame et al., 1968). Helium is primarily doubly

ionized in the solar wind and singly ionized in the ionosphere.
The He^3/He^4 ratio has been measured in the auroral zone by Buhler
et al. (1972) by exposing foils to the auroral ions in two rocket
experiments and subsequently mass analyzing the collected ions in
the laboratory. They conclude that the ions were of solar wind
origin. Other abundance ratios within the magnetosphere, such as
O^+/H^+ (Shelley et al., 1972) and He^{++}/H^+ (Whalen and McDiarmid, 1972)
at lower particle energies and He^+/H^+ (Krimigis, 1973), and C^+/O^+
(Mogro-Compero, 1972) at higher particle energies, have also been
used to infer the origin of the particles.

Processes responsible for the entry of particles into the upper
magnetosphere from the ionosphere or from the solar wind may also be
dependent on the mass or charge of the particles. Solar wind access
to the magnetosphere through diffusive processes (Axford and Hines,
1961; Axford, 1970; Tverskoy, 1969) which depend on electric and
magnetic field turbulence in the region of the magnetopause, may be
dependent on the mass and charge of the particles, whereas access
by merging of the geomagnetic and interplanetary fields on the dayside would at least initially provide access for all the solar wind
particles. Energetic ions in the magnetosphere resulting from field-
aligned flow of ionospheric ions into the magnetosphere as a result
of field-aligned electric fields (Whalen et al., 1974) would be expected to reflect the composition of the ionosphere at the altitude,
latitude, local time, etc., where the electric field process was
operative, whereas population of the magnetospheric tail region by
polar wind ions and subsequent convection deeper into the magnetosphere (Axford, 1970) would reflect the composition of the polar
wind which contains H^+ and He^+ ions (Hoffman et al., 1974). Shelley
et al. (1974b) have used satellite measurements of the H^+ and He^{++}
characteristics to investigate the convection electric field in the
dayside cusp region at low altitudes.

The processes responsible for the energization, transport, and
loss of energetic magnetospheric plasmas may often be concurrent and
strongly coupled. The ability to investigate these processes as a
function of the mass and charge of the plasma components whould be
helpful in identifying the conditions, if any, under which the processes are weakly coupled and in investigating the processes separately or collectively. Cornwall (1972) has shown that a comparison
of the radial diffusion rates of helium ions and protons in the magnetosphere would be a sensitive test of the diffusion process assumed.

In comparison to the extensive observations that are available
on the electron component of the hot magnetospheric plasmas and on
the total of the ionic component (usually assumed to be protons),
observations which can identify the ionic species are scarce and
are limited in scope. This situation has resulted from the relative
complexity of the instrumentation required to make unambiguous measurements of the separate ionic components, particularly those of

low abundance. Until the recent observations of the relatively intense O^+ component in the hot plasma during magnetic storms (Shelley
et al., 1972, 1974a; Sharp et al., 1974a), the principal emphasis
at these lower energies has been on measurements of the helium ions
(Reasoner, 1973). To date the only ionic species in the hot ($<$ 50
keV) magnetospheric plasmas which have been identified and reported
are H^+, He^+, He^{++}, and O^+. The O^+ identification is made in part
on an ionospheric abudnance argument since the mass-per-unit-charge
determination of 16 ± 2 AMU cannot rule out a contribution from N^+
ions (Shelley et al., 1972).

In view of the evidence that some of the components of the hot
magnetospheric plasmas are of ionospheric origin (Shelley et al.,
1972, 1974c; Sharp et al., 1974a; Johnson et al., 1974), the recent
observations of Hoffman et al. (1974), on the cold plasma composition near 1400 km during the large magnetic storm on 4 August 1972,
should be noted. At magnetic latitudes above 55°, they report unusually high abundances of O^+, N^+, O_2^+, N_2^+, and NO^+. The N^+ ion
became the dominant species and the O_2^+, N_2^+, and NO^+ concentration
were near 10^3 ions/cm^3, which is comparable to the O^+ concentrations
during less disturbed times. Thus, a wide range of masses of ionospheric ions at relatively high altitudes are at times available for
injection into the upper magnetosphere, and the need for higher mass
resolution as well as higher detection sensitivities is indicated
for future energetic ion mass spectrometer measurements in the magnetosphere.

OBSERVATIONS

Protons

It is well established that protons are a major component of
the hot magnetospheric plasma. This has been determined from ground-
based observations of the Doppler-shifted hydrogen line emissions in
auroral optical emission spectra (Meinel, 1951), from quantitative
comparisons of ground-based measurements of the auroral hydrogen
emissions with satellite measurements of the precipitating ionic
component (without mass discrimination) (Romick and Sharp, 1967;
Reasoner et al., 1968), and from direct measurements of the protons
with energetic ion mass spectrometers on rockets (Whalen et al.,
1971) and on satellites (Shelley et al., 1972; Sharp et al., 1974a).
Whether protons are the dominant ionic component of the hot magnetospheric plasma at all times and all locations in the magnetosphere
is now subject to question as a result of the satellite observations
by Shelley et al. (1972) that the relatively large precipitating
fluxes of O^+ ions in the energy range 0.7 to 12 keV sometimes exceeded the precipitating fluxes of protons in the same energy range
during the 17 December 1971 magnetic storm. These observations are
discussed in further detail in a later section of this paper.

The polar wind is a flow of ionospheric ions into the tail region of the magnetosphere and satellite measurements of the polar wind ions have been made using ion mass spectrometers (Hoffman et al., 1974). Protons, and frequently He$^+$ ions, are observed streaming outward in the high-latitude regions at velocities of several kilometers per second as determined from phase shifts in the roll modulation maximums of the sensor response between light and heavy ion species. Axford (1970) has shown that the polar wind flux is more than sufficient to provide the auroral proton flux if a magnetospheric process were operating to convect the polar wind ions from the tail region deeper into the magnetosphere and to energize them.

He$^+$ Ions

Observations of precipitating He$^+$ ions with energies up to 1.4 keV have been reported by Johnson et al. (1974) from measurements with energetic ion mass spectrometers on two polar-orbiting satellites at altitudes of about 500 and 800 km. The He$^+$ ions, along with protons and O$^+$ ions were observed on two occasions precipitating into the auroral regions in the morning sector. No He^{++} was observed on these occasions. The spatial distributions and relative intensities for one of the cases are shown in Figure 1 (from Johnson et al., 1974). The peak He$^+$ ion intensity was 0.03 erg/cm^2-sec-sr and occurred at L = 7 in the same region as significant fluxes of H$^+$ and O$^+$. The energy distributions of the O$^+$ and He$^+$ for this case are shown in Figure 2. It was shown from these data that the O$^+$ and He$^+$ ions had similar velocity distributions which suggests that the injection and energization mechanism may impart equal velocity to both ion species. This conclusion is highly tentative since it is based on only one event. From limits on the He^{++} flux and considerations of the charge exchange processes for He^{++} in the magnetosphere, the authors also conclude that the He$^+$ ions are most probably of ionospheric origin. As seen in Figure 1, the observations of protons in the same region as the O$^+$ and He$^+$ ions suggests that they might also be of ionospheric origin.

He^{++} Ions

Rocket measurements in the auroral zone of helium ions in the keV range have recently been reviewed in detail by Reasoner (1973) and will be discussed only briefly here. The most recent and most comprehensive rocket experiment to measure ion composition was that of Whalen and McDiarmid (1972) in which an instrument capable of detecting H$^+$, He$^+$, and He^{++} with energies between 2 and 20 keV was flown to an altitude of about 800 km during an auroral breakup. On the basis of the charge state of the observed helium ions and the measured He^{++}/H$^+$ ratios, they concluded that the origin of the ions

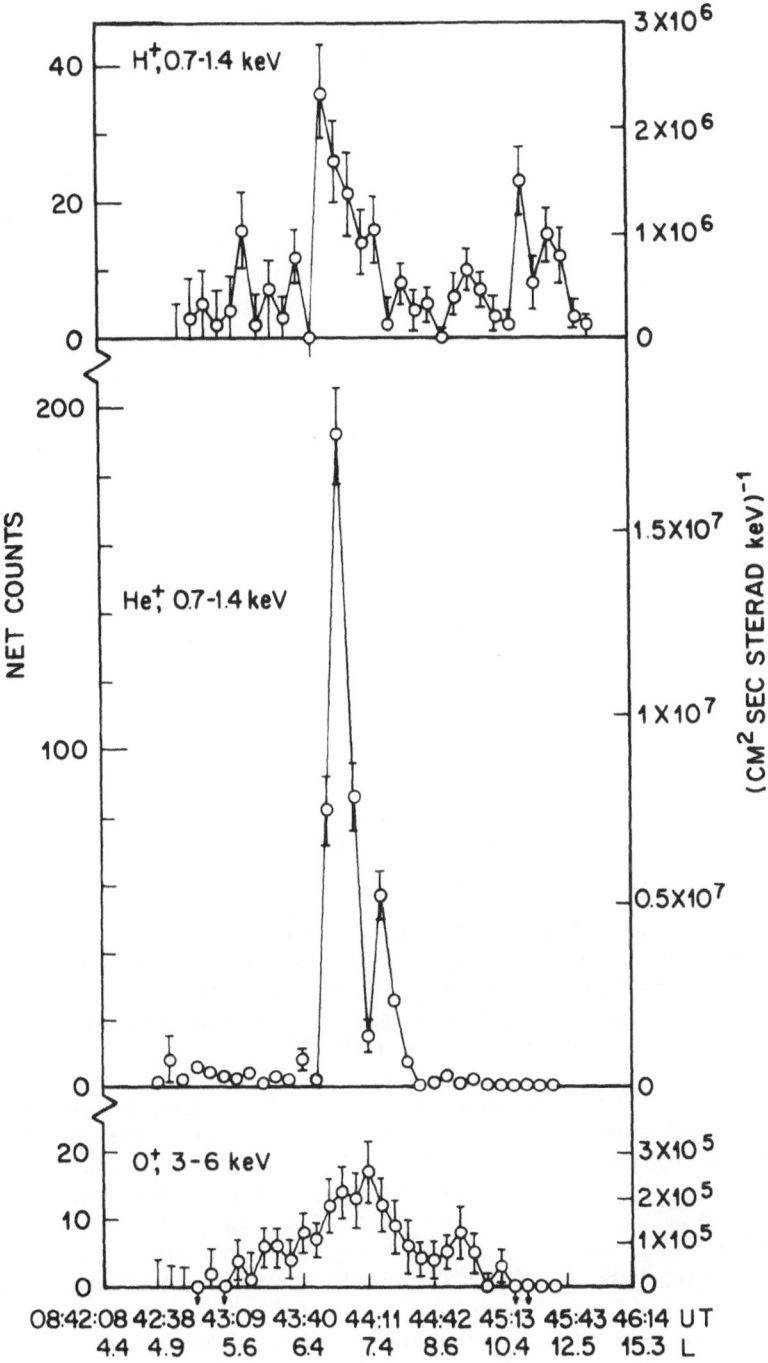

Figure 1.   Latitude distributions of H⁺, He⁺, and O⁺ ions.

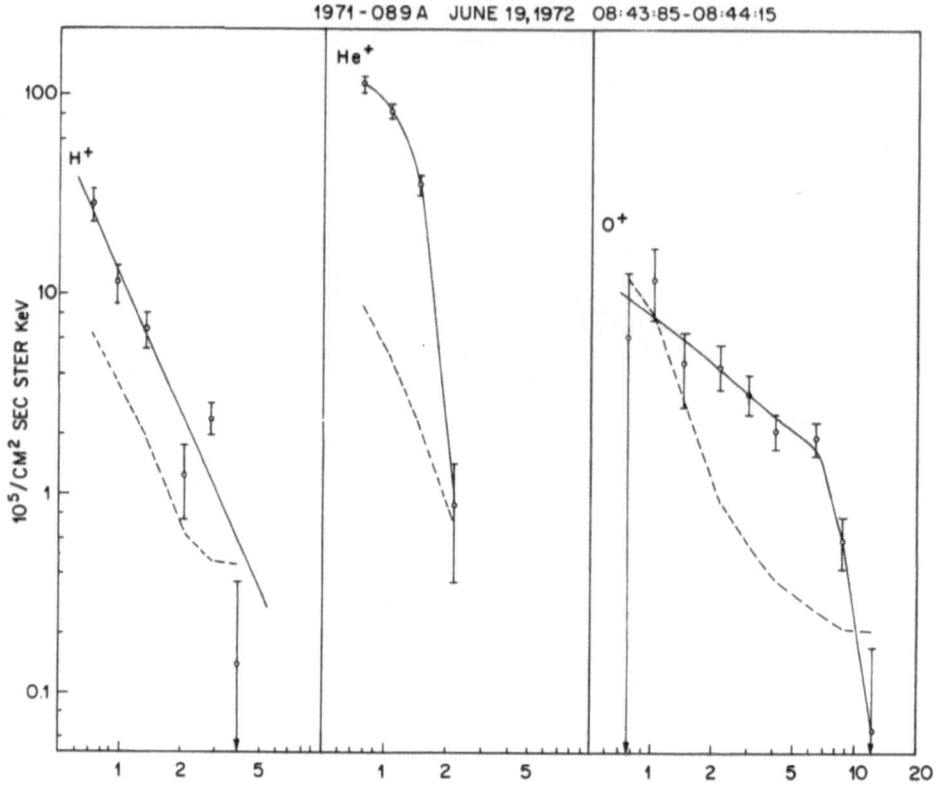

Figure 2. Energy spectrums of H$^+$, He$^+$, and O$^+$ ions.

was the solar wind rather than the ionosphere. On the basis of He^{++} measurements at three energies, they also concluded that the He^{++} and H$^+$ spectrums were peaked at the same energy per unit charge. From this they inferred that the solar wind ions had undergone an electrostatic acceleration of about 6 kV.

Satellite observations of He^{++} have been used to investigate the convection electric field in the low-altitude region of the dayside cusp (Shelley et al., 1974b). The relative positions of the low-latitude cutoffs of the precipitating ion fluxes were found to be approximately inversely proportional to the ion velocity and independent of the ion species. The observations were consistent with a dawn-to-dusk electric field of about 50 mV/meter.

Energetic He^{++} ions have been observed from satellite measurements precipitating into the nightside auroral region during a magnetic storm (Sharp et al., 1974b). The precipitation was observed over a wide latitudinal region from L = 3.8 to L = 7.5 and the aver-

aged He^{++}/H^+ ratio varied from $5.6 \pm 0.7\%$ at $L = 7.3$ to $2.9\% \pm 0.5\%$
at $L = 5.0$. The velocity distributions of the H^+ and He^{++} were surprisingly similar as seen in Figure 3 (from Sharp et al., 1974b).
From considerations of the observed He^{++}/H^+ ratio which is typical
of the solar wind and from the similarity of the velocity distributions, it was concluded that the He^{++} and H^+ originated in the solar
wind and were acted upon within the magnetosphere by adiabatic processes and that no substantial electrostatic acceleration occurred.
It should be noted that in this case the inferred acceleration process is different from the electrostatic acceleration process inferred from the rocket measurements of Whalen and McDiarmid (1972).
Of course, multiple acceleration processes within the magnetosphere
are not excluded.

O^+ Ions

O^+ Ions During Substorms. The precipitating energetic O^+ ions
during the substorms on 12 and 13 December 1971 have been investigated using data acquired with the energetic ion mass spectrometer
aboard the polar-orbiting satellite 1971-089A at an altitude of
about 800 km. The instrument was oriented at $55°$ to the zenith during all the measurements and thus sampled only the precipitating ion
fluxes at the high latitudes of the O^+ data discussed here. It covered the energy range in nine steps from 0.7 to 12 keV and the mass-
per-unit-charge range from 1 to 32 AMU (Shelley et al., 1972).

The magnetic activity on 12 and 13 December 1971 is indicated
by the bottom two curves in Figure 4. Prior to 1500 hours on 12
December, K_p was less than 2 and AE was less than 100 gammas. Two
fairly well isolated substorms are indicated by the AE index. The
second of these, beginning at about 2200 hours, has been investigated by Williams et al. (1974) using Explorer 45 data acquired in
the pre-midnight local time sector. The second substorm was widespread in local time and was observed at Dixon Island (0320 LT) in
the local time sector of the satellite measurements. Investigation
of the details of the first substorm are not yet complete.

The peak intensities of the O^+ ions are shown by the top curve
in Figure 4. During the quiet magnetic period prior to 1500 on 12
December, the O^+ intensities were near or at the limit of the sensitivity of the instrument. Beginning at 1600 hours, the O^+ ions are
observed in the local morning sector during each available satellite
pass for a period of 12 hours and the general correlation with the
substorm activity as indicated by the AE index is quite evident.
Larger intensities were observed in the second substorm, but the
peak fluxes of the first storm may have been missed because of the
limited time coverage provided by the satellite measurement. The
spread in local time of the O^+ observations was from 0200 to 0500
hours.

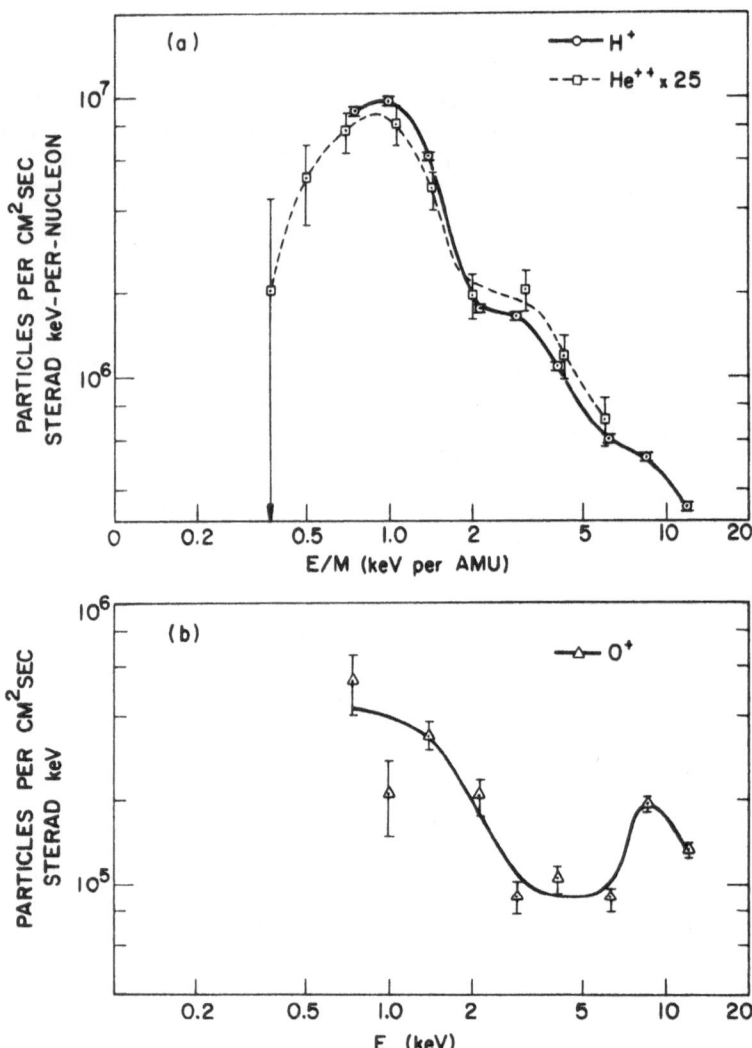

Figure 3.   a) Energy per nucleon spectrums of H$^+$ and He^{++} ions.
            b) Energy spectrums of O$^+$ during the same period.

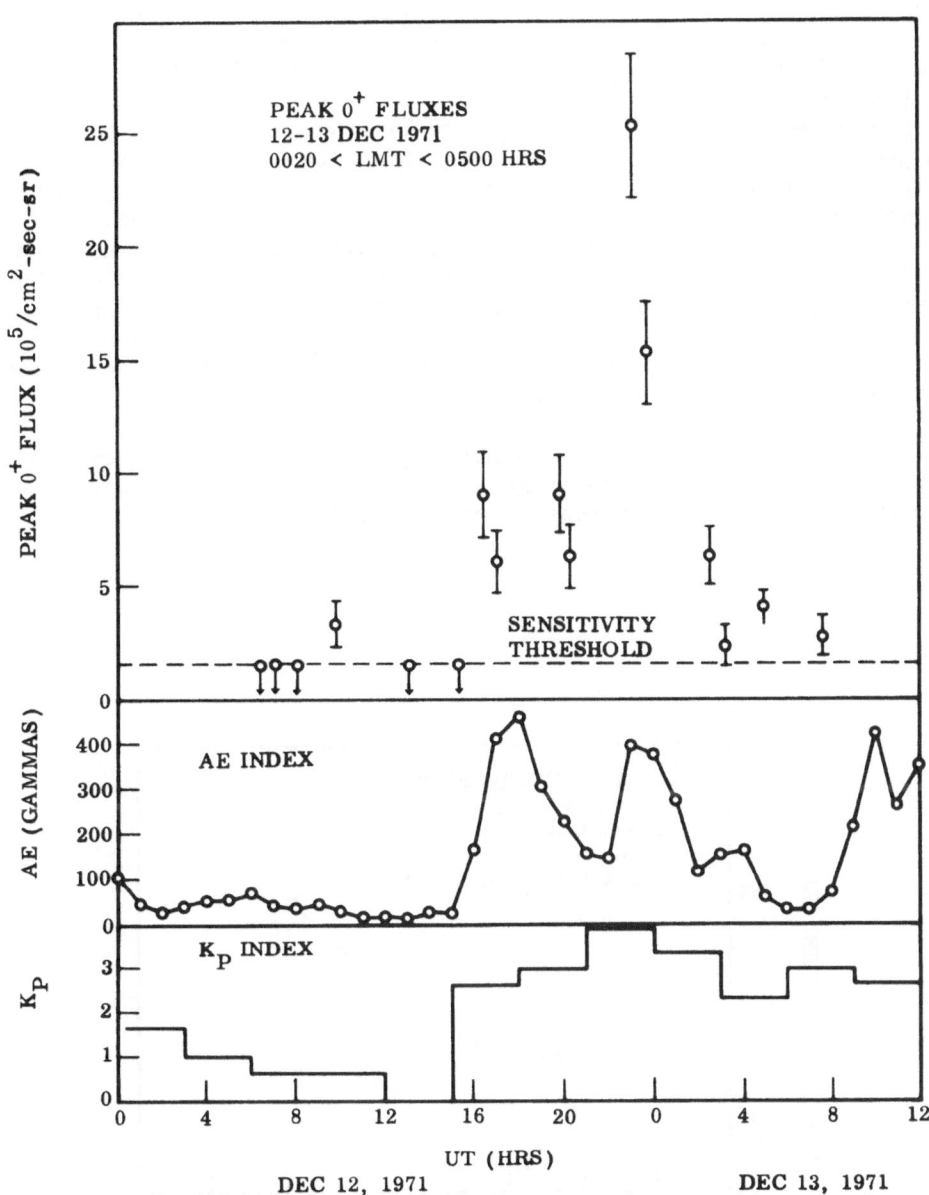

<u>Figure 4</u>.   Peak O^+ fluxes measured during two substorms.

The mean width of the regions of precipitation, defined by the 10% and 90% points of the integral of the number flux as the satellite traverses the region of O^+ precipitation, was found to be 2.5 degrees in magnetic latitude. The location of the mean position of the O^+ region, defined by the 50% point of the above integral is compared in Figure 5 with the location of the lower-latitude edge of the proton flux measured with the same instrument over the same energy range, 0.7 to 12 keV. The low-latitude edge of the protons is generally a distinctive feature of the data and is defined in this study as the lowest latitude at which 50% of the peak proton number flux in the plasma sheet particle region is observed. A strong correlation between these positions is seen in Figure 5, and the mean position of the O^+ fluxes is generally equatorward of the proton edge. During these substorms, the mean position of the O^+ fluxes was also always inside the "cut-off" trapping boundary for energetic electrons as defined by Romick et al. (1974) for electrons with energies greater than 130 keV.

From this study and from synoptic studies of other substorms with less complete data coverage, it is concluded that precipitating O^+ ions are a common feature of magnetic substorms.

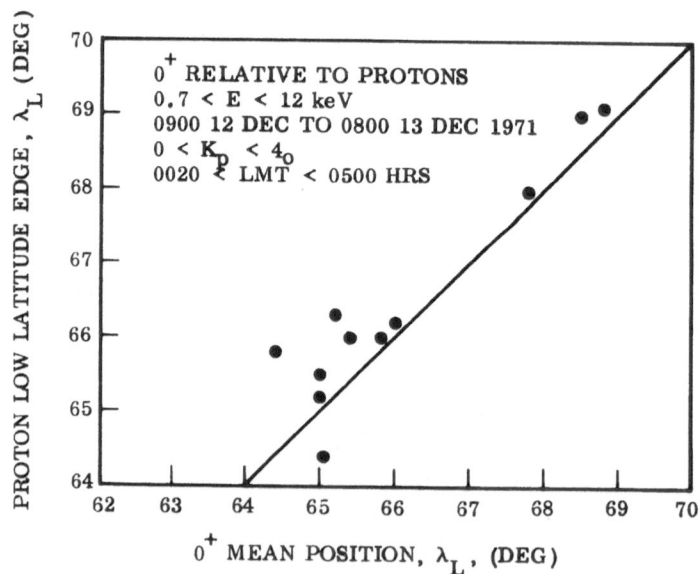

Figure 5. Relative locations of O^+ and H^+ fluxes during the time period shown in Figure 4.

The Morphology of Precipitating O^+ Ions in a Magnetic Storm.
In order to search for clues to the nature of the processes which
accelerate the O^+ ions from the cold ionospheric plasma and to define as closely as possible the morphological parameters which the
theories will have to explain, a detailed study of the morphology
of the O^+ ions during the 17-18 December 1971 magnetic storm is
being conducted and some preliminary results of that study are presently available. This was a rather classic magnetic storm with an
ssc at 1418 UT on 17 December and a large main phase. The peak D_{ST}
was 171γ and the storm lasted until about 2300 UT on the 18th. The
satellite was in the 0300/1500 LT plane during the period of interest and a rather high fraction of data coverage was being obtained,
which provided about as good temporal resolution as can be achieved
for studies such as this with a polar-orbiting satellite. Figure 6
from Shelley et al. (1972) shows the data from 6 consecutive satellite traversals of the nightside (0300 hours LT) auroral and subauroral regions during the main phase of the storm. Data for both H^+
ions (solid curve and closed circles) and O^+ ions (dashed curve and
open circles) are illustrated. The ordinate is the total number of
counts in the mass-per-unit-charge peak for the appropriate species
summed over all nine of the energies sampled in a complete 6-second
cycle of the experiment. It is approximately proportional to the
integral number flux in the $0.7 \leq E \leq 12$ keV range.

The principal features of note in Figure 6 are: 1) the peak
O^+ fluxes are comparable to or greater than those of the protons in
the energy range measured; 2) the O^+ ions are observed over a wide
latitude range implying either an extended source or precipitation
from a trapped population undergoing substantial radial diffusion;
3) the fluxes of both species are highly structured with respect to
latitude and quite variable from one pass to the next with no obvious specific conjugate structure; and 4) the O^+ ions are spatially
displaced with respect to the protons, i.e., they generally extend
equatorward of the protons and the protons generally extend poleward of the O^+ ions.

For the detailed morphological study mentioned above, we have
formulated several different parameters from data such as is illustrated in Figure 6 and have intercompared these parameters with each
other as a function of magnetic latitude and time during the course
of the storm, and with various indices of geomagnetic activity. For
an intensity parameter characteristic of an entire satellite traversal of the auroral and subauroral regions, we computed the integral
energy flux of the O^+ ions over the latitudinal range $44^\circ < \Lambda_L < 80^\circ$.
This quantity is illustrated in Figure 7 for the northern hemisphere
data only. For comparison we have plotted the D_{ST} index on the same
logarithmic scale, but inverted so that increasing ring-current intensity corresponds to an increasing value of the ordinate. One
sees that there is a remarkable quantitative correlation between
these quantities over the period illustrated. The initial rise

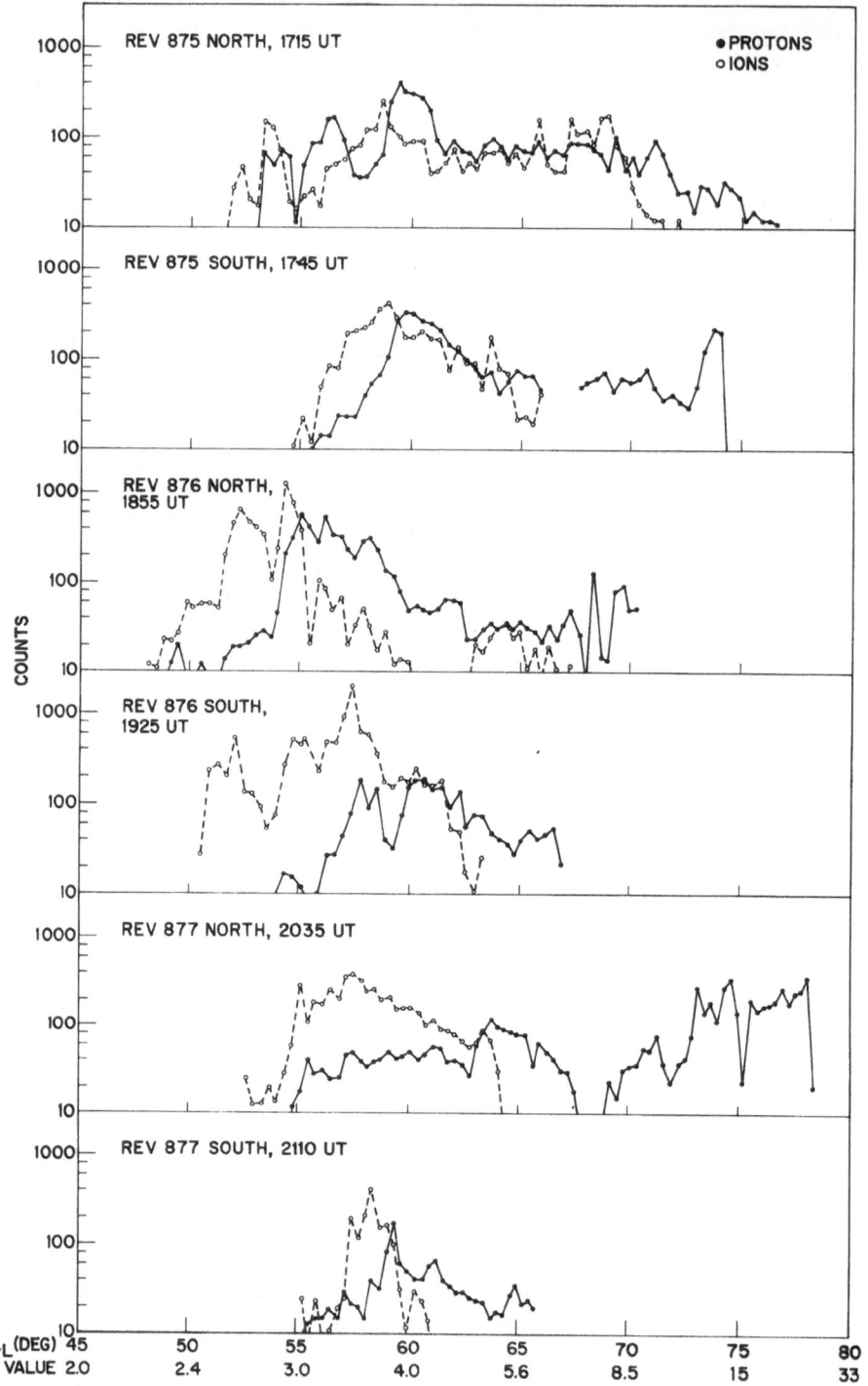

Figure 6. Ion fluxes during the 17 December 1971 magnetic storm.

after the ssc at 1418 is about the same in both the timing and relative amplitudes. The O^+ intensities appear to have a substantially
shorter decay time than the ring current. This trend is also exhibited by the earlier data from remnants of the 16 December storm
(ssc at 1904 UT) which was still in progress at the onset of the
17 December event. The close correspondence of the two illustrated
quantities can be interpreted in several ways. It can be considered
to support the hypothesis of Cladis (1973a,b) that the energy for
the acceleration of the O^+ ions is derived from the ring-current
particles. Alternatively, one could formulate a model in which a
significant fraction of the ring current itself is in the form of O^+

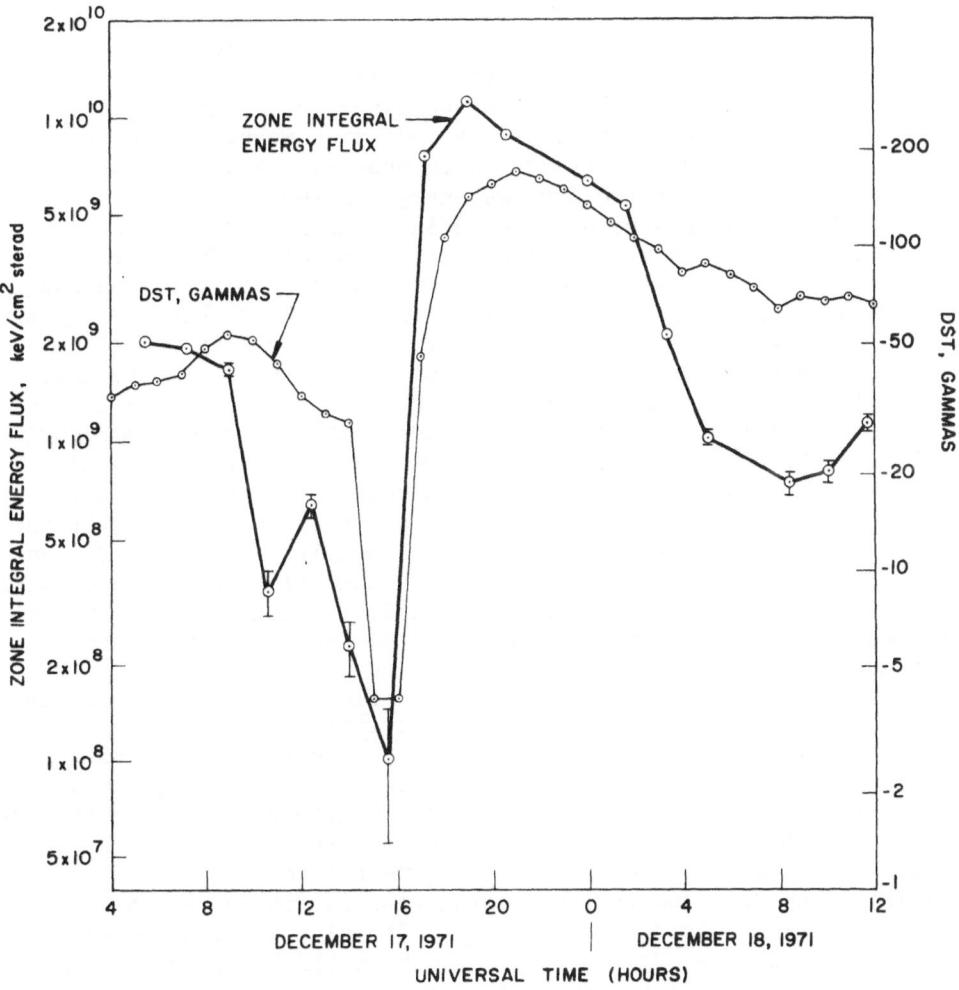

Figure 7. Integral O^+ intensity compared to D_{ST}.

ions and the time dependence of the observed precipitation reflects
the varying flux intensity of the parent trapped population.

Figure 8 from Shelley et al. (1972) shows some representative
energy spectrums from one of the passes illustrated in Figure 6
(Revolution 876 - South). The data are averaged over the time intervals indicated. The error bars represent counting statistics
only. The spectrums vary from a monotonically steeply falling one
at low latitudes to a hard spectrum with a peak at about 4 keV at
high latitudes. Peaks in the few-keV range are a common feature of
the O^+ spectrums which have been examined up to this time and, as
will be seen below, the general trend of a spectral hardening with
increasing latitude has been found to be typical for the storm in
question.

Data from another nightside pass at 1714 UT illustrating this
trend are shown in Figure 9. This pass marked the initial flux increase after the ssc and thus might be expected to have a spectral
signature most representative of the injection mechanism before
distortions due to drift effects could obscure any discernible
trends. For this pass the average energy of the O^+ ions in the 0.7
$\leq E < 12$ keV range has been computed in 2° latitudinal intervals.
The error bars represent counting statistics only. The average energy was considered significant and formulated only in latitudinal
intervals in which the average energy flux was $\geq 10^6$ keV/cm^2-sec-sr.
It was computed by taking the ratio of the energy flux to the number flux, each of which had been averaged over the 2° latitudinal
interval, and thus tends to be weighted by the more intense events
within the interval. An alternative computation was performed by
computing the average energy from those individual (six-second)
spectral measurements that were considered statistically significant and then taking the average value of these quantities over the
2° interval, weighting each measurement equally. The significance
criterion chosen was that the integral number flux in the individual
spectrum had to be $\geq$ twice its standard deviation. The latitudinal
dependence of this parameter computed from the northern nightside
data throughout the time period 0532 UT on 17 December to 1146 UT
on 18 December (cf Figure 7) is given in Figure 10. The error bars
in this figure represent the standard deviations of the means of
these individual average energies and not counting statistics. A
similar plot prepared from the average energy values computed according to the definition utilized in Figure 9 give similar results, indicating that the observed trend is not the result of biasing due to
a specific weighting technique. Thus, we see that the rather surprising trend of a spectral hardening with increasing latitude is
characteristic of this entire period. Radial diffusion processes
such as discussed by Nakada et al. (1965) would be expected to produce an opposite effect. Hopefully, this will be a useful morphological feature with which to test specific theoretical models.

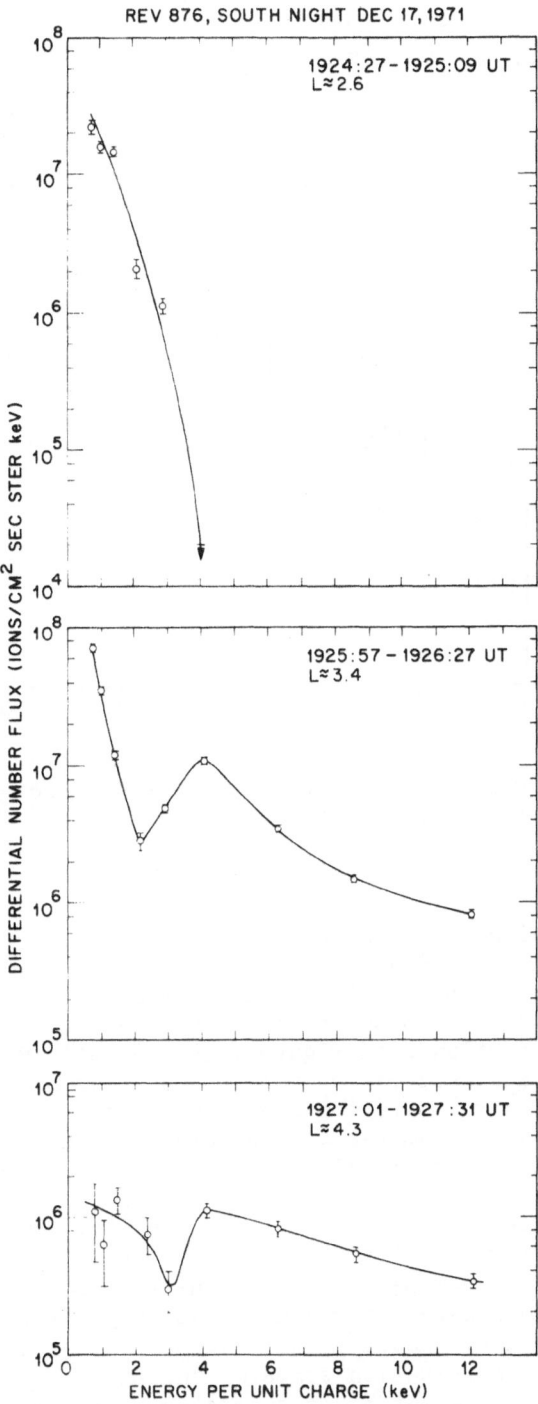

Figure 8.   Energy spectrums of O^+ ions.

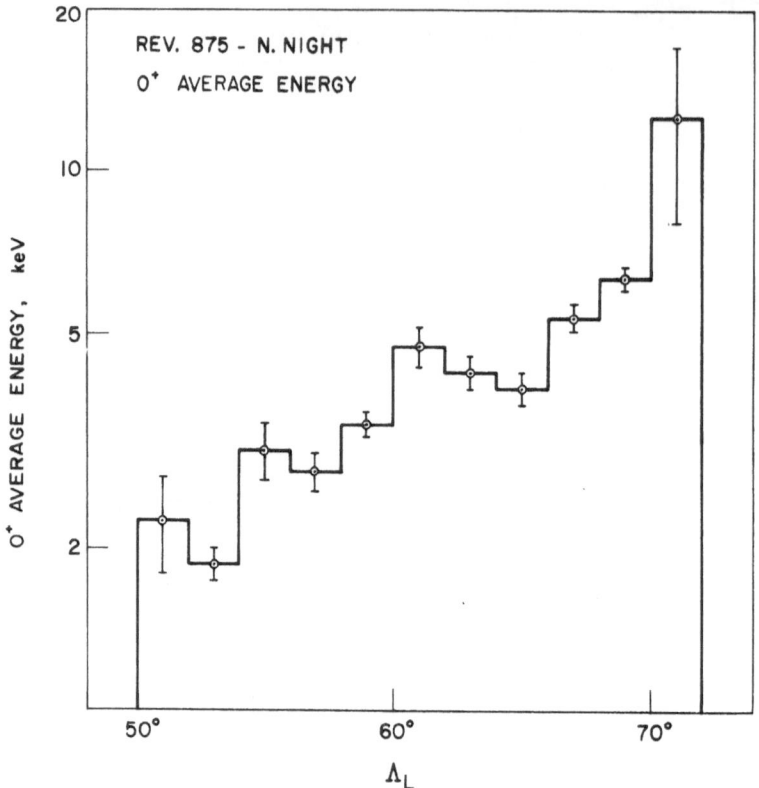

Figure 9. Latitudinal variation of the average energy of O⁺ ions on a single pass.

    The latitudinal variations of the O^+ and H^+ precipitation pat-
terns during this period also exhibit some interesting characteristics. In Figure 11, the center of the observed precipitation region,
measured by the 50% point in the zone integral intensity parameter
described in connection with Figure 7 is indicated with a circle
for H^+ ions and a square for O^+ ions. The horizontal bars represent
the extent of the precipitation region measured by the 10% and 90%
points in this same parameter. In several instances during this
period, the proton precipitation extended well into the polar cap.
The satellite orbit on some passes did not extend higher than 80°
invariant latitude so in order to form a uniform data base, we trunc-
ated the zone integrals at that latitude. Thus, for those points in
Figure 4 in which the upper bar is in the vicinity of 78°-80°, that
point does not in general mark the extreme upper latitude of the
precipitation region. Despite this distortion, we see a remarkable
tracking of the precipitation zones of the two ions with a latitudinal displacement of about 5° as suggested by the data in Figure 6.
This detailed tracking illustrates the intimate relationship between

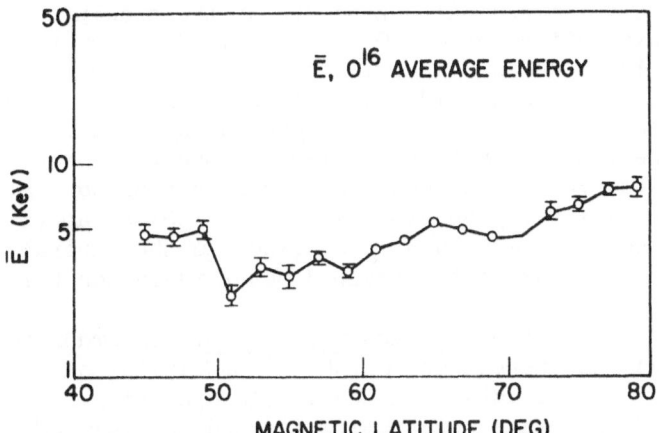

Figure 10. Latitudinal variation of the average energy of O^+ ions during the time period shown in Figure 7.

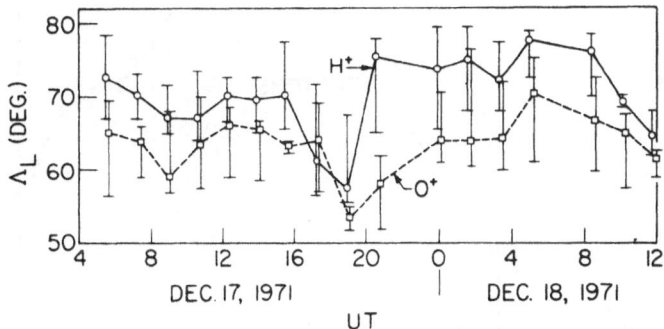

Figure 11. Locations of the precipitation zones of O^+ and H^+ ions.

the O⁺ ions and the magnetospheric proton population in the energy
range measured.

The overall latitudinal dependence of the precipitation intensity during this entire period (0532 UT on 17 December to 1146 UT on
18 December) is shown in Figure 12. We see most clearly in this
figure how the O⁺ ions provide the dominant part of the precipitating energy flux in this energy range over a substantial latitudinal
interval. The latitudinal displacement of the two zones is evident
and as indicated above one sees that the high-altitude extent of
the H⁺ precipitation region extends into the polar cap (i.e., $\Lambda_L >$
80°), while the O⁺ is more nearly confined to the auroral zone and
the subauroral regions of the trapped outer radiation belt particles.

Synoptic Study of O⁺ in Magnetic Storms. In order to investigate the local time dependence of the O⁺ precipitation, a synoptic
study was made of the data from one year's operation of the energetic
(0.7 to 12 keV) ion mass spectrometer aboard the 1971-089A satellite
(Shelley et al., 1974a). During this period the satellite precessed

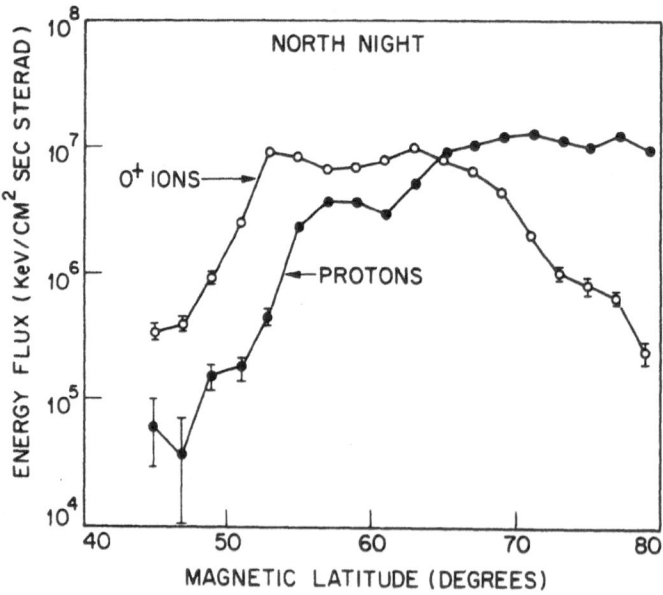

Figure 12.   Latitudinal variation of the energy flux of O⁺ and H⁺
             ions during the time period shown in Figure 7.

through the entire range of local times. Eleven of the principal
magnetic storms in the period from December 1971 to November 1972
were utilized. Data from three orbits were examined for each storm,
selected from the beginning, middle, and end of the storm being
studied. O^+ ions were observed precipitating into the atmosphere
during every storm.

Figure 13 shows the latitudinal extent of the O^+ precipitation
regions. The dot indicates the position of the maximum flux intensity on each pass and the lines indicate the portion of the pass during which the flux was greater than the spectrometer sensitivity
threshold of $\approx 2 \times 10^5$ ions/cm^2-sec-sr.

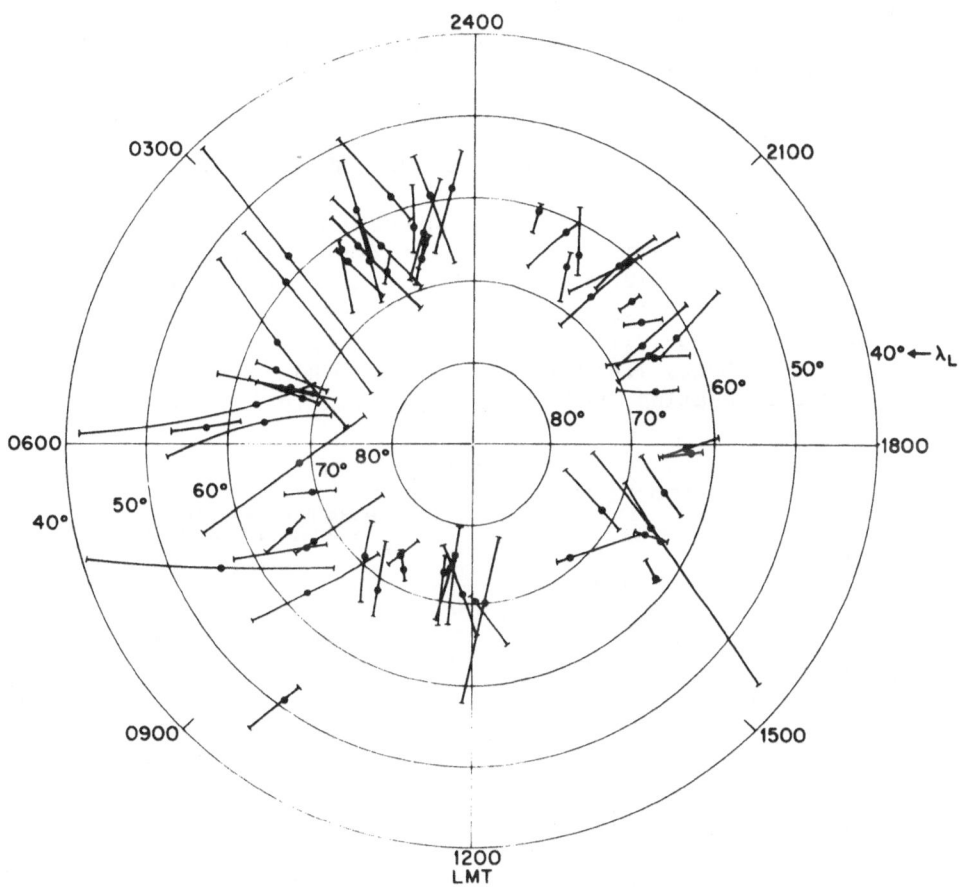

Figure 13. Polar plot in invariant latitude and magnetic local time
            of the regions of observed O^+ precipitation during 11
            major storms.

The peak flux intensities are shown in Figure 14. The solid horizontal bars indicate the median values in each three-hour sector of magnetic local time. For comparison with typical auroral proton intensities, we have indicated with a dashed line the flux level which, for an isotropic angular distribution and a 4-keV average energy, would correspond to a precipitating O^+ energy flux of 0.1 erg/cm^2-sec

It should be noted that the 11 storms utilized for this work had varying magnitudes and no attempt has been made to normalize the results to account for this. Each storm contributes to the data in two restricted local time intervals, so the detailed local time variations evident in the figures should not be interpreted as being representative of the local time variations in a single storm. It is significant, however, that extended regions of precipitating O^+ ions were seen in every storm and at all local times and that the median peak intensities on the nightside are roughly an order of magnitude more intense than those on the dayside.

THEORETICAL AND RELATED EXPERIMENTAL RESULTS

The published theoretical work specifically directed toward interpreting the properties of the heavy ion fluxes in the energy range of these measurements is limited. Cladis (1973a,b) has proposed a mechanism whereby ionospheric O^+ ions are accelerated by resonant interactions with ion cyclotron waves generated by the ring-current protons. Palmadesso et al. (1974) have considered the ionospheric heating due to the electrostatic ion cyclotron tur-

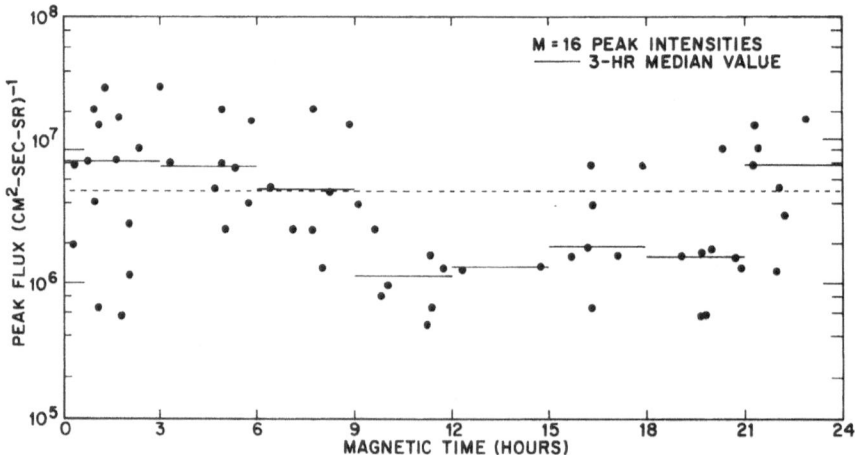

Figure 14. Peak O^+ fluxes observed during 11 major storms as a function of magnetic local time.

bulence which is thought to be associated with the anomalous resistivity leading to field-aligned potential drops in the auroral zone. They suggest that this effect might contribute to the O^+ energization. Brice (1974) has shown how mixtures of heavy ions and protons in the ring current provide conditions where waves at frequencies below the heavy ion cyclotron frequency may be strongly amplified.

Torr et al. (1974) examine the atmospheric effects to be expected from the precipitation of the energetic O^+ ions. They conclude that large upward-moving fluxes of fast neutral oxygen atoms and atmospheric heating above 200 km altitude will result.

Berko et al. (1975) use a self-consistent calculational technique to compare the measured ion fluxes on Explorer 45 with the geomagnetic field deformation caused by these fluxes under the assumption that they are all protons. They set an upper limit of a few percent to the heavy ion contribution to the storm-time ring current at the time of the comparison. Simultaneous observations with the Lockheed energetic-ion mass spectrometer on satellite 1971-089A are not inconsistent with these results. The O^+ precipitation had fallen below the sensitivity threshold of the spectrometer at the time, late in the storm's recovery phase, when the comparison was made.

SUMMARY AND CONCLUSIONS

The results now available on the He^+, He^{++}, and O^+ ions in the hot magnetospheric plasmas provide increasing evidence of the importance of mass and charge composition measurements for investigating the complex electrodynamic processes within and at the boundaries of the magnetosphere. The present measurements on these ions are still greatly limited in energy range, in mass range and resolution, and in detection sensitivities. The data have thus far been acquired only at low altitudes so there are not data in the equatorial plane to assess their importance to understanding processes in that region. Although still limited, the results from the energetic oxygen and helium measurements are beginning to provide sufficient definition of the characteristics of these particles that current theoretical models of the processes which act on them can be more realistically evaluated, but clearly more detailed theoretical work is needed.

Instrumentation is presently under construction for the GEOS and ISEE spacecraft and these instruments will have greatly improved mass resolution and sensitivity which will allow studies to be made of even rarer ionic constituents in the plasmas. The need for large-scale simultaneous observations of the energetic oxygen ions is evident and techniques for this type of observation should

be pursued. Perhaps observations of the Doppler-shifted emissions
from the precipitating ions could be used to differentiate them
from the emissions of the ambient atmospheric constituents. All
of the new techniques which can provide information on the composition of the hot magnetospheric plasma can be expected to be of
increasing importance in future magnetospheric/ionospheric research.

REFERENCES

Axford, W. I., "Helium in the Atmosphere, Aurora, and Solar Wind,"
    in Atmospheric Emissions, B. M. McCormac, ed., p. 317, Van
    Nostrand-Reinhold, 1969.

Axford, W. I., and C. O. Hines, "A Unifying Theory of High-Latitude
    Geophysical Phenomena and Geomagnetic Storms," Can. J. Phys.,
    39, 1433, 1961.

Axford, W. I., "The Origin of Radiation-Belt and Auroral Primary
    Ions," in Particles and Fields in the Magnetosphere, B. M.
    McCormac, ed., p. 46, D. Reidel, Dordrecht, Holland, 1970.

Banks, P. M., and T. E. Holzer, "The Polar Wind," J. Geophys. Res.,
    73, 6846, 1968.

Bame, S. J., A. J. Hundhausen, J. R. Asbridge, and I. B. Strong,
    "Solar Wind Ion Composition," Phys. Rev. Letters, 20, 393,
    1968.

Berko, F. W., L. J. Cahill, Jr., and T. A. Fritz, "Protons as Prime
    Contributors to the Storm-Time Ring Current," J. Geophys. Res.,
    80, 1975 (in press).

Brice, N., "Wave-Wave Coupling in Multiple Ion Plasma," J. Geophys.
    Res., 70, 2520, 1974.

Buhler, F., W. I. Axford, H. J. A. Chivers, K. Marti, P. Eberhardt,
    and J. Geiss, "Rare Gas Isotopes in Auroras," EOS Trans. Am.
    Geophys. Union, 53, 1092, 1972.

Cladis, J. B., "Effect of Magnetic Field Gradient on Motion of Ions
    Resonating with Ion Cyclotron Waves," J. Geophys. Res., 78,
    8129, 1973a.

Cladis, J. B., "Interpretation of Energetic Heavy Ion Fluxes Ob-
    served during the Magnetic Storm of December 17, 1971," Radio
    Science, 8, 1029, 1973b.

Cornwall, J. M., "Radial Diffusion of Ionized Helium and Protons:
    A Probe for Magnetospheric Dynamics," J. Geophys. Res., 77,
    1756, 1972.

Hoffman, J. H., W. H. Dodson, C. R. Lippincott, and H. D. Hammack,
    "Initial Ion Composition Results from the Isis Satellite," J.
    Geophys. Res., 79, 4247, 1974.

Johnson, R. G., R. D. Sharp, and E. G. Shelley, "The Discovery of
    Energetic He$^+$ Ions in the Magnetosphere," J. Geophys. Res., 79,
    3135, 1974.

Krimigis, S. M., "The Charge Composition Aspects of Energetic Trapped
    Particles," Proceedings of the Solar Terrestrial Relations Con-
    ference, held at the Univ. of Calgary, Calgary, Alberta, Canada,
    Aug. 28 - Sept. 1, 1972, D. Venkatesan, ed., p. 207, 1973.

Meinel, A. B., "Doppler-Shifted Auroral Hydrogen Emission," Astrophys.
    J., 113, 50, 1951.
Mogro-Compero, A., "Geomagnetically Trapped Carbon, Nitrogen, and
    Oxygen Nuclei," J. Geophys. Res., 77, 2799, 1972.
Nakada, M. P., J. W. Dungey, and W. N. Hess, "On the Origin of Outer
    Belt Protons," J. Geophys. Res., 70, 3529, 1965.
Palmadesso, P. J., T. P. Coffey, S. I. Ossakow, and K. Papadopoulos,
    "Topside Ionosphere Ion Heating Due to Electrostatic Ion Cyclo-
    tron Turbulence," Geophys. Res. Letters, 1, 105. 1974.
Reasoner, D. L., "Auroral Helium Precipitation," Rev. Geophys. Space
    Phys., 11, 169, 1973.
Reasoner, D. L., R. H. Eather, and B. J. O'Brien, "Detection of
    Alpha Particles in Auroral Phenomena," J. Geophys. Res., 73,
    4185, 1968.
Romick, G. J., W. L. Ecklund, R. A. Greenwald, B. B. Balsley, and
    W. L. Imhof, "The Interrelationship between the > 130 keV
    Trapping Boundary, the VHF Radar Backscatter, and the Visual
    Aurora," J. Geophys. Res., 70, 2439, 1974.
Romick, G. J., and R. D.Sharp, "Simultaneous Measurements of an In-
    cident Hydrogen Flux and the Resulting Hydrogen Balmer Alpha-
    Emission in an Auroral Hydrogen Arc," J. Geophys. Res., 72,
    4791, 1967.
Sharp, R. D., R. G. Johnson, E. G. Shelley, and K. K. Harris, "Ener-
    getic O$^+$ Ions in the Magnetosphere," J. Geophys. Res., 79, 1844,
    1974a.
Sharp, R. D., R. G. Johnson, and E. G. Shelley, "Satellite Measure-
    ments of Auroral Alpha Particles," J. Geophys. Res., 79, 5167,
    1974b.
Shelley, E. G., R. G. Johnson, and R. D. Sharp, "Satellite Observa-
    tions of Energetic Heavy Ions during a Geomagnetic Storm," J.
    Geophys. Res., 77, 6104, 1972.
Shelley, E. G., R. G. Johnson, and R. D. Sharp, "Morphology of Ener-
    getic O$^+$ in the Magnetosphere," in Magnetospheric Physics, B.
    M. McCormac, ed., p. 135, D. Reidel, Dordrecht, Netherlands,
    1974a.
Shelley, E. G., R. D. Sharp, and R. G. Johnson, "Dayside Convection
    Electric Field Deduced from Ion Measurements in the Low-Altitude
    Cusp," EOS Trans. Am. Geophys. Union, 56, 1175, 1974b.
Shelley, E. G., R. D. Sharp, and R. G. Johnson, "The Ionosphere as
    the Source of Ring-Current Particles," EOS Trans. Am. Geophys.
    Union, 55, 1015, 1974c.
Torr, M. R., J. C. G. Walker, and D. G. Torr, "Escape of Fast Oxy-
    gen from the Atmosphere during Geomagnetic Storms," J. Geophys.
    Res., 79, 5267, 1974.
Tverskoy, B. A., "Main Mechanisms in the Formation of the Earth's
    Radiation Belts," Revs. Geophys., 7, 219, 1969.
West, H. I., Jr., "Advances in Magnetospheric Physics 1971-1974:
    Energetic Particles," Rev. Geophys. Space Phys., 1975 (in press).
Van Allen, J. A., "Dynamics, Composition and Origin of the Geomag-
    netically Trapped Corpuscular Radiation," Trans. Int. Astron.
    Union, XIB, 99, 1962.

Whalen, B. A., J. R. Miller, and I. B. McDiarmid, "Evidence for a
    Solar Wind Origin of Auroral Ions from Low Energy Ion Measure-
    ments," J. Geophys. Res., 76, 2406, 1971.
Whalen, B. A., and I. B. McDiarmid, "Further Low-Energy Auroral Ion
    Composition Measurements," J. Geophys. Res., 77, 1306, 1972.
Whalen, B. A., D. W. Green, and I. B. McDiarmid, "Observations of
    Ionospheric Ion Flow and Related Convective Electric Fields
    in and Near an Auroral Arc," J. Geophys. Res., 79, 2835, 1974.
Williams, D. J., J. N. Barfield, and T. A. Fritz, "Initial Explorer
    45 Substorm Observations and Electric Field Considerations,"
    J. Geophys. Res., 79, 554, 1974.

MAGNETOSPHERIC PLASMA REGIONS AND BOUNDARIES

Walter J. Heikkila

The University of Texas at Dallas
P. O. Box 688
Richardson, Texas   75080

ABSTRACT

Several magnetospheric regions and their boundaries are discussed in critical terms. Serious questions are raised by recent
observations that suggest that the magnetosheath plasma on the dayside penetrates deep into the region of closed magnetic field lines.
This deduction, and the evidence that the polar cap field lines
interconnect with the interplanetary magnetic field lines, suggest
that the magnetopause be defined as a plasma boundary, rather than
in terms of open or closed magnetic field lines. The magnetopause
is here defined as that surface where there is an abrupt reduction
in the phase space density from the values typical of the magnetosheath plasma. On the night side this definition, coupled with
the evidence of closed field lines poleward of the auroral oval,
requires a modification to Dungey's model of the magnetosphere in
which there is a gap or space between the end of the plasma sheet
and the X-line. These considerations imply that the usual model of
magnetic merging is not applicable to the magnetosphere. This
implication is supported by the failure to detect the energy
dissipation at the dayside magnetopause that would be a consequence
of the tangential component of the electric field characteristic of
the merging process. Careless use of words or phrases such as
interconnection and generation is discussed. It is suggested that
the resolution of these serious questions can be achieved only by
turning to a more primitive description that avoids the shortcomings
of the fluid theory.

1. INTRODUCTION

Magnetospheric plasma is separated into several regions with distinctly different properties. It is important that we understand the physical nature of the boundary surfaces between them, and that we identify the governing physical processes. While the existence of such surfaces has been known for some time, nevertheless our knowledge of them is in many ways rather poor.

Five important boundaries are indicated in Figure 1. These are the bow shock, magnetopause, outer boundary of the plasma sheet, inner boundary of the plasma sheet, and trapping boundary for energetic particles. In discussing these boundaries and the associated plasma regimes, we need to note that low energy particles (the ones constituting the bulk of the hot magnetospheric plasma and also the primary auroral particles) are influenced greatly by both magnetic and electric fields, while energetic particles are controlled primarily by the magnetic field alone. In essence this constitutes a definition of low and high energy particles for our purposes; the division comes at about 10 kev as a rough guide.

There is reason to believe that the motions of both low and high energy particles in the magnetosphere are to a large extent adiabatic, conserving at least the first two adiabatic invariants. The guiding centers then drift with the transverse velocity

$$\underline{V}_d = \frac{\underline{E} \times \underline{B}}{B^2} + \frac{W \sin^2 \alpha \ (\underline{B} \times \nabla B)}{qB^3} + \frac{2 \ W \cos^2 \alpha (\underline{B} \times \underline{R})}{qB^2R^2} \qquad (1)$$

where the energy W, pitch angle α, and charge q specify the particle, and $\underline{R}$ is the vector from the particle to the center of curvature of the magnetic field line (Falthammar, 1973). We use SI units. The first term is the electric drift velocity; it is also the velocity of the magnetic field lines in the frozen field approximation. The second and third terms specify the gradient and curvature drifts. These latter two drifts are not directly affected by the presence of an electric field. However, if there is a component of the electric field in the direction of these drifts, as there usually is in the magnetotail, then the particles may gain energy from this field; their subsequent gradient and curvature drifts are thus secondarily affected by the electric field. If there is also a component of the electric field parallel to the magnetic field then negative and positive particles will of course be efficiently accelerated in opposite directions.

   In order to understand the essential differences in the
behaviour of particles in the different magnetospheric regions
and at their boundaries we consider electric field effects explicitly.
This means that we do not make use of the familiar concept of frozen
field convection of magnetic field lines. That this is a reasonable
approach may be appreciated upon noting the fact that these three
types of drift may be comparable for the energetic particles with
which we are concerned. It is not wise to treat any one of them as
dominant over the others until that is explicitly justified by a
more fundamental or basic treatment (Hines, 1963). It is especially
in regions such as the various boundary layers, where the scale
lengths may be quite small, that great care must be exercised in
the application of the more approximate fluid theory.

   We consider only the steady state, or quasi-steady state.
This means that we ignore the rate of change of energy stored in
the electromagnetic fields, and that we neglect induced electric
fields. This same approximation is made in all the theories of
magnetic field line merging or reconnection, since curl $E = -\partial B/\partial t$
is taken equal to zero in them. Substorm effects in which temporal
changes are important require separate consideration. Finally, we
will concentrate on primary auroral processes, i.e. those processes
which are necessary for the creation of auroras. Secondary processes
would be those which determine various details of the auroras, such
as small scale structure, and we will make only passing reference
to some of them.

2. BOW SHOCK AND MAGNETOSHEATH

   Solar wind and magnetosheath plasma can behave like a fluid
because of the existence of propagating wave modes, the Alfven and
magnetoacoustic waves (Heikkila, 1975, to be published). Because
of these waves information can be transmitted to the individual
particles so that they can behave in a coordinated manner, for
example flow around an obstacle. The solar wind plasma flow speed
exceeds these wave speeds, and it is therefore not possible to
match the boundary conditions in the linear reversible theory.
Nature's way out of the dilemma is to create a thin non-linear
regime, the bow shock, in which the plasma and flow conditions are
changed abruptly. The main influence of the bow shock on magnetospheric and auroral phenomena is the determination of the magnetosheath plasma characteristics, such as the proton and electron gas
temperatures.

   The post-shock flow in the magnetosheath can be understood in
terms of the linear fluid theory. Because the flowing plasma has
enough energy and momentum it can establish exactly the right
distribution of fields and currents so that each particle goes in
the right direction. To a large extent this flow is without

dissipation, and can be understood without reference to internal
magnetospheric and auroral phenomena. We therefore do not consider
it further here.

3. MAGNETOPAUSE

The magnetosphere has been rather loosely defined as that
region of space in which plasma processes, including distribution
and flow, are controlled primarily by the geomagnetic field (Gold,
1959). In keeping with the convention adopted for naming the outer
boundaries of atmospheric regions as pauses, the outer boundary
of the magnetosphere is called the magnetopause.

In the early theories of the magnetosphere (Chapman and
Ferraro, 1931; Johnson, 1960; Axford and Hines, 1961) the interplanetary magnetic field (IMF) was ignored, and the magnetopause
was consequently the outermost surface of closed field lines, in
effect by definition. All the magnetic field lines in this surface
radiated out from a neutral point in the southern high latitude
dayside region, and reentered back into the magnetosphere through
a similar neutral point in the north. Since the magnetic field
strength vanishes at such singular points so does the magnetic
control of the plasma, and it was conjectured that solar wind or
magnetosheath plasma might be able to penetrate down to low
altitudes through these neutral points. Although there is considerable validity to these ideas, nevertheless the existence of an
IMF implies some essential modifications to this picture.

First let us consider certain topological features of the net
magnetic field in the vicinity of the magnetopause. We can get
some idea of these from the calculations of Cowley (1973) and
Stern (1973), who added a constant external magnetic field, with
arbitrary orientation, to a dipole field. Admittedly the magnetospheric part of the total field is much more complicated than a
dipole field, but nevertheless the basic topology is not really
different; for example, there is symmetry about the equatorial
plane. From such considerations we can deduce that the polar cap
field lines interconnect with those in the interplanetary medium.
Such interconnection is verified by the observation that energetic
solar flare particles have essentially direct access to the polar
caps (Evans and Stone, 1972; Morfill and Scholer, 1973). A
particularly convincing bit of evidence is the observation that
anisotropies of the particle fluxes in space are preserved in the
form of differences between the fluxes over the two polar caps.

Plasma observations far downstream, for example in lunar
orbit, show that the plasma regimes familiar to us from nearer
the earth can still be recognized. The bow shock surrounds what

is clearly the magnetosheath. Going closer to the antisolar
direction the plasma density and flux are abruptly reduced, to
values typical of the magnetotail (Rich et al., 1973). It thus
seems reasonable to accept the idea that the magnetopause continues
on downstream. In doing so we are forced to accept a definition
which permits the high latitude or polar cap field lines to cross
the magnetopause.

We therefore suggest that the magnetopause should be defined
in terms of the plasma characteristics rather than in terms of the
magnetic field. Specifically, the magnetopause should be defined
as that surface where the phase space density of the plasma is
abruptly decreased from its value in the magnetosheath. This
surface is the limit of free penetration of the magnetosheath
plasma; by free penetration we mean any flow for which Liouville's
theorem is obeyed by most of the plasma particles. Acceleration
and energization are not ruled out, provided only that the process
conserves the phase space density. Diffusion is ruled out. This
definition makes no mention of the magnetic field, and it is thus
applicable to both the dayside and downstream regions.

Another feature of the interconnected magnetic fields is that
there is a singular line or separatrix (not a field line) which
defines the boundary of closed field lines. There are three
distinct types of field lines, which can be described as those
having either two, one, or no feet on the ground. These are
indicated in Figure 1 by solid, dashed and dot-dashed lines
respectively. For the special case of a strictly southward IMF the
separatrix is an X-type neutral line, or more accurately an X-ring,
completely around the magnetosphere. The magnetic field strength
at the line vanishes, and the field lines in the plane normal to
the neutral line form an X geometry; this geometry is indicated
by the open X in both the noon and midnight meridian planes in
Figure 1. This is sometimes referred to as the reconnection
geometry, since the process of merging or reconnection of magnetic
field lines takes place at such a neutral line when there is an
electric field parallel to it (Vasyliunas, 1975). In the more
general case where the IMF has an arbitrary orientation there are
two neutral points, more or less on opposite sides of the magnetosphere; these are joined by two separatrices, which are now X-lines
with a tangential component of magnetic field along them. The
projection of the field lines on a plane normal to the X-line still
shows X-type geometry. The essential feature of an X-line is that
there is no component of magnetic field perpendicular to it, and
therefore no Lorentz force acts on a charged particle moving along
it. Cowley (1973) has concluded that the addition of a tangential
component along the X-line does not basically alter the merging
process. We will limit our considerations here to this special
case, since the situation with the southward IMF is the easiest
one to visualize, and since it also appears on both theoretical

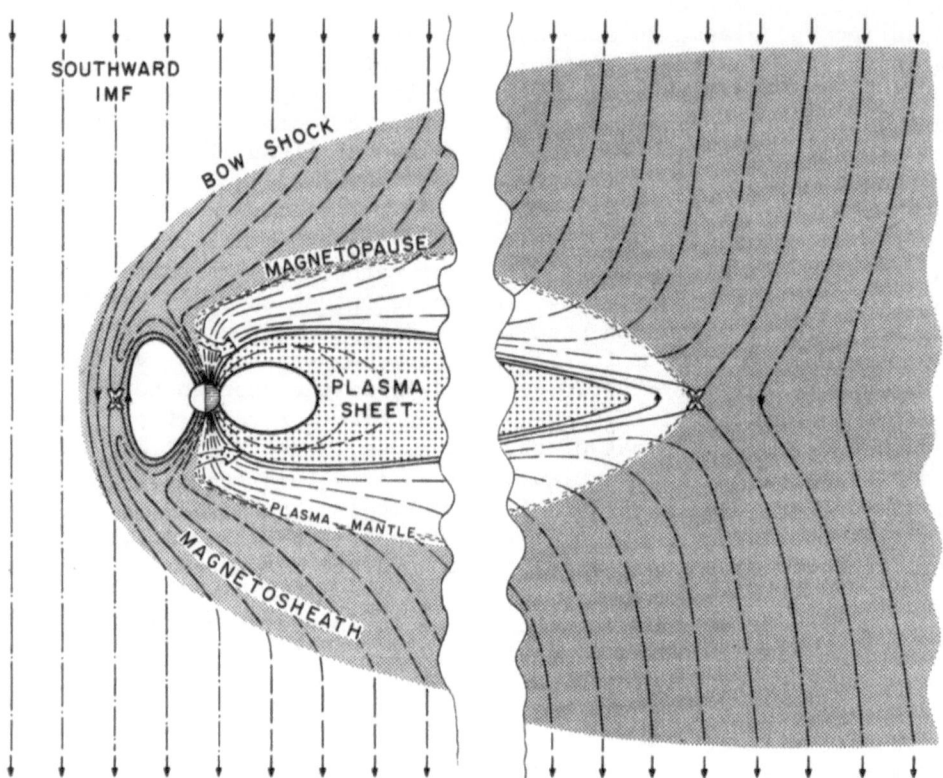

Figure 1. Magnetic field topology associated with interconnected
interplanetary and geomagnetic field lines. Closed field lines
are shown solid, lines with one foot on the ground are dashed, and
lines with no feet on the ground are dot-dashed. The essential
difference from Dungey's model is the gap between the end of the
plasma sheet and the X-line, which is located at the downstream
magnetopause. Magnetosheath particles enter the plasma sheet by
diffusion from the cleft, followed by convection along the auroral
oval. The trapping boundary is within the plasma sheet.

and observational grounds to be the one most favorable for magnetospheric and auroral activity. This special case is more complicated
than the general one only in mathematical details, not in basic
physics. We will use the term X-line, in preference to neutral
line, in order to suggest the general applicability of our reasoning.

It should be noted that the two neutral points in the general case with an IMF of arbitrary orientation are not the same neutral points as those in the closed model of the magnetosphere referred to above. The addition of the external magnetic field means that the field strength at these high latitude regions no longer vanishes, and the word neutral is not appropriate. With the confirmation that magnetosheath plasma does indeed penetrate down to low altitudes through these regions came the designations dayside or polar cusp (Heikkila and Winningham, 1971; Frank, 1971). Further observations showed that the cross-sectional area of the penetrating plasma column is rather large, being some 1° to 4° in north-south extent (Winningham, 1972); accordingly a term such as cleft is more accurately descriptive (Heikkila, 1972a). Unfortunately there is still a lingering tendency toward the use of the term neutral point; this term is archaic, and its use to denote the polar cusp or cleft should be discontinued. Neutral point should be used only for a singular point where the magnetic field strength vanishes.

There is also a regrettable tendency to regard the terms interconnection and reconnection as essentially equivalent (Vasyliunas, 1974). However, we should reserve interconnection to denote an open magnetic topology, without in any way implying anything about an electric field, or about frozen field convection of magnetic field lines. It is the fact of interconnection (not reconnection) that is demonstrated by the easy entry of energetic particles to the polar caps, since Mev particles are not sensitive to weak electric fields with potential differences of only some tens of kilovolts over the length scales involved.

If there is an electric field then the process of merging or reconnection of magnetic field lines may take place. Specifically, the merging rate is proportional to the component of the electric field along the X-line. Thus, a statement that there is magnetic merging or reconnection is a statement about the electric field that may be associated with the open magnetic field. It should be noted that this is more than a statement that an electric field exists; it specifically refers to the component parallel to the neutral or X-line. If the electric field were orthogonal to the X-line the merging process would cease to act.

Now there must be an electric field in the interplanetary medium, given by the Lorentz transformation $\underline{E}' = \underline{E} + \underline{V} \times \underline{B}$, where the prime indicates a quantity measured in a frame of reference moving with the plasma so that $\underline{V}' = 0$. Since the plasma is a good conductor we may set $\underline{E}' = 0$, and thus in our earth-fixed non-rotating frame of reference we would measure a field $\underline{E} = -\underline{V} \times \underline{B}$. For a southward IMF this is directed from dawn to dusk. Assuming that the magnetic field lines are equipotential lines, this field would be mapped down to the polar cap along the open magnetic field lines. A surprising feature of the interconnection is the

fact that the magnetic field lines are twisted in such a way that
the resulting polar cap electric field is always in the dawn-to-
dusk direction, for any orientation of the IMF. From this it is
generally concluded that the magnetospheric electric field (which
is always dawn-to-dusk) is simply due to this mapping.

The process of merging is supposedly involved here, and the
magnetospheric electric field is said to be due to merging. There
is, however, a difficulty with this conventional picture, as pointed
out by Heikkila (1975). Because of the necessary continuity of the
tangential component of the electric field required by Maxwell's
equations, the existence of the merging electric field would imply
that an electric field exists also along the magnetopause. If
this were the case, then the magnetopause current would dissipate
energy; the total energy dissipated by the current of about 10^7
amperes flowing through a typical magnetospheric potential difference
of 10^5 volts would be 10^{12} watts. This is a very large dissipation
rate, exceeding the estimated dissipation due to all auroral processes
on the night side. The energy should reappear in the form of
energized particles flowing out of the dayside magnetopause region.
Such energization is not observed on satellite crossings of the
magnetopause region; in fact the plasma flowing in the magnetotail
boundary layer or plasma mantle carries less, not more, energy than
the adjacent magnetosheath flow. Unless the implied energy dissipation is accounted for it must be concluded that the magnetopause
current does not flow through this potential drop, and that to a
first approximation the magnetopause is an equipotential surface.
This conclusion has serious ramifications for magnetospheric theory
in general; it implies that the model of the electric field upon
which merging theory is based is wrong, and that the mathematical
theory is therefore irrelevant to magnetospheric physics.

While recognizing the need to reconsider the merging model
of the electric field we must also appreciate that there are
difficulties with the alternative, namely an equipotential X-line.
Specifically, if there is no electric field along the X-line then
it seems to be necessary to admit a component of the electric
field along magnetic field lines somewhere. Such parallel electric
fields may perhaps be associated with anomalous resistivity
(Fredricks et al., 1973), or with double layers (Block, 1972). A
parallel electric field is no more questionable, or inadmissable,
than is an electric field along a neutral line. In fact, in some
sense the latter is more difficult to understand, since the current
along a neutral line must be severly limited if the X-type geometry
(as opposed to O-type) is to be preserved. There is no such
limitation to allowed currents along magnetic field lines.

It may also be difficult for some to accept the idea that the
magnetic field might not be convected through an X-line, even
though the more distant parts of the field line can be described

in terms of frozen field convection. However, there is really no basic difficulty, since the field lines have no identity in the sense of string or spaghetti; the magnetic field lines can be identified only by means of the particles that are on them, and the convection of particles near a neutral or X-line is not describable by the usual $\underline{E} \times \underline{B}$ terms. To understand what is going on we must go back to a more primitive description in terms of individual particles in electric and magnetic fields. The merging theory then follows with the a priori assumption that magnetic field lines are equipotential lines.

It is appropriate to point out that an electric field at an equipotential surface has only a normal component. If the magnetic field is nearly tangential to it, then the convection velocity $\underline{V} = \underline{E} \times \underline{B}/B^2$ is also tangential to the surface. This is in agreement with the requirements of fluid flow, namely that the lines of flow should bypass the obstacle, rather than intercept it. The internal magnetospheric circulation can then also have a return flow just inside the magnetopause; such a flow has been observed by Kavanagh et al. (1968). On the other hand, a tangential field would imply a direct interaction between most of the particles in the oncoming flow and the magnetopause, a feature not at all in agreement with a fluid picture.

The confirmation that magnetosheath plasma can penetrate down to low altitudes through the cleft was quickly taken as evidence that the cleft magnetic field lines are open, connected directly with the IMF (Heikkila and Winningham, 1971). However, McDiarmid et al. (1975a, b) have concluded, on the basis of the pitch angle distributions of energetic particles (E > 40 kev), that much of the cleft is on closed field lines. Energetic particles are not greatly affected by the magnetospheric electric field; they follow the magnetic field lines rather closely, provided that the magnetic field is not highly structured on length scales comparable with the gyroradii of the particles. Such energetic particles in at least the lower latitude half of the cleft have a pitch angle distribution which is peaked at 90°, a so-called pancake distribution. This is easily explained by the assumption that the particles in question are trapped on closed magnetic field lines, the particles with pitch angles near 0° and 180° being lost at the two ends of the field lines; it is difficult to find any other simple explanation for it. Thus the use of these energetic particles as tracers provides rather convincing evidence that a large fraction of the cleft is on closed field lines.

At this point we face a serious difficulty in explaining how the magnetosheath plasma can have such easy access deep into the region of closed field lines. One night be tempted to appeal to some process like diffusion; however, such diffusion would have to

to be remarkably fast to virtually wipe out any gradient in the
density or flux of the plasma. Furthermore, the process would
then have to stop abruptly, in view of the observation that the
equatorward boundary of the cleft is generally sharp. Finally,
the inward diffusion would be opposed by the outward convection
due to a dawn-dusk magnetospheric electric field, if one exists
near the magnetopause.

There is the possibility that the magnetosheath plasma does not
follow magnetic field lines, and so there may be a normal component
of the magnetic field at the dayside magnetopause. Such a normal
component would be associated with a rotational discontinuity which
is implied by the fluid or MHD theory, and also by merging theories
(Sonnerup and Ledley, 1974; Vasyliunas, 1975). We can estimate
its magnitude by using reasonable values for the dimensions of
the cleft and magnetopause. Taking an area of 200 km by 2000 km
for the closed part of the cleft at low altitudes we get a flux
of 2.4×10^6 maxwells. If this comes out through an area 7 R_e
wide and extending 9 R_e from the equatorial plane (see Heikkila,
1975, for justification), the result is a normal component of 6 nt.
This is of the same order of magnitude as that deduced by Sonnerup
and Ledley (1974) for two carefully chosen magnetopause crossings
by Ogo 5. However, such clear examples of rotational discontinuities
are rare, and most crossings indicate a tangential discontinuity,
with no measurable normal component. Thus it is not clear that the
observations of McDiarmid and his co-workers can be explained in
this way. At the present time this problem remains unexplained.
Perhaps the solution will be found along with the explanation for
the electric field in this region.

We need also to reconsider the downstream portion of the
magnetopause, and especially its relationship with the X-line.
Dungey's (1961) model of the magnetosphere is incomplete, since
it does not include any mechanism for reducing the phase space
density as the plasma is convected from the magnetosheath into
the magnetotail. Vasyliunas (1975) has pointed out that the
reconnection process should accomplish that by means of the standing
shocks that radiate out from the X-line. If that is the case,
then we must place the X-line at the magnetopause, as shown in
Figure 1. No other mechanism is included in the merging model for
modifying the plasma. Unfortunately, we now face a serious
difficulty; there is no physical mechanism available in this model
for absorbing the momentum of the magnetosheath flow. There is
nothing to turn the flow around at the X-line, and the flow must
therefore continue in the antisolar direction. This downstream
component of flow is about 400 km/sec, far greater than the component
of flow in toward the X-line given by the merging model. Consequently,
the flow symmetry assumed in the merging model with a stagnation
point at the X-line does not exist. There seems to be no possibility of maintaining the symmetrical flow pattern at a stationary
X-line, as assumed in the merging theories.

Perhaps there is no steady state solution possible, and the
X-line moves downstream with the plasma velocity. If the magnetotail is a few hundred R_e long the X-line would leave the system
in one or two hours, presumably to be replaced by a new X-line
closer to the earth. Such a process does indeed seem to take
place during substorms (Nishida and Hones, 1974), and the period
between substorms is often of the order of a few hours. However,
there are also long periods when there are no substorms, although
the solar wind always continues to flow, and thus this explanation
does not appear very promising. Furthermore, the development of
the new X-line remains to be explained.

Taking a cue from the dayside region, where the magnetopause
seems to be approximately an equipotential surface, we might
consider that the magnetopause in the distant downstream portion
of the magnetotail is also an equipotential surface. In that case,
the merging theory is again not applicable.

4. PLASMA SHEET AND ITS OUTER BOUNDARY

There is no doubt that most of the plasma sheet is in the
region of closed magnetic field lines. Direct measurements of the
magnetic field show a northward component, except perhaps briefly
during the expansive phase of a substorm (Nishida and Hones, 1974).
However, it is not clear how far the plasma sheet extends, and in
particular whether it reaches out to the limit of closed field
lines as in the merging model, and whether the outer boundary cuts
across magnetic field lines.

Related to this is the question of the immediate source of
auroral particles. Again there is little doubt that the primary
particles causing the diffuse or continuous auroras at mid-latitudes
are on closed field lines, and come from the plasma sheet region.
However, there have been suggestions that the particles causing
the discrete arcs at higher latitudes may be on open field lines,
or at least at the boundary of open and closed field lines (Frank
and Gurnett, 1971; Eather, 1973). In the merging model, on the
other hand, these high latitude particles might be associated
with the slow shocks of Petschek's (1966) model (see Sonnerup,
1974). If so, then they would be precipitated a short distance
equatorward of the last closed field line.

Observations of energetic particles near auroral forms provide
evidence that in fact the discrete auroral forms do occur on closed
field lines. McDiarmid and his co-workers (see e.g. Venkatarangan
et al. 1975; Whalen and McDiarmid, 1972) have observed that a
weak flux of energetic (E > 40 kev) electrons persists beyond the
trapping boundary, and that it often shows a pancake distribution

in pitch angle even poleward of the discrete auroral forms. Just
as on the day side, this observation is most easily interpreted
as evidence for closed field lines. Therefore, we adopt the view
that the auroras in the nighttime oval region (as distinct from the
polar cap) do occur on closed field lines.

As already pointed out, this finding may not be inconsistent
with the merging picture, if the auroral particles are precipitated
in the slow shock rather than along magnetic field lines, provided
that more energetic particles can travel freely through the shock.
However, the shock is a discontinuity in the magnetic field, and
it should act as a scattering center for energetic particles. It
would take very little scattering to fill in the loss cone, since
this is only about one degree in half-angle in the distant part
of the magnetotail. The fact that the loss cone is preserved in
the bounce motion is convincing evidence that pitch angle scattering
is not an important process in this region (at least under the
steady state conditions that we are considering), and it suggests
that there is no sharp discontinuity which the energetic particles
must cross.

In view of our doubt as to the applicability of the merging
theory to the magnetosphere, we propose the model shown in Figure 1.
Here we assume that the end of the plasma sheet occurs on closed
field lines, some distance earthward from the X-line. The inclusion
of this space or gap, with a low density of particles up to the
X-line which is located at the magnetopause, is the most significant
difference between Dungey's model and the present one.

In this model the entry of magnetosheath particles into the
plasma sheet does not take place over the polar cap region; instead,
the plasma diffuses from the cleft onto field lines that map back
into the sides of the plasma sheet. A diffusion process (not
explainable in terms of frozen field convection except through
some modification like "foot-dragging') is indicated by the
observed decrease in phase space density (Vasyliunas, 1972; Heikkila,
1972b; Hill, 1974). The dotted arrows curving back from the cleft
in Figure 1 are meant to indicate this diffusive entry followed
by convection along the oval (Hill and Dessler, 1971; Frank and
Gurnett, 1971).

Let us consider some important processes in the plasma sheet,
in order to see if this model is consistent. As the name implies
the plasma sheet is a region in which there is a relatively high
density of hot plasma, higher than that in the high latitude lobes
of the magnetotail, but lower than in the magnetosheath (Akasofu
et al., 1973). We assume that it is the plasma sheet particles
that are energized and precipitated to produce auroras. No matter
what the mechanisms may be, or what language is used to describe
those mechanisms, the immediate energy source must be a cross-tail

electric field. There is ample evidence for such a field, existing
at all times, with a dawn-dusk polarity (Gurnett, 1972; Heppner,
1972; Mozer and Lucht, 1974).

If this field is to deliver electromagnetic energy the integral
of the Poynting vector over a closed surface surrounding the plasma
sheet must be negative, where

$$\oint_s (\underline{E} \times \underline{H}) \cdot d\underline{S} = -\int (\underline{H} \cdot \frac{\partial \underline{B}}{\partial t} + \underline{E} \cdot \frac{\partial \underline{D}}{\partial t}) \, dT - \int \underline{E} \cdot \underline{J} \, dT \qquad (2)$$

The first term on the right is the rate of change of the stored
electric and magnetic energies within the volume T; in keeping
with the merging theories in which curl $\underline{E} = -\partial \underline{B}/\partial t$ is set equal to
zero, we will consider only quasi-steady state conditions, and
ignore this term. The remaining term represents a sink of energy
if the integral of $\underline{E} \cdot \underline{J}$ is positive; this energy must then reappear
as kinetic energy of the plasma sheet particles. In the reconnection
theories $\underline{E} \cdot \underline{J}$ is necessarily positive (Sonnerup, 1970), and thus it
has the correct sign for particle energization. In our model,
however, reconnection could not energize auroral particles since
the X-line does not occur in the plasma sheet.

Regrettably the phrase 'electric field created by merging' is
often used in the literature; that phrase implies that the merging
process is a generator, but the positive sign is appropriate to a
load (Heikkila, 1975). If energy is dissipated in a load, then
it must be generated somewhere. The need for such a generator,
and its location and mechanism, have not been discussed in the
literature on merging. It's implied location is in the solar wind;
however, the existence of the bow shock precludes that possibility,
by definition.

In order that there can be particle energization in the plasma
sheet there must be a flow of electromagnetic energy into it.
This requires an electric current as well as an electric field, as
shown by equation 2 (see also the short article by Vasyliunas,
1968). According to Ohm's law,

$$\underline{J} = \sigma_{\parallel} \underline{E}_{\parallel} + \sigma_p \underline{E}_{\perp} + \sigma_{\parallel} \underline{B}/B \times \underline{E} \qquad (3)$$

with the usual notation for conductivities and other electromagnetic
quantities. Then

$$\underline{E} \cdot \underline{J} = \sigma_{\parallel} E_{\parallel}^2 + \sigma_p E_{\perp}^2 > 0 \qquad (4)$$

In the steady state $\nabla \cdot \underline{J} = 0$ and $\underline{E} = -\nabla\phi$, and hence $\underline{E} \cdot \underline{J} = -\nabla \cdot (\underline{J}\phi)$.
Then

$$\int \underline{E} \cdot \underline{J} \, dT = -\int \nabla \cdot (\underline{J}\phi) \, dT = -\oint \underline{J}\phi \cdot d\underline{S} > 0 \qquad (5)$$

This can be of the correct sign for a load only if current
enters (is negative) at regions of high polarity, and leaves
it at low. Laboratory experience with resistive circuits is
of course in harmony with this conclusion. A region in which
electromagnetic energy is created out of some other form of
energy, such as kinetic energy of plasma particles, has the current
directions reversed; this may be appreciated by considering a
battery.

Such basic considerations show that no matter what the
mechanisms might be, energy dissipation in auroral and magnetospheric
processes must be describable in terms of an electric circuit,
with a current generator feeding a load. We have only to identify
the source (not necessarily the mechanism) for the energy, as well
as the sink, in order to be able to draw a complete circuit diagram
(Heikkila, 1974). Of course the detailed nature of the processes
must be explained before we can claim a complete understanding of
the situation. However, it is regrettable that in many theoretical
discussions of auroral and magnetospheric processes the distinction
between generators and loads, which is essentially a distinction
between cause and effect, is often confused, if it is made at all.

We can thus conclude that the energization of plasma sheet
particles, whatever the mechanism may be, requires an electric
current in the same direction as the electric field. In effect,
the plasma sheet particles can be energized by any mechanism which
permits them to fall in the electrostatic field, so that they can
gain energy from it; in doing so the particles will tend to reduce
the strength of the electric field, and so it must be maintained
by the generator. Two such mechanisms are curvature and gradient
drift of particles on closed field lines, as shown by equation 1.
These lead to longitudinal and transverse energization respectively,
of both negative and positive particles (Heikkila, 1974). The
existence of these two mechanisms is the cause of two kinds of
auroras, the high latitude discrete arcs and the lower latitude
continuous aurora.

This dawn-dusk current is largely responsible for creating
the reversal in the magnetic field from antisolar in the southern
lobe to sunward in the northern lobe of the magnetotail. This
current is spatially continuous with the ring current closer to
the earth, although the current strength will vary with particle
density as a function of radial distance. In effect this westward current inflates the magnetosphere, stretching the plasma
sheet into the long tail-like shape.

Suppose that somewhere there is an end to the plasma sheet.
Then there is also an end to the sheet current, and it is easy
to verify that just beyond this end the sheet current contributes

a northward magnetic field. We can think of this neutral sheet
current and the magnetopause currents as forming a solenoid; the
resulting magnetic field in effect emerges from the southern part
of the solenoid, turns around, and returns to the earth inside
the northern part.

The magnetic field strength just beyond the end of the plasma
sheet must be somewhat larger than inside the plasma sheet, since
the diamagnetic effect of the plasma is not present. This increase
can also be seen by a pressure balance argument. This region may
be considered as a magnetic bottle containing the plasma sheet.
This model is therefore quite consistent with the requirements for
containment of the plasma sheet.

Still further out the contribution of the magnetotail currents
to the total magnetic field will decrease. If no other local sources
of magnetic field were present then as the higher order terms fall
off the field would become approximately dipolar. To this is
added the IMF, producing an X-line in the manner demonstrated by
Cowley (1973) and Stern (1973). No local electric current is
required for the formation of the X-line. If there are some local
currents, such as the tail end of the magnetopause current, then
the structure of the magnetic field would be correspondingly
modified.

We can consider the plasma sheet currents from another point
of view, namely that of providing a $\underline{J} \times \underline{B}$ force on the plasma.
Throughout the plasma sheet this force points toward the earth,
and is just the force required to compress the plasma, and to
prevent the plasma from expanding and escaping. The only other
force available is the Coulomb force, but that is in general
negligible compared with the Lorentz force. This way of describing
the force on the plasma is completely equivalent to saying that
the magnetic pressure (tensor) balances the plasma pressure
(Schmidt, 1966).

The plasma sheet current arises largely from the gradient
drift of the particles. Again it is easy to see that this is in
the correct direction, since $\underline{B} \times \nabla B$ is westward. There will also
be a contribution from particles with smaller pitch angles under-
going curvature drift, and again in the westward direction.

Similarly, the gradient and curvature drifts on the open field
lines beyond the X-line are also westward, and the resulting
$\underline{J} \times \underline{B}$ force is antisunward as appropriate to outward plasma flow.
These relationships are also embodied in the reconnection model.

5. HIGH LATITUDE LOBES

Early observations suggested that the high latitude lobes of
the magnetotail were devoid of plasma, but more recently the
existence of weak fluxes of low energy particles has been demonstrated (Akasofu et al., 1973; Winningham and Heikkila, 1974).
It does not seem likely that these particles play any major role
in plasma sheet and (oval) auroral processes. Nevertheless, they
are of interest in their own right as indicators of magnetospheric
structure and plasma processes.

The most common type of particle flux observed at low altitudes
is a very weak flux of electrons with the same spectral shape as for
the cleft plasma; this has been called polar rain (Winningham and
Heikkila, 1974). While the spectral shape suggests a magnetosheath
origin for the polar rain, the low intensity indicates that the
manner of entry is not simply flow or convection; some kind of
stochastic or diffusive process must be involved in the entry,
just as for the plasma sheet particles. Looking at Figure 1 we can
make the guess that the magnetosheath electrons are scattered
along the downstream magnetopause; electrons with sufficiently
small pitch angles can then follow the field lines down to low
altitudes, being convected to some extent across the polar cap at
the same time by the polar cap electric field. Magnetosheath
protons have a much more directed velocity than do the electrons,
and would not therefore be scattered as easily; this may explain
the absence of protons in the polar rain.

It is significant that the polar rain intensity usually
decreases a short distance poleward of the nighttime oval auroras.
This observation is consistent with the gap between the end of the
plasma sheet and the limit of closed field lines, as shown in
Figure 1. It is not consistent with precipitation along a slow
mode shock from the X-line.

Auroral and particle observations show that a second type of
flux is present over the polar caps during moderately quiet
conditions; this is the 'polar shower' responsible for the sun-
aligned auroral arcs (Winningham and Heikkila, 1974). Again
there are no protons accompanying the electrons. The electrons
are more energetic than the cleft electrons. Assuming that the
origin of the shower electrons is the magnetosheath plasma, then
some kind of energization mechanism must be operating; it is one
which raises the spectral peak up to a few kev at times, and also
produces a field aligned pitch angle distribution. All of these
observations are consistent with acceleration by a parallel
electric field. That such a parallel field is not normally present
over large regions of the polar cap is shown by the lack of change
in the polar rain spectrum, as compared with the cleft or magnetosheath spectrum, and also by the escape of atmospheric photoelectrons

(Winningham and Heikkila, 1974). The location and cause of such a parallel electric field in the shower regions has not been explained. It may, perhaps, be of interplanetary origin, unlike the field inside the plasma sheet which we believe to be generated within the magnetospheric system.

A third type of polar cap flux, called a polar squall (Heikkila, 1972b; Winningham and Heikkila, 1974), is observed only during geomagnetic storms. It will not be considered here.

Recent observations by the Vela, Heos, and Hawkeye satellites have shown the existence of a boundary layer or mantle of plasma just inside the magnetopause (Hones et al., 1972; Rosenbauer et al., 1975; Frank, 1975, private communication). This plasma is apparently of magnetosheath origin in view of the similarities in spectral shapes and flow velocities. However, the mantle plasma does usually have lower temperature, density, and flow velocity than the nearby magnetosheath plasma. This again is indicative of a diffusive process of entry probably at the high latitude edge of the cleft (Rosenbauer et al., 1975).

At the present time there is no indication of what (if any) role this mantle plasma might have in plasma sheet and auroral processes. There is a possibility that this plasma somehow gets into the plasma sheet; it is a low beta plasma, and its flow can be controlled by the magnetic and electric fields arising from other causes. It is also possible that the mantle plasma somehow manages to leave the magnetotail near the X-line.

Somewhere in this boundary region there may also be older magnetospheric plasma circulating inside the magnetosphere (Kavanagh et al., 1968). The dawn-dusk electric field would convect this plasma back out to the dayside magnetopause, from where it must exit somewhere. According to the merging theory this removal would occur over the lobes of the magnetotail in the slow mode shocks. If instead of a tangential electric field at the dayside magnetopause there is a perpendicular one, then the removal would be out along the equatorial boundary region. It is thus possible that the high and low latitude portions of the boundary layer have different origins, and thus comparison of observations taken simultaneously in these two regions should provide clues for resolving some of the basic physics involved.

6.   LOWER LATITUDE BOUNDARIES

There is a pitch angle dependent trapping boundary for energetic particles indicated by the dashed lines in the plasma sheet in Figure 1 (Roederer, 1969). This trapping boundary is the outer

limit of stably trapped particles. It will also depend on the
structure of the electric field, to an extent that depends on the
energy and pitch angle of the particles in question. While this
is an interesting phenomenon, it has little to do with primary
auroral processes which take place on higher L-shells.

Similarly the inner boundary of the plasma sheet is of
secondary importance to the creation of auroras. It is associated
with the removal of particles energized by plasma sheet processes
taking place further out. It has been discussed extensively in
the literature, and while many interesting details remain to be
resolved there does not appear to be any basic difficulty involved.

7. SUMMARY

Serious questions must be faced as a result of recent observations (McDiarmid et al., 1975a, b; Venkatarangan et al., 1975)
regarding the position of the last closed field line in relation
to plasma boundaries such as the dayside magnetopause and the
outer boundary of the plasma sheet. These findings are embodied
in our revision of Dungey's model of the magnetosphere, shown in
Figure 1. The apparent large separation in both regions is not
compatible with current ideas based on the theory of merging
or reconnection of magnetic field lines.

Another key question, that of the implied power dissipation
at the dayside magnetopause (Heikkila, 1975) comes from consideration
of $\underline{E} \cdot \underline{J}$. That this question has not been raised previously is
indicative of a weakness in the magnetohydrodynamic or fluid
formulation of the theory. In the MHD theory $\underline{E}$ is replaced by
$-\underline{V} \times \underline{B}$, and $\underline{J}$ is replaced by $\nabla \times \underline{B}/\mu_o$. Thus the simple quantity
$\underline{E} \cdot \underline{J}$ is replaced by the more cumbersome expression $- \underline{V} \times \underline{B} \cdot \nabla \times \underline{B}/\mu_o$
and its significance is lost.

The fluid theory fosters a tendency to regard the electric
field as secondary, rather than primary, in importance. For example,
the phrase "electric field due to merging", often used in the
literature, suggests a lack of attention to causality. Similarly,
the phase "electric field due to convection" is appropriate to a
dynamo region but not to a load like the magnetosphere with
positive $\underline{E} \cdot \underline{J}$.

While the fluid theory is often adequate for magnetospheric
discussions it is nevertheless an approximation and as such it
should be used with caution (Hines, 1963; Heikkila, 1973). The
resolution of such serious questions as those discussed here needs
to be sought in more primitive terms.

REFERENCES

Akasofu, S.-I., E. W. Hones, Jr., S. J. Bame, J. R. Asbridge, and
    A. T. Y. Lui, Magnetotail and boundary layer plasmas at a
    geocentric distance of ∿ 18 R_E: Vela 5 and 6 observations,
    J. Geophys. Res., 78, 7257, 1973.

Axford, W. I., and C. O. Hines, A unifying theory of high-latitude
    geophysical phenomena and geomagnetic storms, Can. J. Phys.,
    39, 1433, 1961.

Block, L. P., Potential double layers in the ionosphere. Cosm.
    Electrodyn., 3, 349, 1972.

Chapman, S., and V. C. A. Ferraro, A new theory of magnetic storms,
    Terr. Magn. Atmos. Elect., 36, 77, 1931.

Cowley, S. W. H., A qualitative study of the reconnection between
    the Earth's magnetic field and an interplanetary field of
    arbitrary orientation, Radio Science, 8, 903, 1973.

Dungey, J. W., Interplanetary magnetic field and the auroral zones,
    Phys. Rev. Letters, 6, 47, 1961.

Eather, R. H., The auroral oval - A reevaluation, Rev. Geophys.
    Space Phys., 11, 155, 1973.

Evans, L. C., and E. C. Stone, Electron polar cap and the boundary
    of open geomagnetic field lines, J. Geophys. Res., 77, 5580,
    1972.

Fälthammar, C.-G., Motion of charged particles in the magnetosphere,
    Chapter 9, Cosmical Geophysics, Eds. A. Egeland, O. Holter,
    and A. Omholt, Universitetsförlaget, Oslo, Norway, 1973.

Frank, L. A., Plasma in the earth's polar magnetosphere, J. Geophys.
    Res.,76, 5205, 1971.

Frank, L. A. and D. A. Gurnett, Distributions of plasmas and electric
    fields over the auroral zones and polar caps, J. Geophys. Res.,
    76, 6829, 1971.

Fredricks, R. W., F. L. Scarf and C. T. Russell, Field-aligned
    currents, plasma waves, and anomalous resistivity in the
    disturbed polar cusp, J. Geophys. Res., 78, 2133, 1973.

Gold, T., Motions in the magnetosphere of the earth, J. Geophys.
    Res., 64, 1219, 1959.

Gurnett, D. A., Electric field and plasma observations in the magnetosphere, Critical Problems of Magnetospheric Physics, Ed. E. R. Dyer, published by the IUCSTP Secretariat, National Academy of Sciences, Washington, D. C., 1972.

Heikkila, W. J., Is there an electrostatic field tangential to the dayside magnetopause and neutral line, accepted for publication in Geophys. Res. Letters, April, 1975.

Heikkila, W. J., Outline of magnetospheric theory, J. Geophys. Res. 79, 2496, 1974.

Heikkila, W. J., Critique of fluid theory of magnetospheric phenomena, Astrophys. Space Sci.,23, 261, 1973.

Heikkila, W. J., The morphology of auroral particle precipitation, Space Res. XII, 1343, 1972a.

Heikkila, W. J., Penetration of particles into the polar cap and auroral regions, in Critical Problems of Magnetospheric Physics, Ed. E. R. Dyer, pp. 67-82, published by the IUCSTP Secretariat, National Academy of Sciences, Washington, D. C., 1972b.

Heikkila, W. J., and J. D. Winningham, Penetration of magnetosheath plasma to low altitudes through the dayside magnetospheric cusps, J. Geophys. Res., 76, 883, 1971.

Heppner, J. P., Electric fields in the magnetosphere, Critical Problems of Magnetospheric Physics, Ed. E. R. Dyer, published by the IUCSTP Secretariat, National Academy of Sciences, Washington, D. C., 1972.

Hill, T. W., Origin of the plasma sheet, Rev. Geophys. Space Phys., 12, 379, 1974.

Hill, T. W. and A. J. Dessler, Plasma-sheet structure and the onset of magnetospheric substorms, Planet. Space Sci., 19, 1275, 1971.

Hines, C. O., The energization of plasma in the magnetosphere: Hydromagnetic and particle-drift approaches, Planet. Space. Sci., 10, 239, 1963.

Hones, E. W., Jr., J. R. Asbridge, S. J. Bame, M. D. Montgomery, S. Singer, and S.-I. Akasofu, Measurements of magnetotail plasma flow made with Vela 4B, J. Geophys. Res., 77, 5503, 1972.

Johnson, F. S., The gross character of the geomagnetic field in the solar wind, J. Geophys. Res., 65, 3049, 1960.

Kavanagh, L. D., Jr., J. W. Freeman, Jr., and A. J. Chen, Plasma flow in the magnetosphere, J. Geophys. Res., 73, 5511, 1968.

McDiarmid, I. B., J. R. Burrows, and E. E. Budzinski, Average characteristics of magnetospheric electrons (150 eV to 200 keV) at 1400 km, J. Geophys. Res., 80, 73, 1975a.

McDiarmid, I. B., J. R. Burrows, and E. E. Budzinski, Particle properties in the dayside cleft, to be published in J. Geophys. Res., 1975b.

Morfill, G. E. and M. Scholer, Study of magnetosphere using energetic solar particles, Space Sci. Rev.,15, 267, 1973.

Mozer, F. S. and P. Lucht, The average auroral zone electric field, J. Geophys. Res., 79, 1001, 1974.

Nishida, A. and E. W. Hones, Jr., Association of plasma sheet thinning with neutral line formation in the magnetotail, J. Geophys. Res., 79, 535, 1974.

Petschek, H. E., The Solar Wind, Ed. R. J. Mackin, Jr. and M. Neugebauer, Pergamon Press, New York, New York. p. 257, 1966.

Rich, Frederick J., David L. Reasoner, and William J. Burke, Plasma sheet at lunar distance: Characteristics and interactions with the lunar surface, J. Geophys. Res.,78, 8097, 1973.

Roederer, J. G., Quantitative models of the magnetosphere, Rev. Geophys., 7, 77, 1969.

Rosenbauer, H., H. Grunwaldt, M. D. Montgomery, G. Paschmann, and N. Sckopke, Heos 2 plasma observations in the distant polar magnetosphere - the plasma mantle, accepted for publication in J. Geophys. Res., 1975.

Schmidt, G., Physics of High Temperature Plasmas, Academic Press, 1966.

Sonnerup, B. U. O., The reconnecting magnetosphere, Magnetospheric Physics, 23-33, D. Reidel Publ. Co., Dordrecht-Holland, 1974.

Sonnerup, B. U. O., Magnetic-field re-connexion in a highly conducting incompressible fluid, J. Plasma Phys.,4, 161, 1970.

Sonnerup, B. U. O., and B. G. Ledley, Magnetopause rotational forms, J. Geophys. Res., 79, 4309, 1974.

Stern, D. P., A study of the electric field in an open magnetospheric model, J. Geophys. Res.,78, 7292, 1973.

Vasyliunas, V. M., Magnetospheric structure, field line merging and plasma convection, contained in article by Martin Walt, Magnetospheric Physics, 3-19, Ed. B. M. McCormac, 1974. D. Reidel Publ. Co., Dordrecht-Holland, 1974.

Vasyliunas, V. M., Theoretical models of magnetic field line merging, Rev. Geophys. Space Phys., 13, 303, 1975.

Vasyliunas, V. M., Entropy, Liouville's theorem and particle acceleration, paper presented at the Solar Terrestrial Relations Conference, Univ. of Calgary, Calgary, Alberta, Canada, August 28 to September 1, 1972.

Vasyliunas, V. M., Discussion of paper by Harold E. Taylor and Edward W. Hones, Jr., 'Adiabatic motion of auroral particles in a model of the electric and magnetic fields surrounding the earth', J. Geophys. Res., 73, 5805, 1968.

Venkatarangan, P., J. R. Burrows and I. B. McDiarmid, On the angular distributions of electrons in 'inverted V' substructures, J. Geophys. Res., 80, 66, 1975.

Whalen, B. A. and I. B. McDiarmid, Observations of magnetic-field-aligned auroral-electron precipitation, J. Geophys. Res., 77, 191, 1972.

Winningham, J. D., Characteristics of magnetosheath plasma observed at low altitudes in the dayside magnetospheric cusps, Earth's Magnetospheric Processes, 68-80, Ed. B. M. McCormac, D. Reidel Publ. Co., Dordrecht-Holland., 1972.

Winningham, J. D. and W. J. Heikkila, Polar cap auroral electron fluxes observed with ISIS-1, J. Geophys. Res., 79, 949, 1974.

This work was supported by NASA Grants NGR 44-004-150 and NGR 44-004-124, and by AFCRL Contract F-19628-75-C-0032.

AURORAL ELECTRON BEAMS NEAR THE MAGNETIC EQUATOR

Carl E. McIlwain

Physics Dept., University of California, San Diego

La Jolla, California 92037 USA

ABSTRACT

Intense beams of electrons travelling parallel to the local magnetic field have been observed at a magnetic latitude of 11° and a radial distance of 6.6 R_e. The distribution function for electrons travelling within 8° of the field line direction is typically flat or slightly rising up to a break point beyond which it decreases as v^{-5} to v^{-10}. The energy corresponding to the break point velocity is usually between 0.1 and 10 keV. These beams are found to occur on closed field lines at the inner edge of the plasma sheet and thus at the root of the earth's magnetotail. Beams with break point energies greater than 2 keV seem to occur only within the first 10 minutes after the onset of hot plasma injection associated with a magnetospheric substorm. Although the origin and destiny of these electrons is as yet unknown, considerations of the total energy and the number of particles transported guarantee that they must play a dominant role in many key magnetospheric processes.

INTRODUCTION

Before any direct particle measurements, ground-based observations of auroral arcs had already indicated that most of the auroral light was caused by energetic electrons entering the atmosphere from higher altitudes (Lenard, 1911; Störmer, 1955; Omholt, 1959). Since the first measurement of intense electron fluxes in and above an auroral arc (McIlwain, 1960), there have been a large number of such observations, including some at altitudes of up to 2000 km. The implications of the energy and angular distributions of these electrons are discussed in the contribution by D. Evans to these proceedings.

Many measurements have been made in the more distant regions
of the magnetosphere with instruments easily capable of sensing the
strong auroral electron fluxes, but no definite detection has been
reported. This has led to the conclusion (1) that they occur only
relatively near the earth, which in turn led to the tentative conclusion (2) that the acceleration processes must also occur only
near the earth. Data presented in the present paper show that the
first conclusion was incorrect, but that the second conclusion may
still be correct.

The previous non-observation is easy to explain in retrospect.
The high intensities occur only within a small (< .03 steradian)
solid angle centered on the magnetic field direction. A sensor
with poor angular resolution can thus underestimate the flux by
more than a factor of ten, and a sensor with good angular resolution
will rarely, if ever, be looking in the correct direction. In particular, there were no observations by the instruments on ATS-5
(DeForest and McIlwain, 1971) even though there must have been
thousands of opportunities in the first 3 years of operation. The
reason is simple. When auroral electron beams are present, the
magnetic field is always tilted between 20 and 70 degrees with
respect to the satellite's spin axis so that none of the sensors,
which are pointed parallel and perpendicular to the spin axis, is
ever properly aligned with the magnetic field.

The present observations tend to confirm the speculation by
Hones et al (1971) that the bumps sometimes found in the azimuthal
dependence of electrons measured by the Vela satellites at 18 earth
radii are due to field aligned electron beams. They assume that the
bumps occur when the magnetic field direction is included in the 6°
by 110° electron acceptance fan, and conjecture that a high resolution detector would measure 20 to 30 times higher fluxes.

THE AURORAL PARTICLES EXPERIMENT ON ATS-6

ATS-6 was launched in geosynchronous orbit (6.62 earth radii)
on May 30, 1974 and is being kept near 94° west longitude for about
1 year, after which it is to be moved to 35° east. The orbital
inclination is less than 2 degrees. At 94° west, the average magnetic latitude is $+10.5^{\circ}$, and it is estimated that variations in
the geographic latitude and the changing aspect to the solar wind
can make the distance from the magnetic equator as small as 6°
and as large as 15°.

The scientific package on this satellite, the "Environmental
Measurements Experiment", includes a good array of particle sensors
and an excellent magnetometer, but, unfortunately, no electric
field or high frequency wave sensors.

The University of California, San Diego (UCSD) auroral particles experiment on ATS-6 consists of 5 electrostatic analyzers.
These measure both ions and electrons over the range of 1 to
80,000 eV with an energy resolution of about 0.2E plus 2 eV full
width at half maximum. The analyzers are unique in two ways:
the analyzing plates are ovoidal, i.e. have different curvatures
in the parallel and perpendicular directions, and there is a
short focal length electrostatic lens preceding the 0.3 cm diameter Bendix Spiraltron sensors. The result is a large geometric
factor (over 0.001 cm^2 ster) in spite of the small sensor area
and, at the same time, good angular resolution: 2.8° by 7° for a
flat spectrum.

The spacecraft is three axis stabilized with the scientific
package mounted behind the 10-meter diameter parabolic antenna
which is normally kept pointed at the earth. To obtain the dependence of the plasma upon pitch angle, one pair of analyzers is
mechanically swept from north up to radially out and on to the
south covering a range of 220°. To obtain the dependence of the
plasma on the azimuthal angle (for flow and gradient determinations)
another pair of analyzers is swept back and forth from east up to
radially out and on to west before reversing. In each case the
motion is at 1.4 degrees per second so that a complete cycle takes
314 seconds. It is also possible to stop the sweeping motion so
that time variations in a particular direction can be studied.
The fifth analyzer is stationary and measures ions travelling
westwards.

The energy range is normally covered in 64 steps taken 4 per
second. In between these 16-second energy scans, the instrument
can be commanded to dwell on any of the 64 energies for 0, 1, 2, 4,
- - -, 128 seconds and to then follow the next scan by a dwell at
an energy step that 0, 1, 2, - - - 32 steps higher and to continue
this sequence 0, 1, 2, - - -, 64 times. This capability has proven
invaluable in studying fast time variations and sharp angular de-
pendences.

The sweeping of the "North/South" pair of analyzers takes the
look direction 13° to the west of north instead of directly through
north in order to avoid the large solar cell panels. At the 35°
east longitude location, the average magnetic field is within a
few degrees of this plane of motion. At the 94° west longitude
location, the field is usually about 10° away from this plane so
that particles with pitch angles less than 10° cannot be measured.
Field-aligned currents do, however, sometimes give an azimuthal
twist to the magnetic field so that smaller pitch angles can be
viewed.

Instantaneous pitch angles are determined using magnetometer
data kindly supplied by Dr. P. Coleman and Dr. R. McPherron of the

University of California, Los Angeles. Studies to determine the
offsets due to spacecraft magnetic fields are still in progress,
but the consistency of the particle data indicates that pitch
angle errors are typically less than 4° using present procedures.

The absolute fluxes given in this paper are provisional. In
particular, a known gradual rise of about a factor of two in the
electron efficiency below 1 keV has not been incorporated in the
present conversion algorithms.

FIRST DETECTION OF FIELD-ALIGNED FLUXES

Within the first day of operation, the UCSD plasma detectors
encountered three events involving intense beams of electrons
travelling along the local magnetic field direction heading toward
the northern auroral zone. The differential number flux measured
by the two electron sensors during one of these events is shown in
Figure 1. For a brief time, the ambient magnetic field was perturbed so that very small pitch angles could be sampled. It is
interesting to note that this fortuitous field perturbation could
be reasonably explained by assuming the spacecraft encountered a
sheet of electrons just like the ones being measured.

The spectrum at large pitch angles has both the shape and
intensity typically found in the midnight region at 6.6 earth radii
following each magnetospheric substorm (DeForest and McIlwain,
1971). The shape is close to Maxwellian with a temperature of
4000 eV, and the intensity corresponds to a density of 0.9 elec-
trons/cm^3. By contrast, the shape of the spectrum at small pitch
angles is unusually peaked, and the absolute differential intensity
at 7 keV is believed to be higher than any previous observation in
this region of space. On the other hand, this spectrum is both
less peaked and less intense than some of the spectra observed
(Albert, 1967; Evans, 1968A, 1968B; Westerlund, 1969; O'Brien and
Reasoner, 1971; Bryant, Courtier, and Bennett, 1972; Arnoldy, Lewis
and Isaacson, 1974) near the earth in association with auroral arcs.
The number flux at small pitch angles is 1.5×10^9 electrons/cm^2
s ster) compared with 1.8×10^8 at large angles. The energy fluxes
are 19. and 2.2 ergs/(cm^2 s ster) respectively. As will be shown,
the beam occupies only a very small solid angle and thus does not
make a large contribution to the omnidirectional fluxes. When
these electrons approach the northern auroral zone, their pitch
angles become much larger so that they would deposit energy at
about the rate of 100 ergs/cm^2 sec provided they are not repulsed
or further accelerated by parallel electric fields. This energy
flux is sufficient to produce bright auroral light emissions.

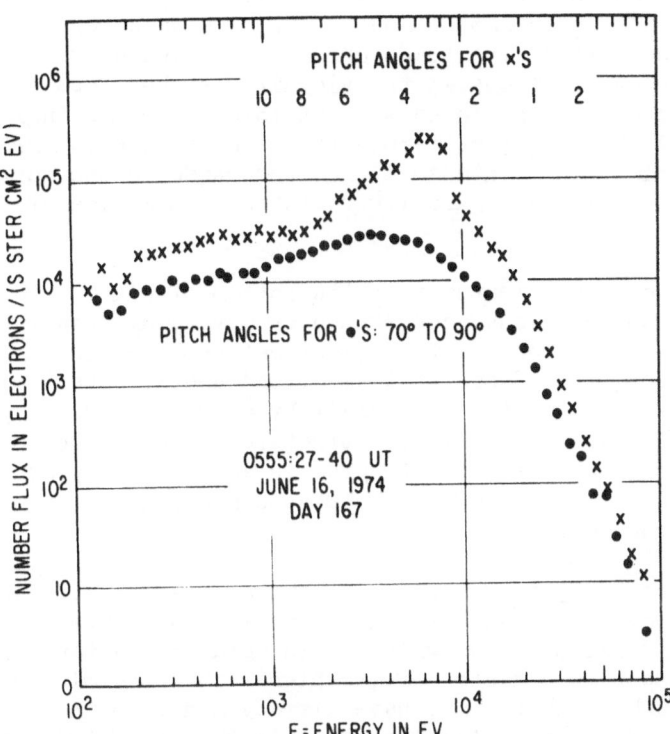

Figure 1 The differential number flux of electrons travelling close to the magnetic field direction measured during the first minutes of a magnetospheric substorm plasma injection.

Figure 2 is an energy-time (and angle) spectrogram of data taken by the pair of detectors sweeping past north and south each 5.23 minutes. The magnetic field is inclined between -40 and -65 degrees to the horizontal plane during this time period (as is normal for this region of space at the root of the geomagnetic tail and at a magnetic latitude of +11 degrees). The detectors thus view particles going along the magnetic field twice each sweep cycle (at about half-way between vertical and south). Particles travelling antiparallel to the field (heading for the magnetic equator and on toward the southern hemisphere) come up from below the spacecraft and cannot be observed. The dark vertical lines in the electron part of the spectrogram are produced by the deficiency in high energy electrons in the vicinity of the "loss cone". Also bright vertical lines can be seen at lower energies indicating the presence of what might be termed the "source cone". Arrows mark the three times that intense fluxes were seen in the source cone. The spectra shown in Figure 1 were taken at the time of the second arrow.

The two bright features at the bottom of Figure 2 are made by low energy ambient ions accelerated by a varying negative spacecraft potential. Most of the lowest energy electrons are believed to be spacecraft produced photoelectrons and secondary electrons that are returned to the spacecraft by a negatively charged sheath (Whipple, 1975). The periodic structure at the bottom of the electron portion of the spectrogram is an artifact produced by a 16-second dwell at 24 keV every 128 seconds (and by an unfinished computer program).

Figures 3 and 4 show the portions of the distribution function corresponding to the spectra in Figure 1. The shape seems to be one which should excite waves which would tend to flatten the peak. The peak is less than a factor of two higher than the minimum, indicating that such processes may have already modified the velocity distribution. The statistical accuracy in the vicinity of the peak is quite good. As will be shown later, the irregularities are probably due to fast time variations and not to fine structure in the instantaneous velocity distribution. It is not impossible that the distribution is in fact flat, with the relative minimum being an artifact of time and angle variations.

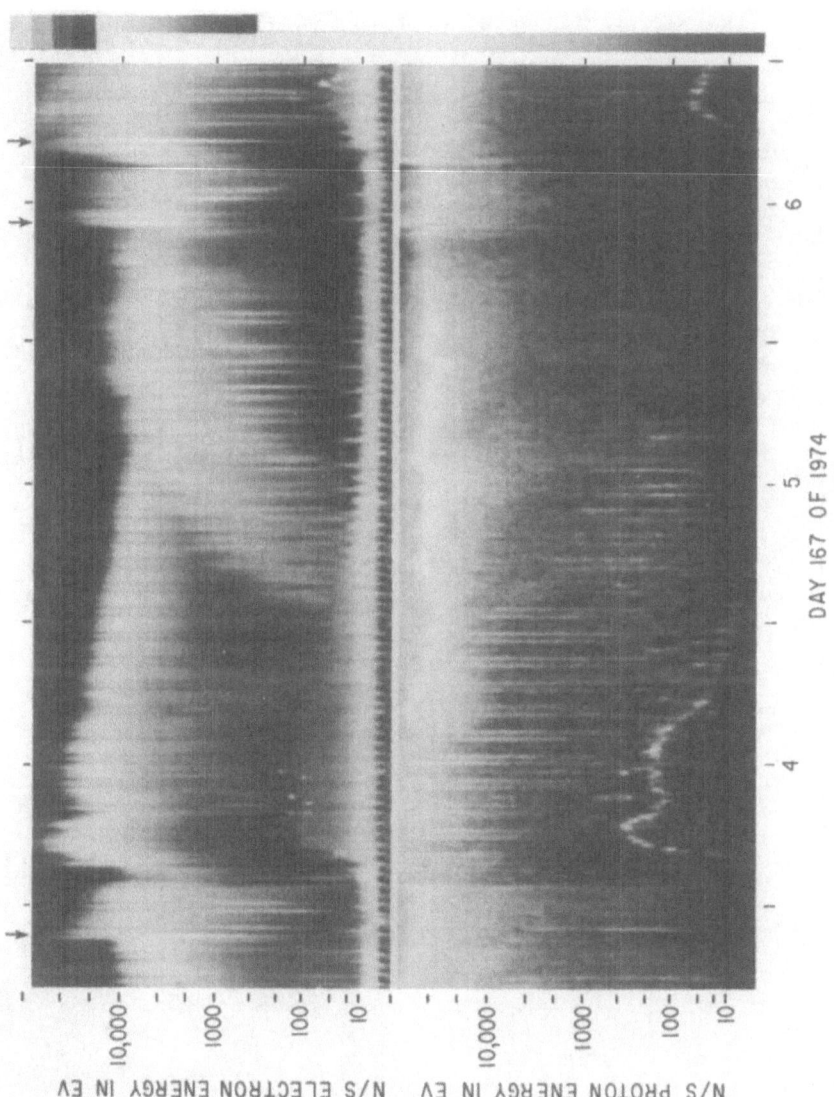

Figure 2 An energy-time spectrogram of 3 hours data obtained on June 16, 1974 by the North/South analyzers of the UCSD experiment on ATS-6. The periodic structure is due to the analyzers mechanically sweeping back and forth over a 220° range each 314 seconds.

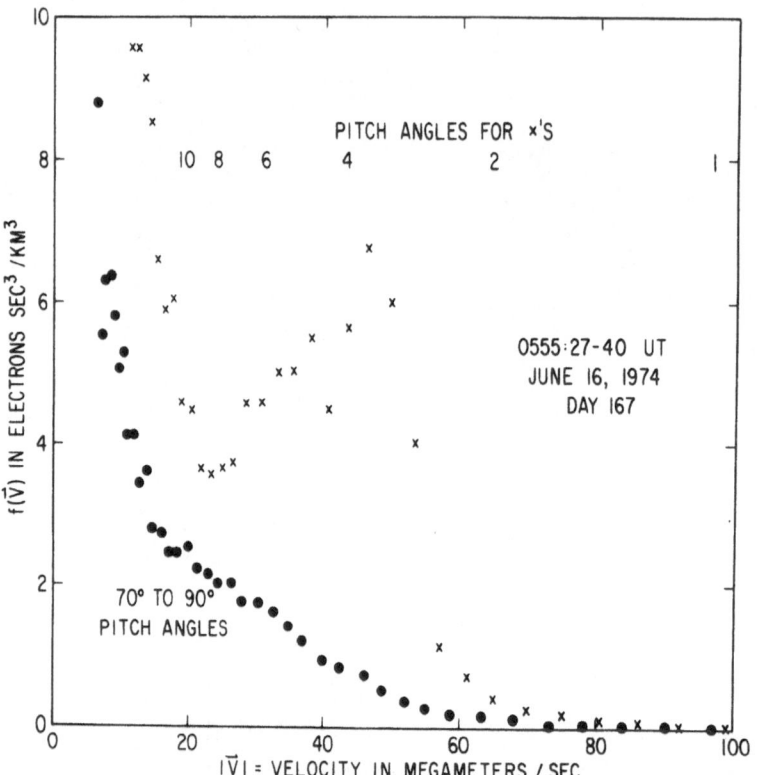

Figure 3 The portions of the distribution function $f(\vec{v})\ d^3x d^3v$, corresponding to the spectra shown in Figure 1.

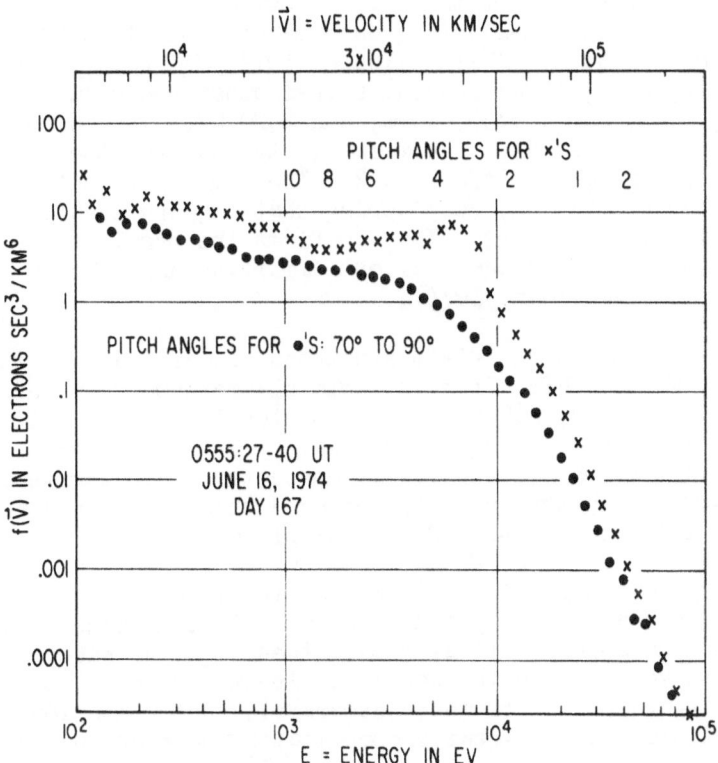

Figure 4   The distribution function replotted using logarithmic scales.

CONTEXT IN WHICH THE BEAMS ARE FOUND

Figure 5 shows the velocity distributions measured during an event on July 20, 1974. Again, the distribution is relatively flat or slightly rising up to a break point, beyond which it decreases very rapidly -- in this case approximately as $E^{-3.5}$ or v^{-7}.

To show the full context of this event, 24 hours of data from both ATS-6 and ATS-5 are shown in Figures 6 and 8. ATS-5 is also in synchronous orbit at a longitude of 105° west, which is 11° west of ATS-6. Analysis of these spectrograms in terms of drifting plasma clouds injected during each magnetospheric substorm (DeForest and McIlwain, 1971; McIlwain, 1972; McIlwain, 1974; Kamide and McIlwain, 1974) indicates that substorms occurred at about 0020, 0440, 0520, 0635, 1010, and 2020 UT. The number and intensity of the first five substorms is typical for the observed K_p magnetic index values of around 2. The similarity between the two data sets is striking, considering that the spacecraft are over a thousand kilometers apart. This confirms the previous conclusion that substorm injections occur almost simultaneously over a wide region of space (McIlwain, 1974).

ATS-6 encountered the first electrons associated with the plasma sheet at 0350 UT. If this "cold plasma" boundary were stationary in local time, ATS-5 would encounter it 4 x 11° = 44 minutes later, at 0434 UT (Mauk and McIlwain, 1974). The ATS-5 spectrogram shows that the boundary was encountered as expected, but that the 0440 UT substorm rapidly replaced this low energy residual of previous injections with a fresh hot plasma.

The very low energy ions visible at the lower left of Figure 6 can be used to measure the plasma flow velocity. Between 0 and 2 hours UT (18 and 20 hours local time), these ions do have a measureable drift velocity in the westward direction (relative to the 3 km/sec eastward motion of the spacecraft). Determination of the magnitude of the flow is greatly complicated, however, by variable large amplitude transverse Alfvén waves which are common at these local times. In this case, their period was 3 to 4 second, and their perturbations of the flow velocity were often comparable to the average flow rate. These waves were presumably driven by the anisotropic energetic ion population which was present at that time.

Figure 7 is an expanded version of Figure 6 which shows that the plasma injected by the 0440 UT substorm was almost isotropic over the measured pitch angle range of 17° to 135°. By 0505 UT, however, the high energy electrons had already developed a pronounced deficiency around the loss cone. The 0520 UT event was either localized at some distance from the two spacecraft or was something different than a regular substorm. At 0518 UT, there was a

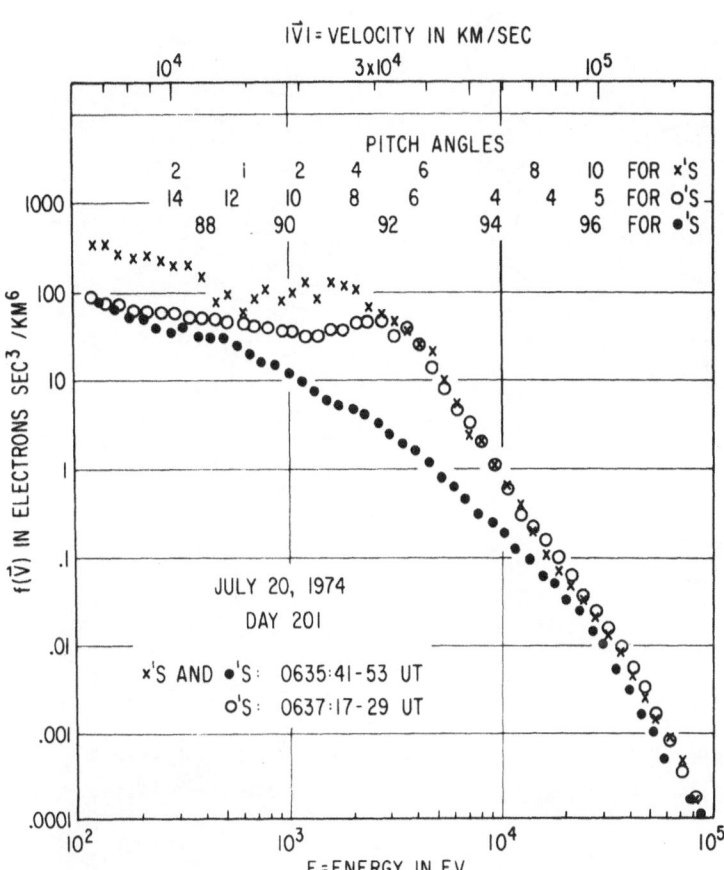

Figure 5 Distribution functions measured shortly after a substorm on July 20, 1974.

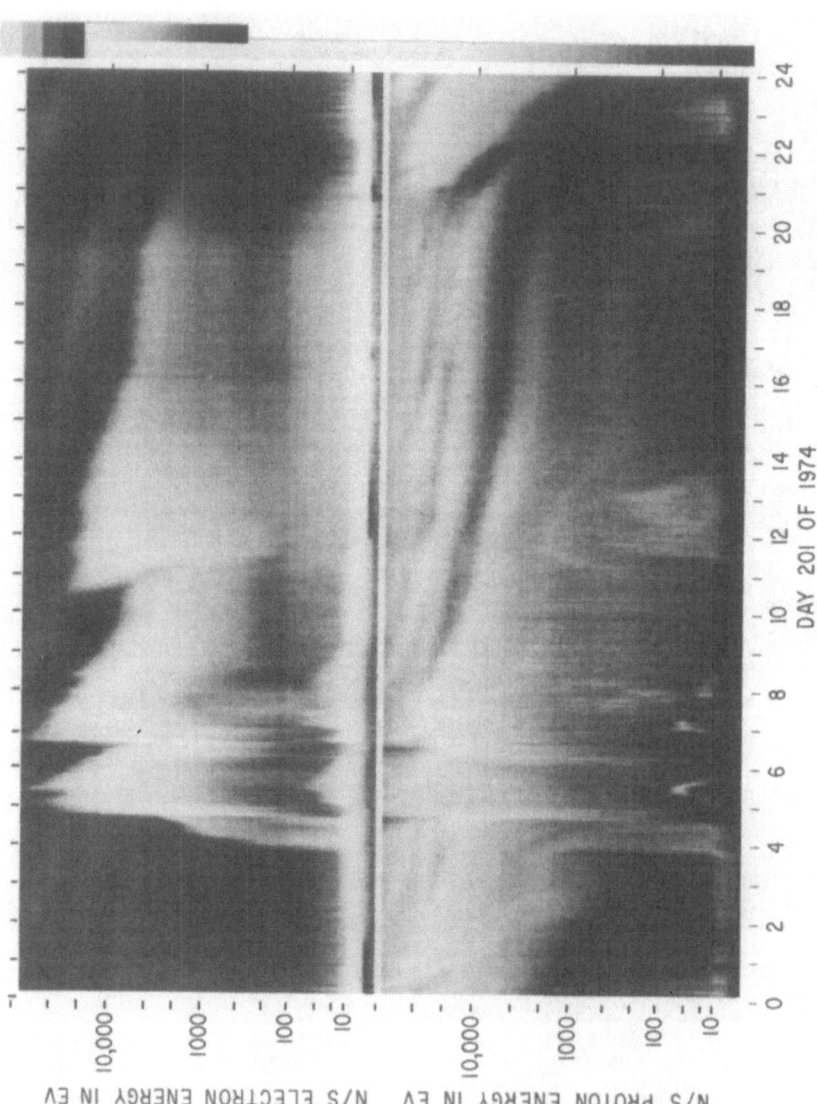

Figure 6. An energy-time spectrogram of 24 hours of data obtained on July 20, 1974 by the North/South analyzers on ATS-6. The structure due to the periodic motion of the analyzers is almost invisible on this time scale.

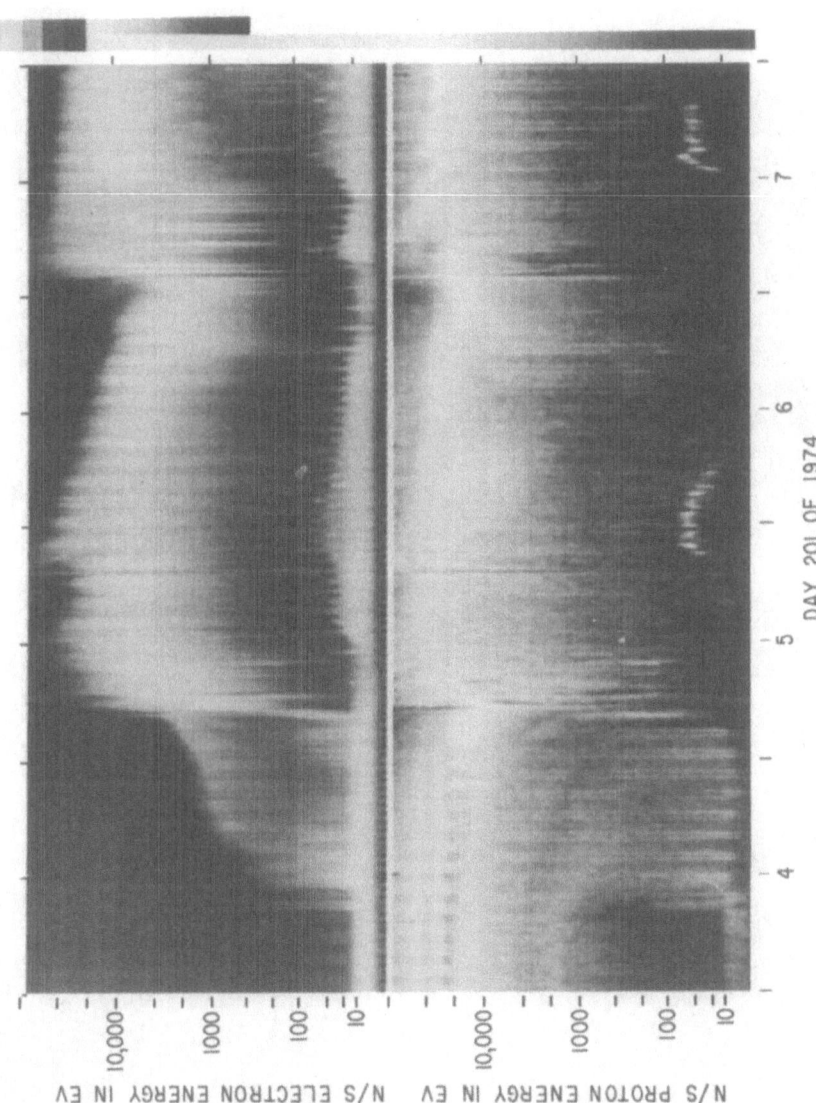

Figure 7 An energy-time spectrogram of 4 hours of data obtained on July 20, 1974 by the North/South analyzers on ATS-6. Details of particle behavior during the 0440, 0520 and 0635 substorms are visible.

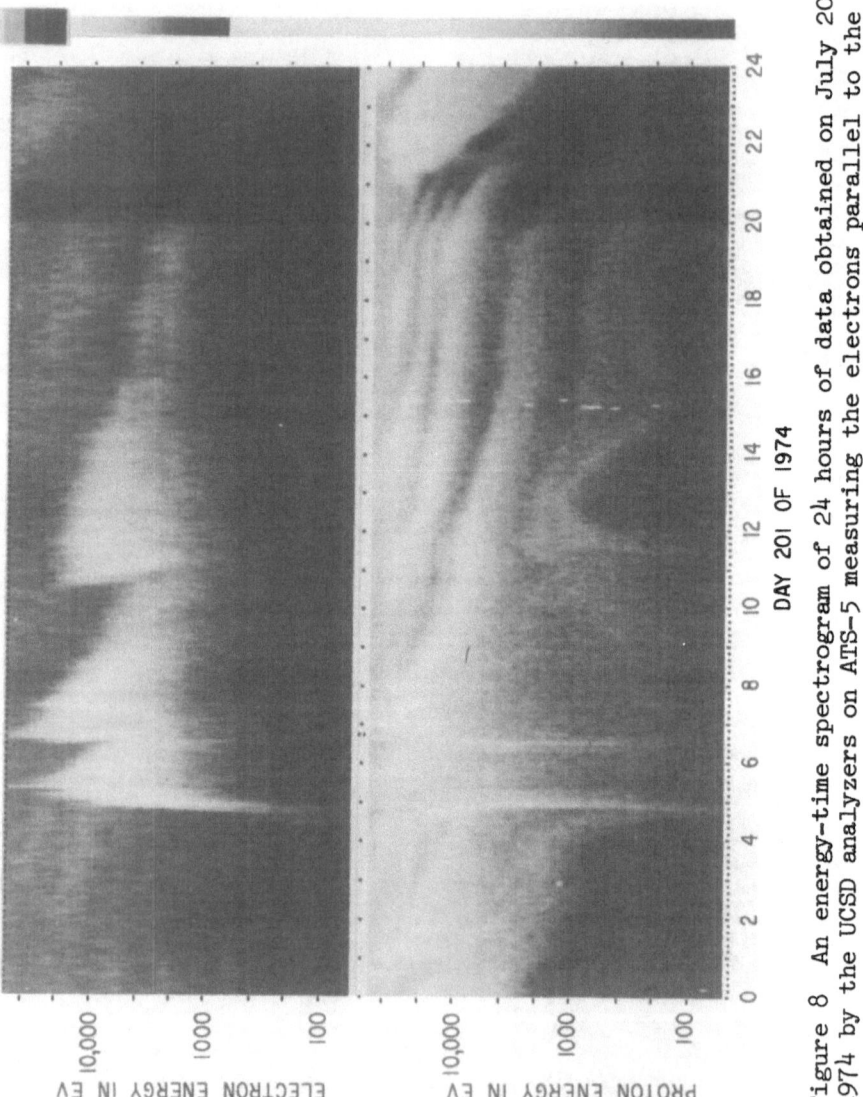

Figure 8 An energy-time spectrogram of 24 hours of data obtained on July 20, 1974 by the UCSD analyzers on ATS-5 measuring the electrons parallel to the spin axis and the ions perpendicular to the spin axis.

momentary factor of two change in both ion and electron fluxes and a simultaneous variation in the magnetic field. This suggests the data should be studied in terms of the passage of a solitary wave (soliton).

As can be seen in Figures 5 and 7, the 0635 event did produce a short lived intense beam of electrons. Within 15 minutes, however, the electron distribution was almost isotropic except for the usual empty loss cone region at high energies. This is about the time expected for the development of a loss cone when there is strong pitch angle diffusion (Schulz, 1974).

It is interesting to note that while the most energetic ions also develop an empty loss cone, the ions with energies less than 10 keV are enhanced at small pitch angles. In other words, their "source cone" region is filled. Field-aligned ion fluxes have also been observed near the ionosphere (Hultqvist et al, 1971).

Referring back to Figure 2, it can be seen that parallel beams of ions were present on day 167 with a slowly decreasing maximum energy between 0400 and 0500 UT. Around 0500 UT it can be seen that both parallel ions and electrons are present at low energies.

Simultaneous ion and electron beams at high energy, however, do not seem to occur. This strongly suggests that the beams are to be associated with parallel currents. Unfortunately, the net current cannot be measured using data taken at the 94° west longitude location, because particles with pitch angles around 180° are obscured by the spacecraft.

PITCH ANGLE DEPENDENCE AND RAPID TIME VARIATIONS

On July 3, 1974 (day 184), the auroral particles experiment was commanded to dwell successively at 1.6, 1.8, 2.1, and 2.4 keV for 128 seconds between the regular 16-second energy scans. Figure 9 shows a case in which the North/South detectors swept almost directly past the field line direction while dwelling at 1.6 keV. In the absence of a parallel electric field, particles with pitch angles less than 3° go to altitudes of less than 500 km in the northern auroral ionosphere. Ignoring an apparent offset of about 4°, the pitch angle distribution has a width at half maximum of about ± 10°. Figure 10 is a replot of the same data on a linear scale. Large fast fluctuations are present but only within the central ± 5°.

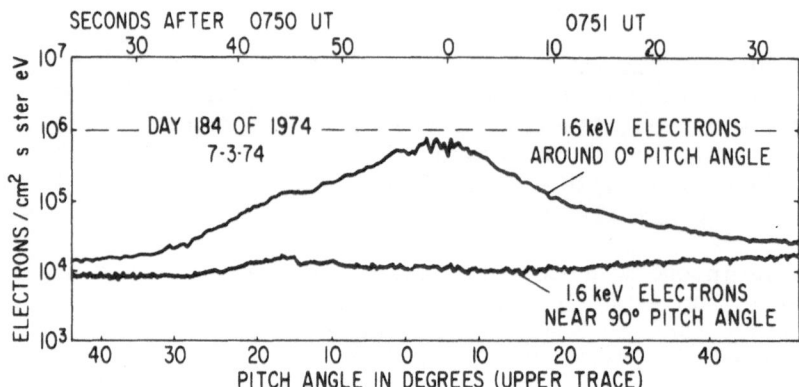

Figure 9   The time and pitch angle dependence of electrons measured during a long dwell at a fixed energy of 1.6 keV.

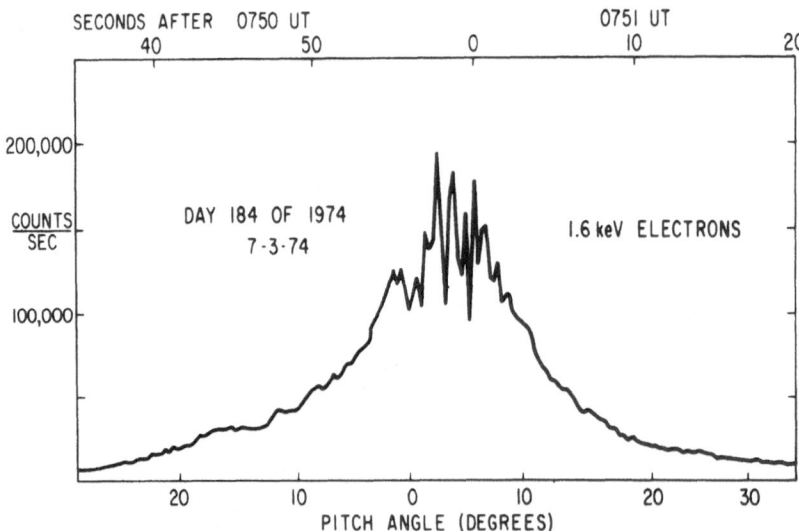

Figure 10   A replot of Figure 9 on linear scale to exhibit the fluctuations in the central ±5°.

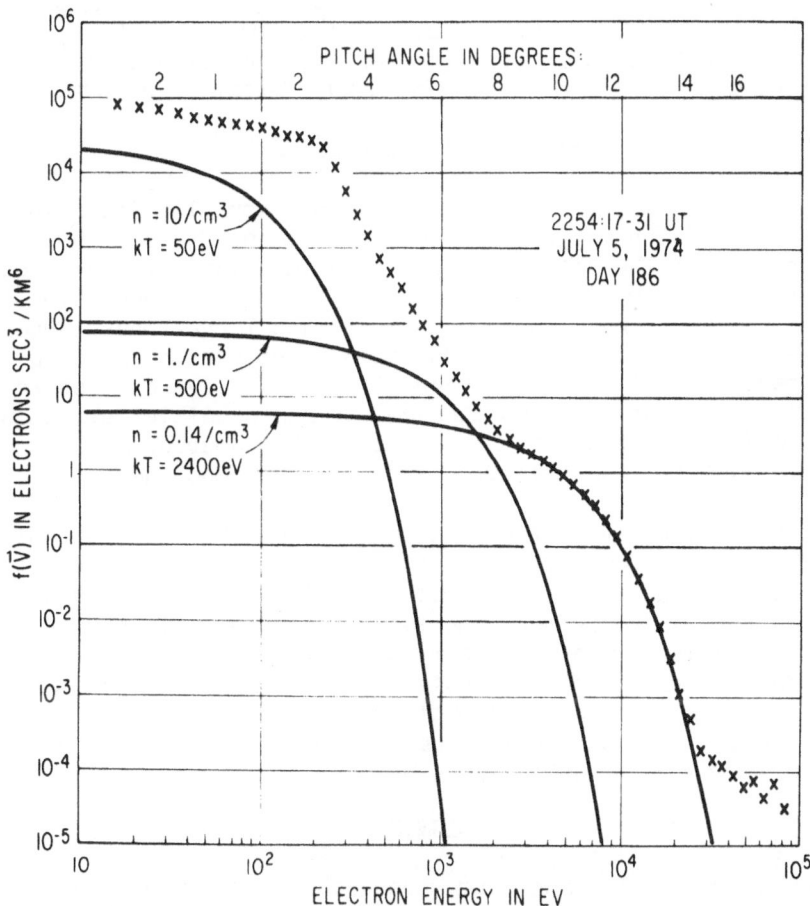

Figure 11 A distribution function measured at 17 hours local time
during a magnetic storm. Also shown are 3 isotropic Maxwellian
distributions. The 10/cm³ and 1/cm³ examples correspond roughly to
the distributions found in the magnetosheath and in the plasma sheet
respectively.

One possibility is to have an initial beam of field-aligned electrons which excites waves, which in turn scatter some of the electrons into neighboring pitch angles (where they can survive for longer periods of time). It seems difficult to account for all characteristics of the observed beams by means of simple convection driven Fermi acceleration (Axford, 1968). The addition of parallel electric fields to this process, however, might yield a viable theory (Fälthammar, 1972; Boström, 1974; Heikkila, 1974) In this case, the electrons in the neighboring pitch angles could be part of the parent population instead of scattered ones.

ORIGIN OF THE ELECTRONS

The distribution function of electrons normally found in the plasma sheet of the magnetotail reaches values of about 100 electrons sec^3/km^6. The distributions shown in Figures 4 and 5 could therefore have been produced by accelerated plasma sheet electrons. In other field-aligned events, however, much larger phase space densities have been observed. One such event was observed in the late afternoon during a large magnetic storm. Figure 11 shows that electrons with energies less than 200 eV had densities exceeding 10,000 sec^3/km^6. There are only two available sources of electrons with this density: the magnetosheath and the ionosphere.

At this point it must be remembered that the residence time of particles on the closed field lines in the 5 to 15 earth radii region can be many hours. This is quite long enough for substantial densities of ionospheric particles to accumulate by means of parallel current driven transport to great heights, and by concurrent wave scattering into trajectories with high altitude mirror points. The ATS-6 data provide strong evidence that this process does in fact occur, and that it may be operating almost continuously over a wide region of space.

CONCLUSION

The observed electron beams presumably carry a net current even though this cannot be proven using the present particle data. It seems probable, however, that the magnetic variations observed during the events can be used to resolve this question. A complete absence of electrons around $180°$ pitch angles is not to be expected because of pitch angle scattering within one-half of a bounce period caused both by waves and by collisions in the ionosphere.

Ignoring the opposing contribution of these electrons, current densities per unit solid angle of 2.5, 8., and 13. $\mu A\ m^{-2}\ ster^{-1}$ are obtained for the spectra observed at 0555 UT on day 167, at 0635 UT on day 201 and at 2254 UT on day 186 respectively. If confined to a 5° half-angle cone, the current densities are only 0.06, 0.2, and 0.3 $\mu A/m^2$. When the beam approaches the ionosphere, the converging magnetic field increases the current density by at least a factor of 100 giving values greater than 6, 20, and 30 $\mu A/m^2$ respectively. These are similar to the current densities observed flowing in and out of the ionosphere (Vondrak, Anderson, and Spiger, 1971; Choy et al, 1971; Cloutier et al, 1973; Armstrong, 1974; Arnoldy, 1974; Zmuda and Armstrong, 1974). There is thus every reason to assume that the equatorial electron beams are part of a Birkeland current system.

While the observed electrons are capable of producing bright auroras, acceleration in or near the ionosphere does still seem to be required to explain the low altitude observations of field-aligned quasi-monoenergetic electrons (Arnoldy, 1974).

It is quite possible, however, that the electron beams observed near the equator in fact establish and maintain the potential differences believed to be responsible for electron acceleration near the ionosphere (Hultqvist, 1971; Evans, 1974).

ACKNOWLEDGMENTS

I thank R. LaQuey, R. Judge, W. Fillius, and E. Strein for their contributions in the design and construction of the ATS-6 Auroral Particles Experiment Detector; G. Peters and L. McDaniel for preparing computer programs; and S. DeForest and B. Mauk for their assistance in the ATS-6 calibrations and data analysis. This research was supported by NASA Grant NGL 05-005-007 and Contracts NAS 5-10364 and NAS 5-21055.

REFERENCES

Albert, R. D., Nearly monoenergetic electron fluxes detected during
    a visible aurora, Phys. Rev. Letters, 18, 369, 1967a.

Armstrong, J. C., Field aligned currents in the magnetosphere,
    Magnetospheric Physics, ed. B. M. McCormac, p. 155,
    D. Reidel Pub. Co.., Dordrecht, Holland, 1974. (Ref. in
    Zmuda, A. J., and J. C. Armstrong, The diurnal flow pattern
    of field-aligned currents, J. Geophys. Res. 79, 4611, 1974).

Arnoldy, R. L., Auroral particle precipitation and Birkeland
    currents, Rev. of Geophys. and Space Phys., 12, 217, 1974.

Arnoldy, R. L., R. B. Lewis, and P. O. Isaacson, Field-aligned
    auroral electron fluxes, J. Geophys. Res., 79, 4208, 1974.

Axford, W. I., Magnetospheric convection, Rev. of Geophys. 7,
    421, 1969.

Boström, R., Ionosphere-magnetosphere coupling, Magnetospheric
    Physics, ed. B. M. McCormac, p. 45, D. Reidel Pub. Co.,
    Dordrecht, Holland, 1974.

Bryant, D. A., G. M. Courtier, and G. Bennett, Electron Intensities
    over auroral arcs, Earth's Magnetospheric Processes, ed.
    B. M. McCormac, p. 141, D. Reidel Pub. Co., Dordrecht,
    Holland, 1972.

Choy, L. W., R. L. Arnoldy, W. Potter, P. Kintner, and L. J.
    Cahill, Field-Aligned particle currents near an auroral arc,
    J. Geophys. Res. 75, 8279, 1971. (Ref. in Zmuda and
    Armstrong, The diurnal flow pattern of field aligned currents,
    J. Geophys. Res. 79, 4611, 1974.)

Cloutier, P. A., B. R. Sandel, H. R. Anderson, P. M. Pazich, and
    R. J. Spiger, Measurement of auroral Birkeland currents and
    energetic particle fluxes, J. Geophys. Res., 78, 640, 1973.

DeForest, S. E., and C. E. McIlwain, Plasma clouds in the magneto-
    sphere, J. Geophys. Res. 76, 3587, 1971.

Evans, D. S., Fine structure in the energy spectrum of low energy
    auroral electrons, in Atmospheric Emissions, eds. B. M.
    McCormac and A. Omholt, p. 107, Van Nostrand Reinhold,
    New York, 1969.

Evans, D. S., Observations of a near monoenergetic flux of auroral
    electrons, J. Geophys. Res. 73, 2315, 1968.

Evans, D. S., Precipitating electron fluxes formed by a magnetic
    field aligned potential difference, J. Geophys. Res. 79,
    2853, 1975.

Fälthammar, C.-G., Magnetospheric processes, Earth's Magneto-
    spheric Processes, ed. B. M. McCormac, p. 16-28,
    D. Reidel Pub. Co., Dordrecht, Holland, 1972.

Hall, D. S., and D. A. Bryant, Collimation of auroral particles
    by time-varying acceleration, Nature 251, 5474, 1974.

Heikkila, W. J., Outline of a magnetospheric theory, J. Geophys.
    Res. 79, 2496, 1974.

Hones, E. W., Jr., J. R. Asbridge, S. J. Bame, and S. Singer,
    Energy spectra and angular distributions of particles in the
    plasma sheet and their comparison with rocket measurements
    over the auroral zone, J. Geophys. Res. 76, 63, 1971.

Hultqvist, B., On the production of a magnetic-field-aligned
    electric field by the interaction between the hot magneto-
    spheric plasma and the cold ionosphere, Planet. Space Sci.
    19, 749, 1971.

Hultqvist, B., H. Borg, W. Riedler, and P. Christophersen,
    Observations of magnetic-field aligned anisotropy for 1 and
    6 keV positive ions in the upper ionosphere, Planet. Space
    Sci. 19, 279, 1971.

Kamide, Y., and C. E. McIlwain, The onset time of magnetospheric
    substorms determined from ground and synchronous satellite
    records, J. Geophys. Res. 79, 4787, 1974.

Lenard, -, Über die Absorption der Nordlichtstrahlen in der
    Erdatmosphäre', S. B. Heidelberg. Adal. Wiss., Math.
    Naturw. Kl. 1911, No. 12.
    (Ref. in International Monographs on Radio, ed. by
    E. Appleton and R. L. Smith-Rose, Oxford University Press,
    London, p. 192, 1955.)

Mauk, B. H., and C. E. McIlwain, Correlation of K_p with the
    substorm-injected plasma boundary, J. Geophys. Res. 79, 3193,
    1974.

McIlwain, C. E., Direct measurement of particles producing visible
    auroras, J. Geophys. Res. 65, 2727, 1960.

McIlwain, C. E., Plasma convection in the vicinity of the geo-
    synchronous orbit, Earth's Magnetospheric Processes, ed.

B. M. McCormac, p. 268, D. Reidel Pub. Co., Dordrecht, Holland, 1972.

McIlwain, C. E., Substorm injection boundaries, Magnetospheric Physics, ed. B. M. McCormac, D. Reidel Pub. Co., Dordrecht, Holland, 1974.

O'Brien, B., and D. L. Reasoner, Measurements of highly collimated, short-duration bursts of auroral electrons and comparison with existing auroral models, J. Geophys. Res. 76, 8258, 1971.

Omholt, A., Studies on the excitation of aurora borealis, 1, the hydrogen lines, Geofys. Publikasjoner, 20 (11), 1-40, 1959.

Schulz, M., Particle lifetimes in strong diffusion, Astrophys. & Space Science 31, 37, 1974.

Shook, G. B., R. D. Sharp, M. F. Shea, R. G. Johnson, and J. B. Reagan, Trans. Am. Geophys. U. 47, 64, 1966.

Störmer, C., The polar aurora, Oxford University Press, London, 1955.

Vondrak, R. R., H. R. Anderson, and R. J. Spiger, Rocket-based measurements of particle fluxes and currents in an auroral arc, J. Geophys. Res. 76, 7701, 1971.
(Ref. in Zmuda and Armstrong, The diurnal flow pattern of aligned currents, J. Geophys. Res. 79, 4611, 1974).

Westerlund, L. H., The auroral energy spectra extended to 45 eV, J. Geophys. Res., 74, 351, 1969.

Whalen, B. A., and I. B. McDiarmid, Observations of magnetic-field-aligned auroral electron precipitation, J. Geophys. Res. 77, 191, 1972.

Whipple, E. C., private communication, 1975.

Zmuda, A. J., and J. C. Armstrong, The diurnal flow pattern of field-aligned currents, J. Geophys. Res. 79, 4611, 1947.

A STUDY OF AURORAL DISPLAYS PHOTOGRAPHED FROM THE DMSP-2 AND

ISIS-2 SATELLITES

S.-I. Akasofu

Geophysical Institute, University of Alaska

Fairbanks, Alaska 99701

ABSTRACT

Recent studies of auroral photographs taken from the DMSP and ISIS-2 satellites are briefly summarized. A particular emphasis is made in describing:
   (1) New large-scale morphological features, such as the distinction between discrete and diffuse auroras.
   (2) The roles of the north-south component of the interplanetary magnetic field on large-scale auroral dynamics, in particular the finding of an important quantity S which controls both the occurrence of magnetospheric substorms and their intensity.
   (3) The relationships between the auroral electrojet and large-scale auroral features.

1.   INTRODUCTION: The permanent nature of the solar wind-magnetosphere dynamo.
The interaction between the solar wind and the magnetosphere constitutes a magneto-hydrodynamic (MHD) dynamo. The solar wind blows across the polar cap geomagnetic field lines which are merged with interplanetary magnetic field lines. The dynamo process takes place near the magnetopause, resulting in a potential of the order of 40-50 kV across the magnetotail. Much of the current generated by the dynamo is discharged across the magnetotail. This portion of the current is called the cross-tail current. However, a small amount of the current is also discharged across the polar cap. This part of the circuit is called the oval circuit and is responsible for the aurora. That is to say, the aurora is a discharge

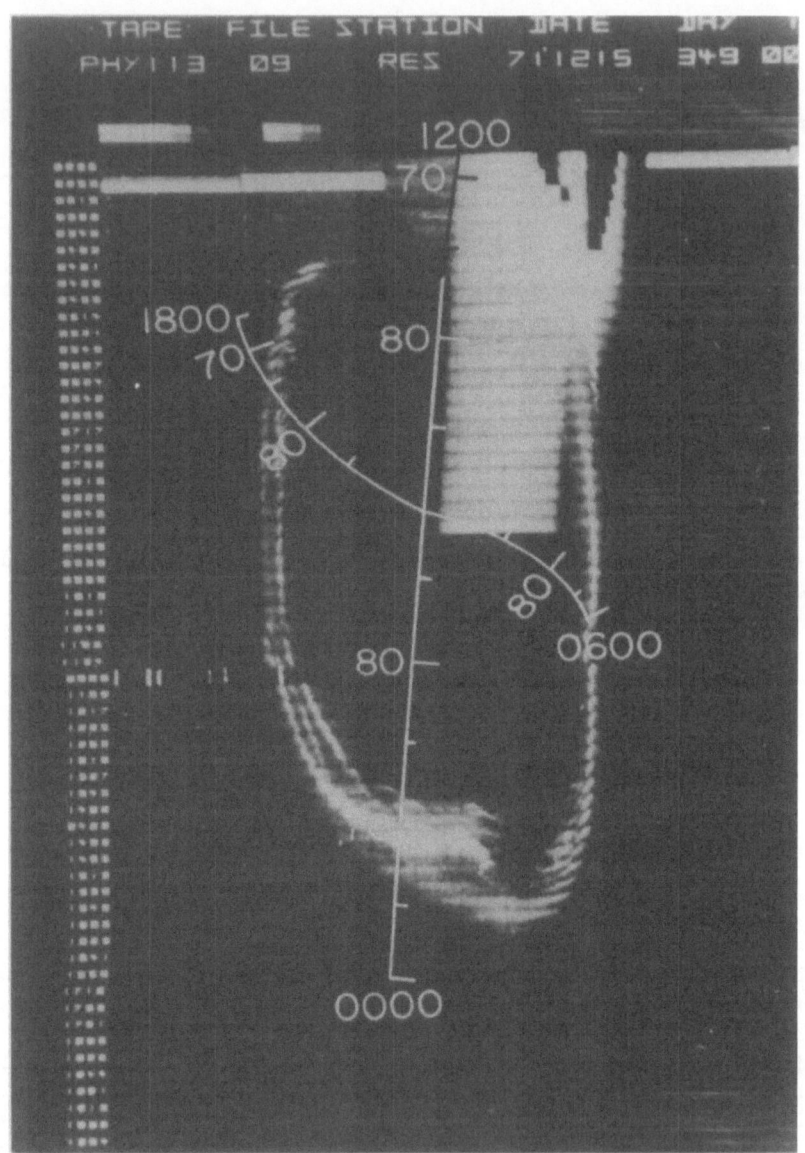

Figure 1 Photograph of the auroral oval taken from the ISIS-2
satellite; 0402 UT, 15 December, 1971 (courtesy of Dr C Anger).

phenomenon associated with the oval circuit. Figure 1 shows an
example of photographs taken from the ISIS-2 satellite. It shows
a nearly complete oval (Lui et al., 1975).

The interaction between the magnetosphere and the solar wind
appears to be a permanent feature. Meng and Anderson (1974) examined extensively magnetotail data at the lunar distance ($60R_E$) for
both low Kp values (Kp$\leq$1+) and high Kp values (Kp$\geq$2-) and showed
that there is no significant difference in the structure of the
magnetotail for both conditions. Thus, there must always be a
finite amount of open flux even during very quiet periods. Indeed, the auroral oval appears to be a permanent feature, although
it became too dim to be observable by the auroral scanning device
aboard the DMSP satellite during a prolonged quiet period. In
section 2.2, we shall see a few examples of small ovals during
very quiet periods.

(Note that even in such a condition, auroras were visible in all-
sky photographs, and auroral particles were detected.) The presence
of the quiet day polar cap daily magnetic variation (S_q^p) is also
an indication that the solar wind-magnetosphere dynamo is a permanent feature, because this particular magnetic variation arises
from electric currents which flow along the oval circuit.

2.   A NEW CLASSIFICATION OF MAGNETOSPHERIC DISTURBANCES
      Magnetospheric disturbances can be classified into two: (1)
reversible or quasi-reversible disturbances and (2) irreversible
disturbances (Akasofu, 1975a).

2.1 Reversible or quasi-reversible disturbances
      A reversible disturbance is defined to be one for which the
magnetosphere returns to a quiet-time configuration after the responsible interplanetary disturbance is removed, with or without
any significant increase of the heat production from the quiet-time
value in the oval. We have so far found three interplanetary disturbances which cause reversible or quasi-reversible disturbances.
      (a) A weak interplanetary shock wave - A weak interplanetary
shock wave simply compresses the magnetosphere. After the solar
wind returns to a pre-shock condition, the magnetosphere expands,
returning to a quiet-time configuration. There is no significant
increase of heat production by this process.
      (b) The east-west component of the interplanetary magnetic
field - Svalgaard (1973) showed that there exists a counterclock-
wise current (observing from above the north geomagnetic pole along
the geomagnetic latitude circle of about 75-80°, when the interplanetary magnetic field has a westward component (i.e., in the
"away" sector) and a clockwise current for an eastward component
(i.e., in the "toward" sector). Heppner (1972) showed that the
electric field distribution becomes asymmetric with respect to the
Sun-Earth line when the interplanetary magnetic field has an east-
west component.

It may be possible that some of these observations can be
understood in terms of the east-west asymmetry of the oval circuit,
instead of the growth of an entirely new current system. If it is
indeed the case, these particular disturbances can also be class-
ified as reversible processes. The oval circuit returns to its
symmetric configuration, without an increase of the heat production,
when the east-west component is removed.

(c) The north-south component of the interplanetary magnetic
field - It has been suggested that the north-south component of
the interplanetary magnetic field controls the merging rate of the
geomagnetic field lines with the interplanetary magnetic field
lines. An increase in the merging rate results in an increase in
area of the polar cap and thus of the auroral oval. Some of the
other observed changes associated with an increased merging rate
are: (i) Earthward shift of the magnetopause; (ii) equatorward
shift of the cusp; (iii) an increase of the magnetic field intensity in the high latitude lobe; (iv) an increase of the flaring
angle of the magnetotail.

The expansion of the auroral oval and thus of the oval circuit
causes a particular type of magnetic variations at magnetic observ-
atories in the polar region, even if there is no change in the total
current intensity of the polar cap circuit. However, the actual
change, identified first by Nishida (1968) as the DP-2 variations,
can be understood in terms of a combination of an expansion of the
circuit and of an appreciable increase of the total current intensity (Akasofu, Yasuhara and Kawasaki, 1973). Thus, the response
of the magnetosphere to the southward component of the interplanetary
magnetic field (and thus to an increase in the merging rate) is not
strictly a reversible process. However, since the increase in heating is rather small compared with what will be described as an irreversible process in the next section, we may call it a quasi-
reversible process.

This dynamical feature of the auroral oval associated with
changes of the north-south component of the interplanetary magnetic
field has only recently been recognized (Akasofu et al., 1973).
Since most of the auroral observations are conducted along the
auroral zone (the dipole latitude of about 60-67°), it is not possible to observe the auroral oval when it contracts poleward, to the
dipole latitude of about 70° or above; this situation occurs when
the interplanetary magnetic field has a large northward component.
It had thus been thought by most that the aurora and the associated
magnetic disturbance does not occur when the interplanetary field
has a northward component. Actually, auroral phenomena are present,
but well beyond the field of view of auroral zone stations.

On the other hand, when the interplanetary magnetic field has
a southward component, the auroral oval descends to the latitude
of the auroral zone in the midnight sector. This has been inter-

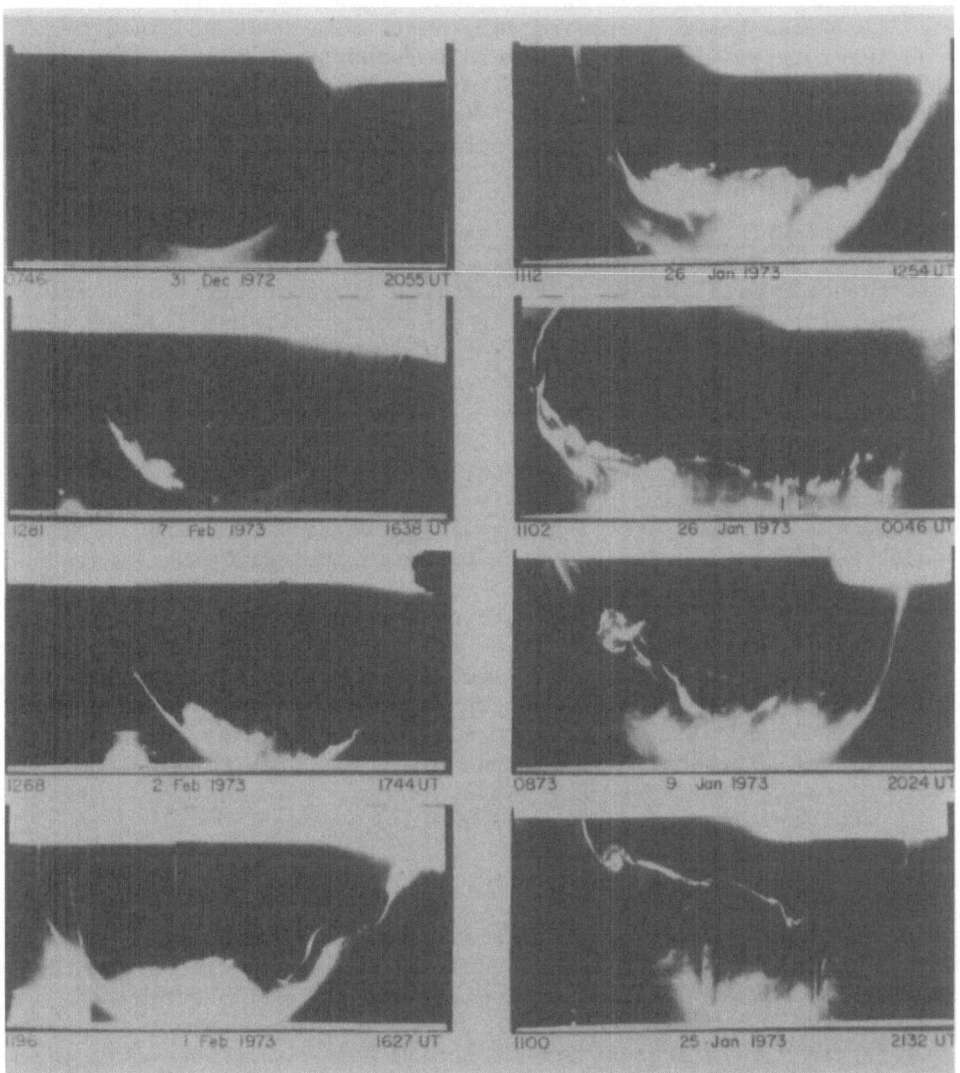

Figure 2   Eight DMSP photographs chosen to show how the expansive phase of the auroral substorm might appear when DMSP photographs could be taken every several minutes apart; note that they were taken on different days.

preted by many as an indication that the aurora can occur only
when the interplanetary magnetic field has a southward component.
For the same reason, typical substorm features can be observed at
the standard auroral zone stations only when the interplanetary
magnetic field has a southward component. Thus, it had incorrectly
been interpreted that the southward component was a necessary condition for the occurrence of substorms. Fortunately, all these
misconceptions have been removed recently by improved auroral
observations, particularly by a meridian chain of all-sky cameras
and satellite photography.

2.2 <u>Magnetospheric substorms as an irreversible disturbance</u>.
     Here, we identify the magnetospheric substorm as an irrevers-
ible phenomenon. Indeed, there occurs a large increase of heating
in the polar cap ionosphere during substorms. Figure 2 shows eight
DMSP photographs which are chosen to show how the expansive phase
of the auroral substorm might appear when DMSP photographs could
be taken every several minutes apart.

     During the last several years, one of the most importan sub-
jects in magnetospheric physics is the roles of the north-south
component B_z of the interplanetary magnetic field on magnetospheric
processes, in particular on magnetospheric substorm processes.

     Let Φ_D be the production rate of merged (or open) field lines
along the dayside X-line. This quantity is also equal to the potential drop along the X-line and also to the potential drop across the
dawn-dusk meridian in the polar cap (Sonnerup, 1974; Gonzalez and
Mozer, 1974). Let us define here also the production rate Φ_N of
reconnected (or closed) field lines along the nightside X-line.

     Consider then the quantity S defined by

$$S = \int_0^T (\Phi_D - \Phi_N) dt$$

where the time t=0 is chosen at the time when the interplanetary
magnetic field begins to decrease after having a large northward
component ($B_z \gtrsim +5\gamma$) for an extended period, say 6-12 hours.

     Figure 3 illustrates schematically time variations of Φ_D,
Φ_N, $(\Phi_D - \Phi_N)$ and $S = \int_0^T (\Phi_D - \Phi_N) dt$ when the north-south component
of the interplanetary magnetic field B_z varies from a large positive
value to a negative value for about 2 hours and then back to a large
positive value again. Note that an extended period of a large positive B_z value is the initial condition.

     It should be noted that the quantity S is equal to the amount
of open magnetic flux at a time reckoned from t=0 (after a prolonged
period of a large positive B_z value).

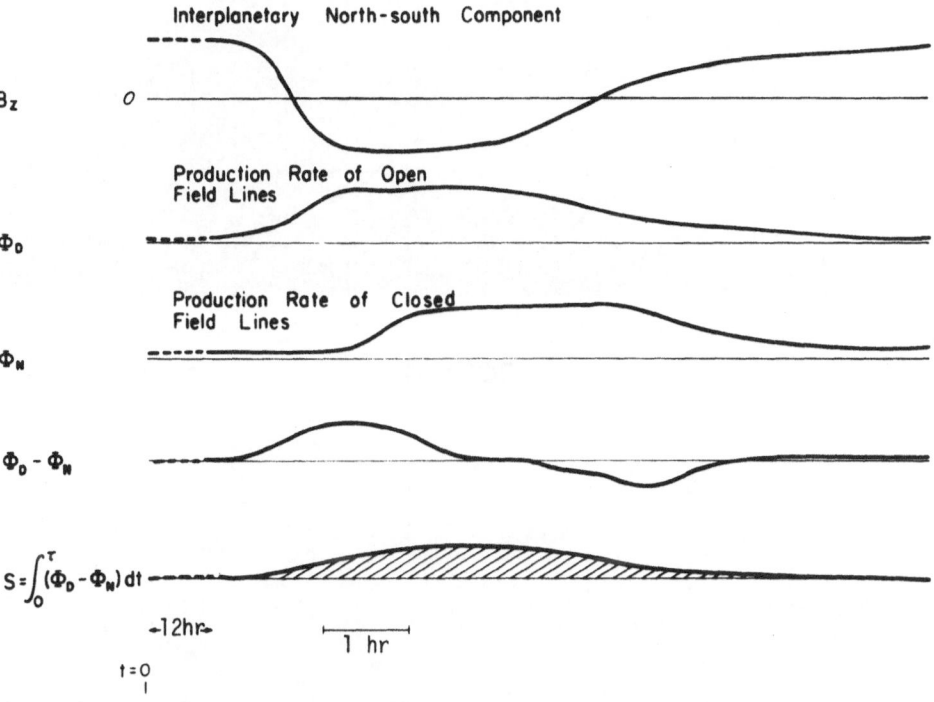

Figure 3 Hypothetical change of the B_z component and the resulting changes of ϕ_D, ϕ_N, $\phi_D - \phi_N$ and $S = \int_0^\tau (\phi_D - \phi_N) dt$.

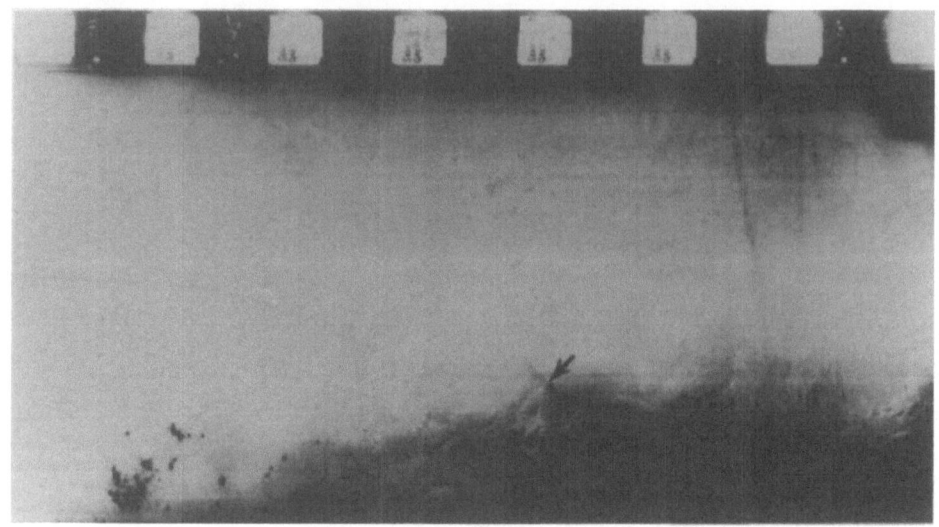

647   **2057 UT**   **24 Dec., 1972**

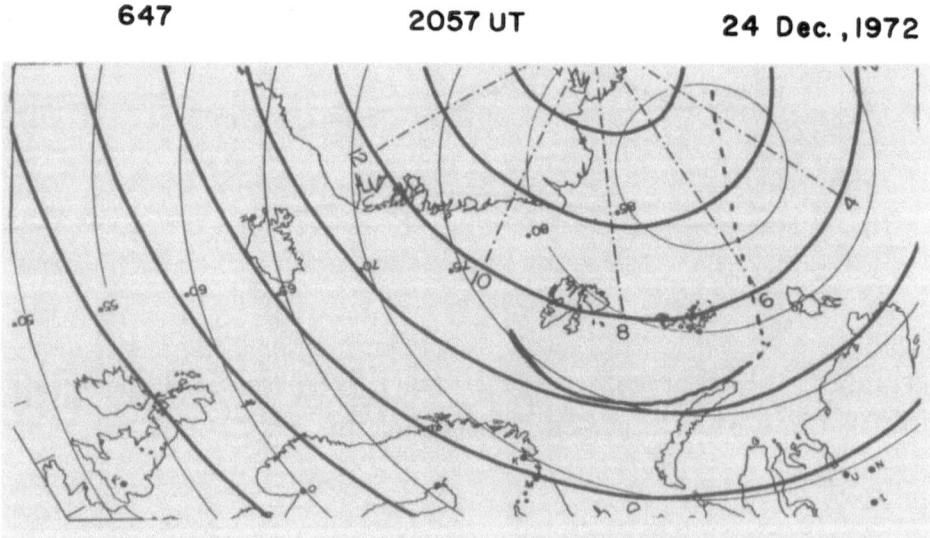

Figure 4   DMSP photograph (in negative) taken at 2057 UT on 24
December, 1972. It shows one of the smallest ovals.

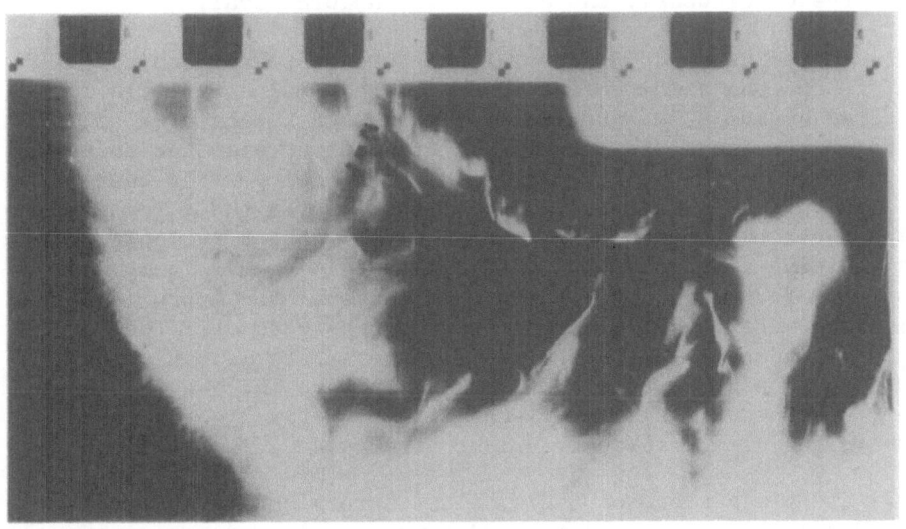

0911   1252 UT   12 Jan 1973

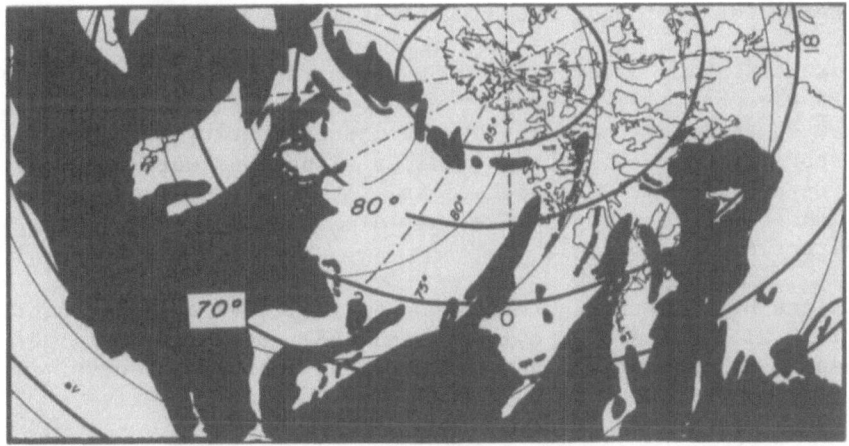

Figure 5   DMSP photograph taken at 1252 UT on 12 January, 1973.
It shows an intense auroral activity along an expanded oval.

$$S = B_p \, (A_1 - A_0)$$

where A_0 and A_1 are the area of the minimum oval and an expanded oval, respectively. Thus, the amount of S can be monitored if one can observe continuously the area of the auroral oval.

An extensive study of auroral photographs taken from the DMSP satellite and the simultaneous interplanetary magnetic field data (monitored by the HEOS satellite) indicates that substorms can occur when the auroral oval is larger than its minimum size and thus when S is positive (Akasofu, 1975b,c). This includes periods when (1) the B_z component is positive, but decreasing, (2) the B_z component becomes negative (the southward turning), (3) the B_z component becomes positive (the northward turning), and (4) the B_z component is positive for a few hours. Only when the oval had approximately the minimum sixe (namely, during prolonged periods of a large positive B_z component), and thus S∿0, were substorms not observed. It is also found that substorms tend to be more intense when S is larger. Here, the intensity of substorms is defined to be proportional to the area covered by bright auroras.

Figure 4 shows one of the smallest auroral ovals observed by the DMSP satellite. It was taken during a prolonged period of a large positive B_z value. Figure 5 shows an intense auroral substorm along an expanded medium size oval; this event occurred when the B_z component was negative. There is a considerable difference in the size of the oval and auroral activity.

Figure 6 shows a series of DMSP photographs which were taken on 24-25 December, 1972. They provide one of the most illuminating examples in studying the roles of the interplanetary magnetic field on substorm processes, although similar situations were also observed on a number of other occasions. At about 08 UT on 24 December the B_z component became positive, and a prolonged period of the northward directed field began.

It can be seen that auroras were quite active at least until about 1410 UT (orbits 640, 641, 642 and 643), in spite of large positive B_z values during that period. The substorm observed during orbit 642 occurred when B_z decreased, but remained positive.

One of the most interesting features of auroral activity during this period is that the size of the auroral oval was gradually decreasing after each substorm. The oval at orbit 643 was considerably smaller than that at orbit 640. The oval became even smaller and quite dim at orbit 645. During the next two orbits (646 and 647), the oval contracted even further polewards. The fact that the oval was very small during orbit 647 suggests that a small decrease of the B_z component after a prolonged period of a large positive value contributes only a small amount to the quantity S. However, during orbit 648, a typical substorm feature was observed; B_z decreased, but remained positive at that time. The substorm observed during orbit 649 occurred when B_z was positive.

The B_z component was positive throughout the day of 26 December, 1972, except for a short period between 1945-2130 UT. The auroral oval was very dim during orbits 650 and 651. Although it is not shown here, the B_z component had small positive values between 04 and 13 UT on that day. A dim contracted oval was seen around 12-14 UT, together with weak substorm activity; its midnight latitude was about 70°. It is likely that S had a small positive value during that period.

A prolonged period of large positive B_z values occurred also on 27 December. Figure 7 shows DMSP photographs, the interplanetary magnetic data and the AE index between 18 UT, 26 December and 24 UT, 27 December. A dim small oval was observed during orbits 674 and 675. Then, a weak substorm activity was observed during orbits 676 and 677, although the AE index had no typical substorm features. The DMSP photograph at orbit 679 shows a bright arc in the afternoon part of the oval; the B_z component was negative for about one hour before that time. After orbit 679, the oval became too dim to study in detail in DMSP photographs during the rest of the day. However, auroras were seen over Sachs Harbour (one of the Alaska meridian chain stations, at inv. lat. ~74°) at least until 12 UT.

It should be emphasized that our conclusion differs significantly from that put forward by McPherron, Russell and Aubry (1974) who claimed that the southward turning of the B_z component would lead to a series of processes (such as an increase of the magnetic flux in the magnetotail and thinning of the plasma sheet (?)) and eventually to the expansive phase of the substorm.

Our observation indicates that so long as S is positive, a completely opposite series of processes (the northward turning of the B_z component, a decrease of the magnetic flux in the magnetotail and expansion of the plasma sheet (?)) leads to the expansive phase of the substorm. Therefore, it is very doubtful that their proposed series of processes has any importance in substorm processes. As Akasofu (1974a) pointed out, their proposed growth phase signatures must simply be effects of the southward turning of the B_z component.

It may be concluded thus:
(1) The magnetosphere is always in a state of 'growth phase' except for prolonged periods of a large positive B_z value.
(2) The substorm can be considered as a process by which the magnetosphere tends to remove the excess energy associated with S. Since the total energy ε in the magnetotail is given (Gonzalez and Mozer, 1974) by

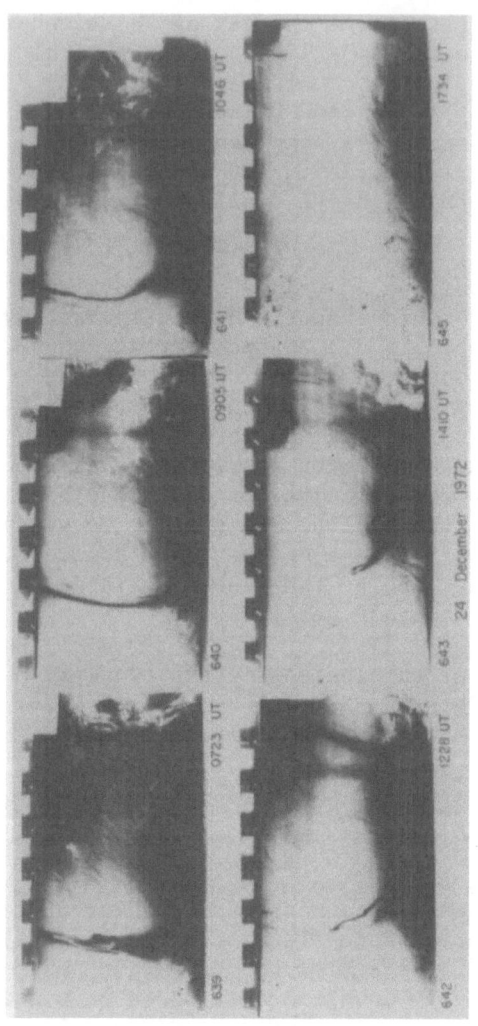

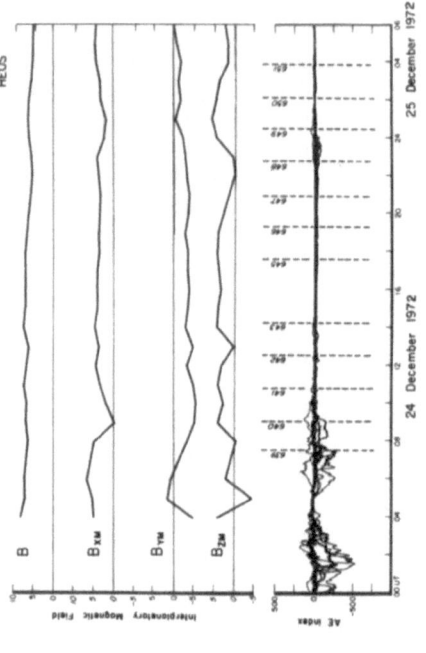

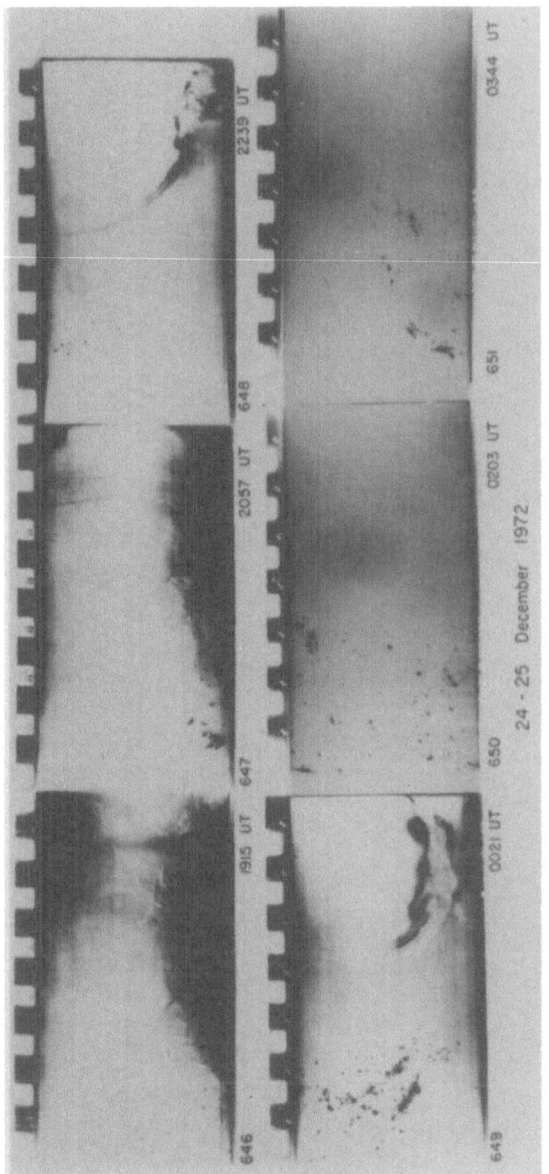

Figure 6 Some of the DMSP photographs taken on 24-25 December, 1972. The interplanetary magnetic field data (HEOS), B, B_{XM}, B_{YM}, B_{ZM}; the AE index.

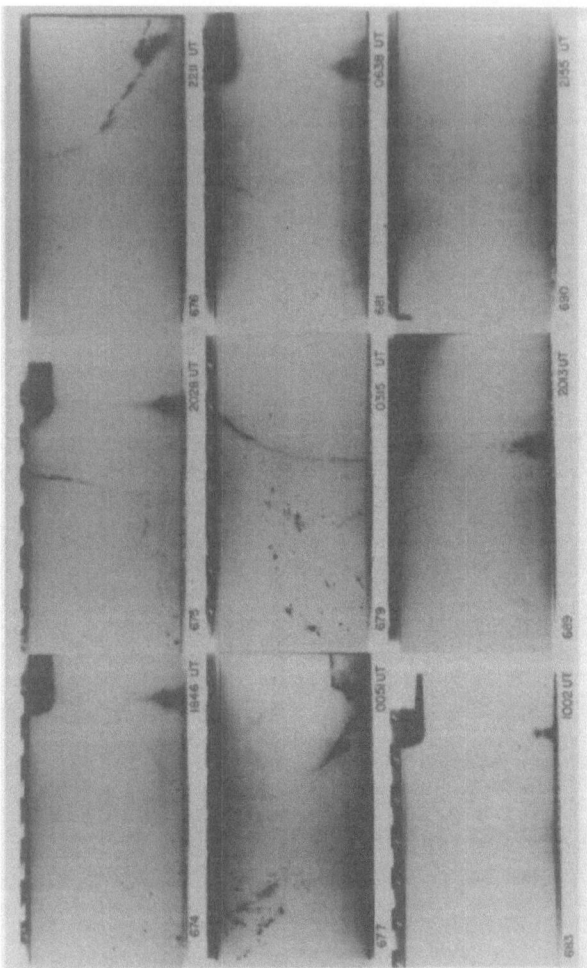

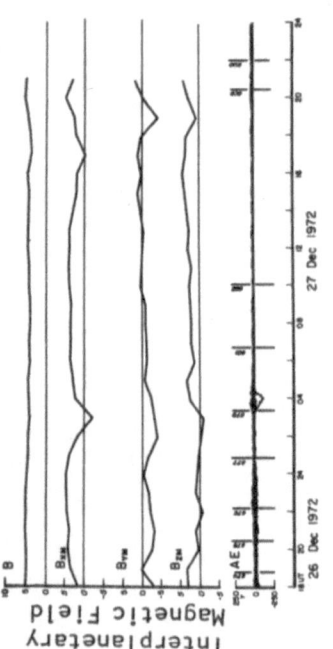

Figure 7 Some of the DMSP photographs taken on 26-27 December, 1972. The interplanetary magnetic field data (HEOS), B, B_{XM}, B_{ZM}; the AE index.

$$\varepsilon \overset{\sim}{=} (d_1 d_2)^2 \ (\mathcal{L} B_p{}^2 / \pi^2 R_T{}^2 \ V_i)$$

where d_1 = the noon-midnight dimension of the polar cap
 d_2 = the dawn-dusk dimension of the polar cap
 V_i = the solar wind speed
 R_T = the radius of the magnetotail
 $\mathcal{L}$ = the length of the magnetotail
substorm energy ε_s for a large $d(=d_1=d_2)$ is given by

$$\varepsilon_s \sim (\mathcal{L} / R_T{}^2 \ V_i \ \pi^4) \ s^2.$$

(3) The sporadic nature of substorms indicates that this
removal process is not a steady process. Akasofu (1975c) suggested
a certain analogy of this process with the cyclogenesis.

3. DISCRETE AURORAS AND DIFFUSE AURORAS
In DMSP-2 photographs, we can tentatively define two types of
auroras (Lui and Anger, 1973; Snyder and Akasofu, 1974):
Discrete Aurora:
A discrete aurora appears as a single strand, separated from
others by a dark space of width of order a few tens of kimometers.
When it is seen from the ground, it has a curtain-like structure.
Diffuse Aurora:
A diffuse aurora appears as a broad band of auroral luminosity
with a width of at least several tens of kilometers. It is not
easily visible from the ground, but covers one half of the sky.

Such a distinction between these two main auroral types cannot
always be made. Figure 8 shows a schematic diagram of auroral distribution which is constructed on the basis of a large number of
DMSP and ISIS-2 photographs. Discrete auroras are indicated by
lines; the brightness is represented by the thickness of the lines.

The equatorward boundary of the belt of auroras can easily be
recognized by a smooth boundary of the diffuse aurora. The diffuse
aurora has a fairly uniform luminosity in the evening sector. However, in the midnight and early morning sectors, it is not a uniform
luminosity. Dark filamentary structures are embedded in a relatively
uniform glow. In general, this tendency becomes more prominent in
the late morning sector where one can recognize individual arcs.
In fact, the diffuse aurora in the late morning sector is, in general,
composed of a large number of arcs which are packed in a rather narrow belt. A patchy structure is also often seen in the diffuse
aurora in the morning sector, particularly near its equatorward
boundary.

The poleward boundary of the diffuse aurora is far more complicated than the equatorward boundary. Often, a series of waves or of
torch-like structure developes during substorms. Some of such
structures have been called the omega band. Further, faint arcs

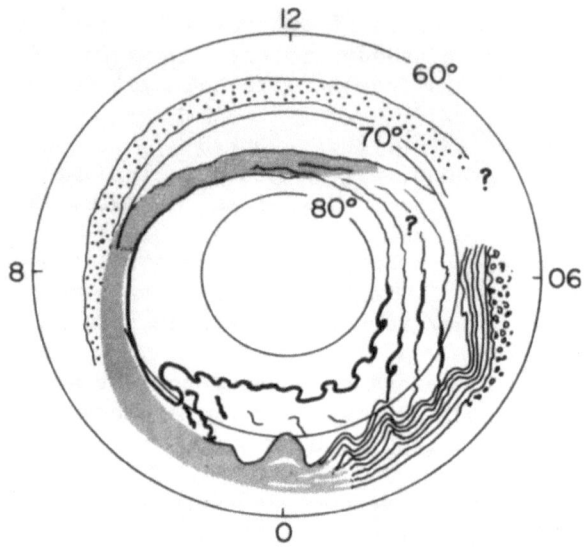

Figure 8   Schematic diagram showing the auroral distribution of different types.

often appear to develop from the tips of these waves and lie in the
morning half of the polar cap. These auroras are called 'polar cap
auroras'.

One of the striking features in DMSP photographs is that there
are distinct differences in the characteristics of auroras between
the evening and morning sectors. Roughly speaking, discrete auroras
are predominantly an evening feature, while the diffuse aurora
occupies the morning sector, as well as the equatorward half of the
whole belt of auroral luminosity. Most of the features pointed out
in the above are seen in the eight photographs in Figure 2. For
details of DMSP photographs, see Akasofu (1974b). Figure 9 shows
an example of photographs taken from the ISIS-2 satellite; it shows
also most of the features mentioned in the above.

4.   THE AURORAL ELECTROJET AND GLOBAL AURORAL FEATURES

A close relationship between polar geomagnetic disturbances
and auroral activity was found in early studies by a large number
of workers. Owing to lack of simultaneous auroral records, however,
polar geomagnetic disturbances were studied largely independently of
auroral morphology.

The IGY All-sky camera program provided the first opportunity
to observe the distribution of luminous aurora over the entire polar
region. However, even the IGY all-sky camera and magnetic data were
not extensive enough to study auroral and polar magnetic substorms
simultaneously over the entire polar region.

This difficulty has been removed to a great extent by photographs taken from the DMSP satellites and the ISIS-2 satellite. It
is thus of great interest to identify the position of large-scale
auroral forms relative to the auroral electrojet and other features
of polar magnetic substorms observed from the ground. Here, an
example of photographs taken from the DMSP-2 satellite on 25 January,
1973, together with the simultaneous ground magnetic data from a
number of polar stations is shown in Figure 10 (Kamide and Akasofu,
1975). In the figure, the auroral photograph and its sketch are
shown in the invariant latitude (at 100 km level) frame, along with
the equivalent overhead current vectors, which are deduced from
ground magnetic disturbance vectors (the actual disturbance vectors
in the horizontal plane can be obtained by rotating them counter-
clockwise by 90°). For reference, the current intensity scale is
given, outside the diagram, in terms of the intensity of magnetic
disturbance. The sketch is reasonably accurate, since most of
available city lights are used in identifying the geographic area.

The westward auroral electrojet is the dominant feature. It
flows along the diffuse aurora in the midnight and morning sectors,
as well as along the vicinity of active westward traveling surges
in the evening sector. The westward electrojet does not end in the

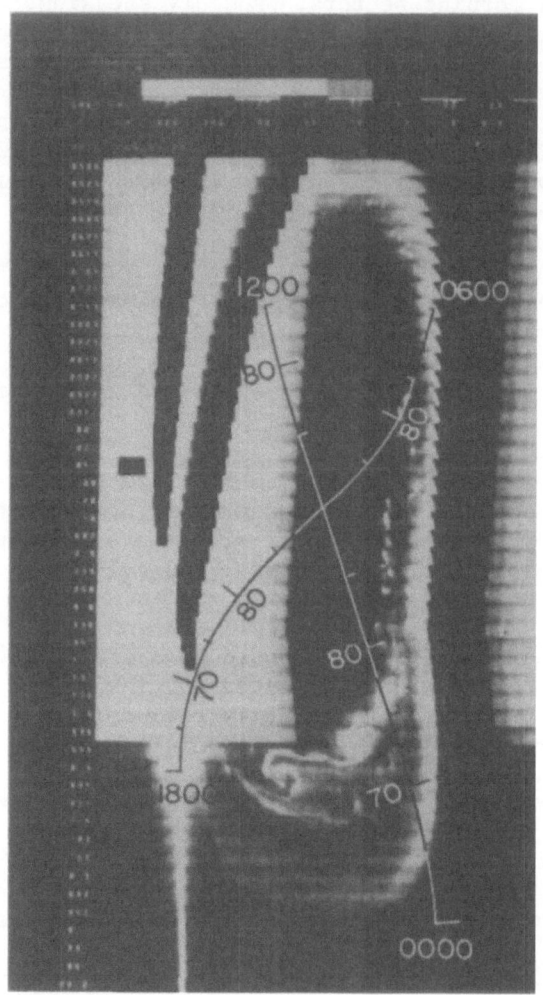

Figure 9   Active auroras photographed by the ISIS-2 satellite:
0205 UT, 17 January, 1972; (courtesy of Dr C Anger).

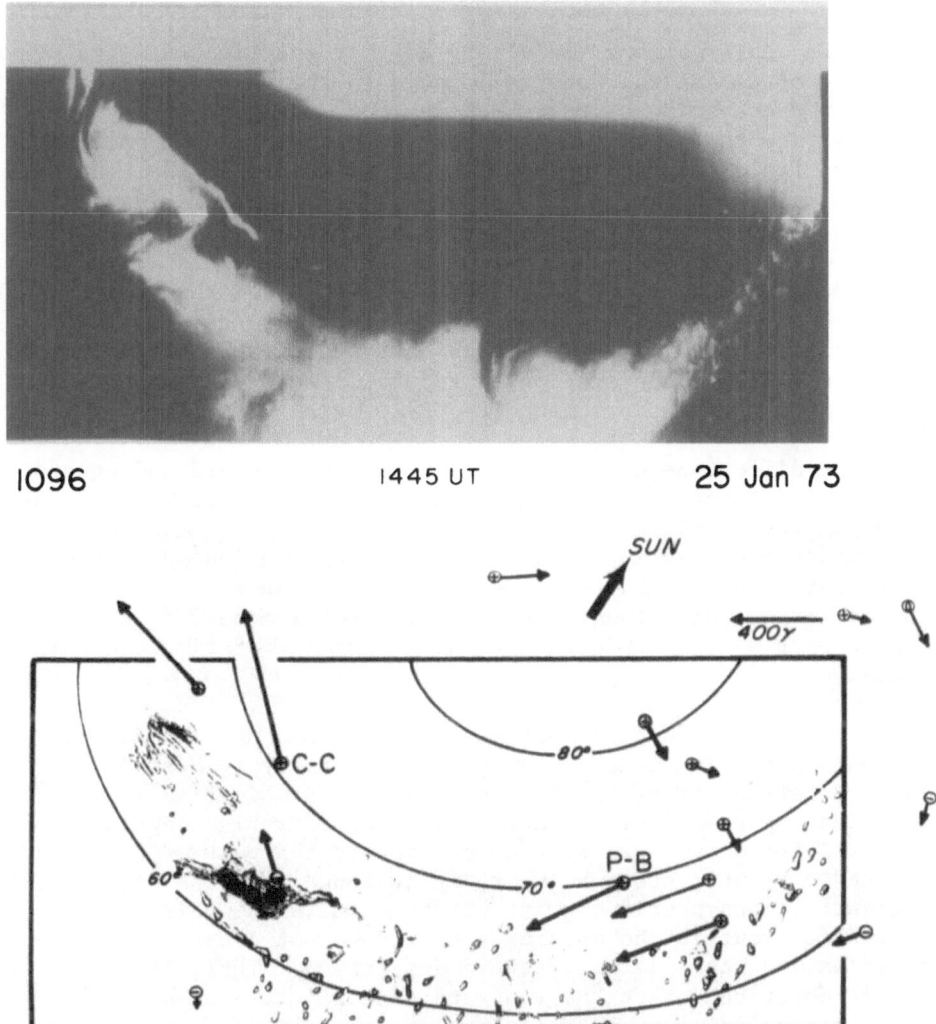

Figure 10   DMSP photograph, together with the equivalent
current vectors at several magnetic stations.

midnight meridian, but extends into the evening sector along the
poleward half of the auroral oval. The eastward current flows mostly
equatorward of discrete auroras (perhaps along the diffuse aurora)
in the evening sector, but sometimes near or in the region of discrete arcs of a stretched or looped form.

5.   FIELD-ALIGNED CURRENTS
     Žmuda and Armstrong (1974) showed that the field-aligned currents
consists of two pairs; one is located in the morning sector and the
other in the evening sector. However, our analysis of magnetic
records from the TRIAD satellite suggests that in each pair the
poleward field-aligned current is more intense than the equatorward current, a typical ratio being 2:1 (Yasuhara, Kamide and
Akasofu, 1975). Armstrong, Akasofu and Rostoker (1975) showed that
in the evening sector, the region of the upward field-aligned current
coincides with the region of discrete auroras and the region of the
downward current with the region of diffuse auroras. So far, such
a correlative study is based on all-sky photographs taken from the
Alaska meridian chain stations. Figure 11 shows an example of DMSP
photographs and the corresponding TRIAD data.

     The difference of the intensity of the poleward and equator-
ward field-aligned currents has a fundamental importance in understanding the coupling between the magnetosphere and the ionosphere.
We demonstrate this importance by computing the ionospheric current
distribution by solving the continuity equation $\vec{\nabla} \cdot \vec{I} = j_{\shortparallel}$ using
the 'observed' distribution of $j_{\shortparallel}$ for several models of the ionosphere with a high conductive annular ring (simulating the auroral
oval).

     It is shown that the actual field-aligned and ionospheric
current system is neither a simple Birkeland type, Boström type nor
Žmuda-Armstrong type, but is a complicated combination of them. The
relative importance among them varies considerably, depending on the
conductivity distribution, the location of the peak of the field-
aligned currents, etc. Further, it is found that the north-south
segment of ionospheric current which connects the pair of the field-
aligned current in the morning sector does not close in the same
meridian and has a large westward deflection. Thus, it has an appreciable contribution to the westward electrojet. One of the model
calculations shows that the entire north-south closure current contributes to the westward electrojet. Figure 12 shows an example of
such calculations.

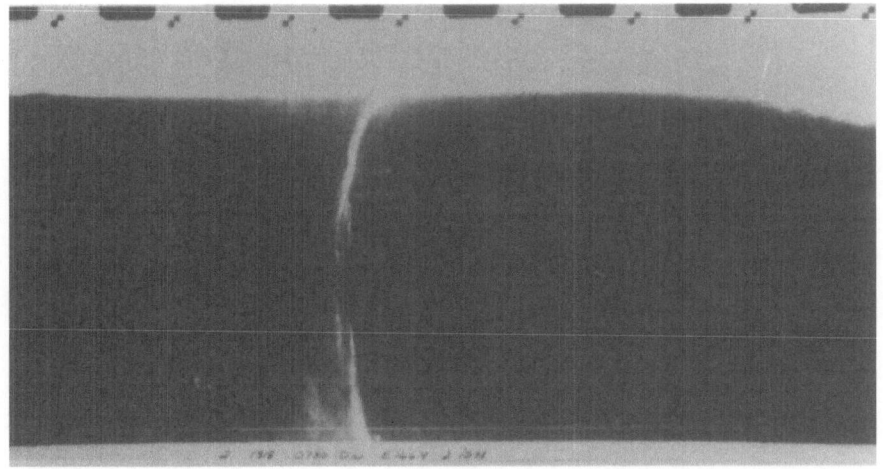

1318   0725 UT   10 FEB. 1973

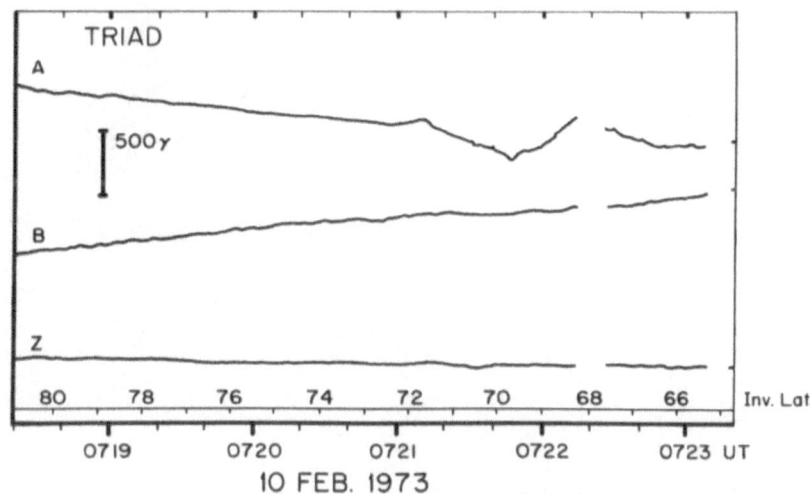

Figure 11   DMSP photograph and the corresponding TRIAD data. The satellite orbit was nearly parallel to a geographic meridian. Prudhoe Bay, Fairbanks, Palmer, Anchorage, Norman Wells can be recognized.

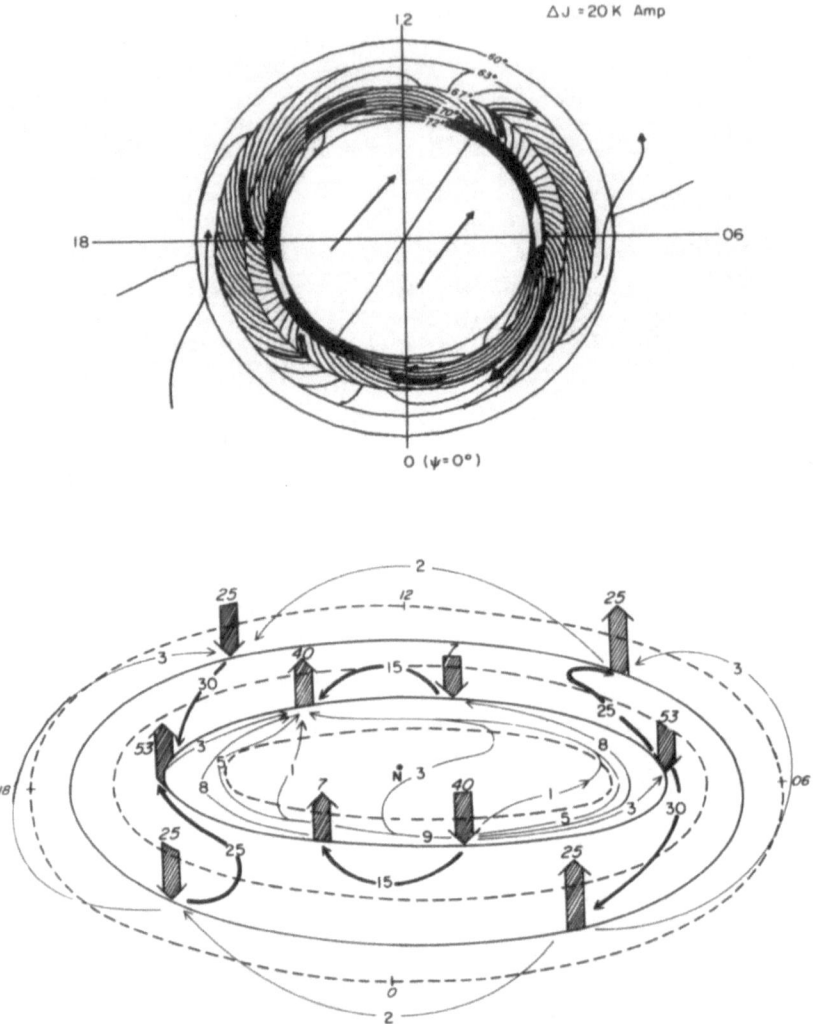

Figure 12 Calculated ionospheric current pattern. The auroral oval
is simulated by a double ring of high conductivities, bounded by
the latitude circles of $\lambda = 72^{\circ}$, 67° and 60°. The poleward field-
aligned currents are located along the latitude scale of $\lambda = 70^{\circ}$,
and the equatorward field-aligned currents along the latitude
circles of $\lambda = 63^{\circ}$.

REFERENCES

Akasofu, S.-I., A study of auroral displays photographed from the DMSP-2 satellite and from the Alaska meridian chain of stations, Space Sci. Rev., 16, 617-725, 1974b.

Akasofu, S.-I., The aurora and the magnetosphere: the Chapman memorial lecture, Planet. Space Sci., 22, 885, 1974a.

Akasofu, S.-I., The roles of the north-south component of the interplanetary magnetic field on large-scale auroral dynamics observed by the DMSP satellite, Nature (in press), 1975b.

Akasofu, S.-I., The roles of the north-south component of the interplanetary magnetic field on large-scale auroral dynamics observed by the DMSP satellite, Planet. Space Sci. (in press), 1975c.

Akasofu, S.-I., The solar wind-magnetosphere dynamo and the magnetospheric substorm, Planet. Space Sci. (in press), 1975c.

Akasofu, S.-I., P. D. Perreault, F. Yasuhara and C.-I. Meng, Auroral substorms and the interplanetary magnetic field, J. Geophys. Res., 78, 7490, 1973.

Akasofu, S.-I., F. Yasuhara and K. Kawasaki, A note on the DP-2 variation, Planet. Space Sci., 21, 2232, 1973.

Armstrong, J. C., S.-I. Akasofu and G. Rostoker, A comparison of satellite observations of Birkeland currents with ground observations of visible auroras and ionospheric currents, J. Geophys. Res., 80, 575, 1975.

Gonzalez, W. D. and F. S. Mozer, A quantitative model for the potential resulting from reconnection with an arbitrary interplanetary magnetic field, J. Geophys. Res., 79, 4186, 1974.

Heppner, J. P., Polar cap electric field distributions related to the interplanetary magnetic field direction, J. Geophys. Res., 77, 4877, 1972.

Kamide, Y. and S.-I. Akasofu, The auroral electrojet and global auroral features, J. Geophys. Res. (in press), 1975.

Lui, A.T.Y. and C. D. Anger, A uniform belt of diffuse auroral emission seen by the ISIS-2 scanning photometer, Planet. Space Sci., 21, 799, 1973.

Lui, A.T.Y., C. D. Anger, D. Venkatesan, W. Sawchuk and S.-I. Akasofu, The topology of the auroral oval as seen by the ISIS-2 scanning photometer, J. Geophys. Res. (in press), 1975.

McPherron, R. L., C. T. Russell and M. P. Aubry, Satellite studies of magnetospheric substorms on August 15, 1968 9. Phenomenological model for substorms, J. Geophys. Res., 78, 3131, 1973.

Meng, C.-I. and K. A. Anderson, Magnetic field configuration in the magnetotail near 60R$_E$, J. Geophys. Res., 79, 5143, 1974.

Nishida, A., Geomagnetic DP-2 flucuations and associated magnetospheric phenomena, J. Geophys. Res., 73, 1795, 1968.

Snyder, A. L., Jr., and S.-I. Akasofu, Major auroral substorm features in the dark sector by a USAF DMSP satellite, Planet. Space Sci., 22, 1511, 1974.

Svalgaard, L., Polar cap magnetic variations and their relationship
    with the interplanetary magnetic sector structure, J. Geophys.
    Res., 78, 2064, 1973.
Yasuhara, F., Y. Kamide and S.-I. Akasofu, Field-aligned and iono-
    spheric currents, Planet. Space Sci. (submitted), 1975.

RECENT OBSERVATIONS RELATING TO THE DYNAMICS AND ORIGIN OF THE

MAGNETOTAIL PLASMA SHEET

E. W. Hones, Jr.

University of California, Los Alamos Scientific Laboratory

Los Alamos, New Mexico 87544, USA

ABSTRACT

This paper reviews the substorm behavior of the plasma sheet and the evidence that a substorm expansive phase onset signals the sudden formation of a magnetic neutral line across a near region of the magnetotail. The neutral line forms at distances less than $\sim 18\ R_E$ in moderate to intense substorms, but at distances greater than $\sim 30\ R_E$ in weak substorms, or substorms detected only at high latitudes (on the contracted auroral oval). Several types of observations are presented as evidence that solar wind plasma enters the plasma sheet via convection along the cleft, through the magneto tail surface layer, through the magnetotail lobes and finally earthward from a magnetotail neutral line. These observations were made, however, during periods of southward interplanetary magnetic field; possibly this is a less effectual route for plasma entry to the plasma sheet when the IMF is northward.

1. INTRODUCTION

The model of Dungey (1961) for the interaction between the solar wind and the earth's magnetosphere has been subjected to a number of tests in the years since its proposal. These tests have mostly been of a statistical nature, many being associations of the occurrence of geomagnetic and auroral activity with the existence of southward interplanetary magnetic field (IMF) (Fairfield and Cahill, 1966; Fairfield, 1967; Arnoldy, 1971; Foster et al., 1971; Akasofu et al., 1973). Most such tests have shown positive correlations to these phenomena and thus have generally supported the view that the diversion of solar wind energy into the magnetosphere is

137

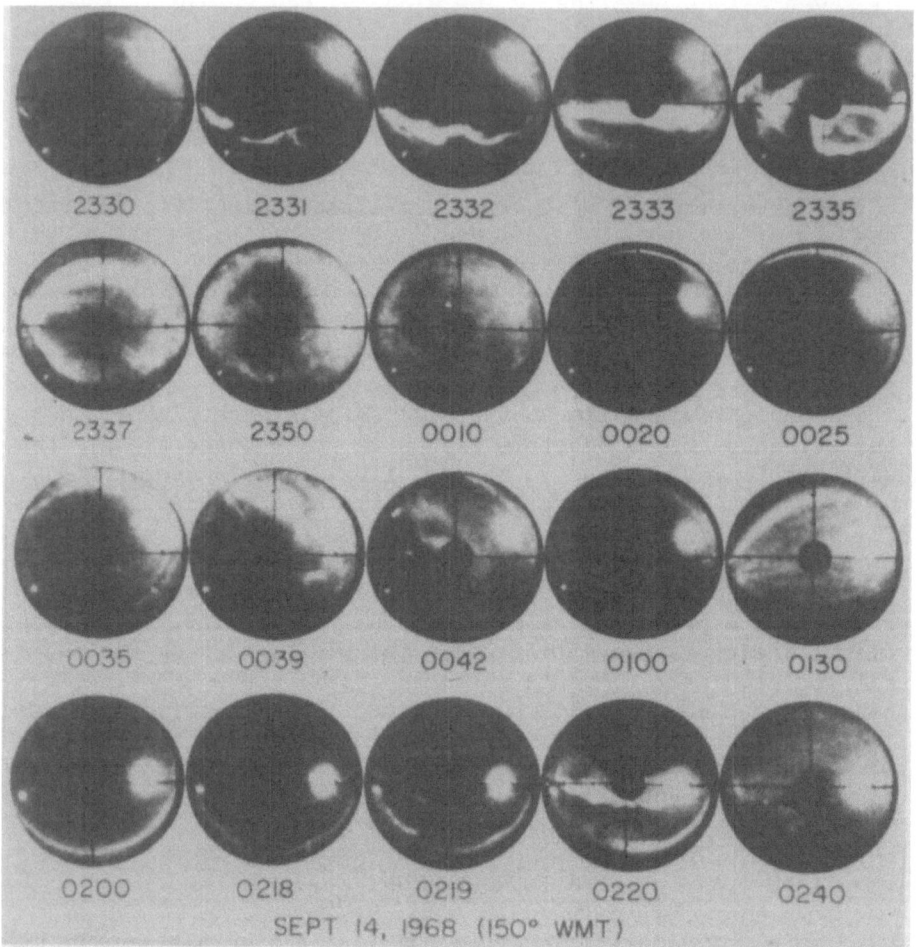

Figure 1 Auroral substorms recorded by an all-sky camera at College, Alaska (magnetic latitude $\approx 65^\circ$). Magnetic local time span of pictures $\approx$ 2130 to 0040.

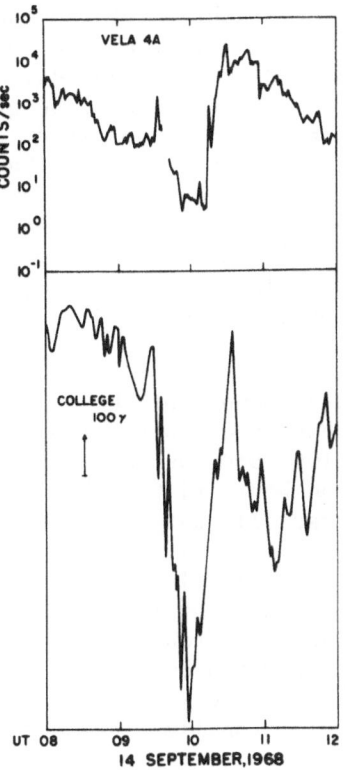

Figure 2 Top: Counting rate of a Geiger counter sensitive to electrons, $E_e \gtrsim 40$ keV on Vela 4A. Bottom: H-component magnetic record from College, Alaska

substantially enhanced by southward turning of the IMF, as Dungey's model would imply.

Essential ingredients of Dungey's model are (a) the merging of magnetic field lines at the magnetopause and in the magnetotail and (b) the convective flow of solar wind plasma into the magnetotail and thence earthward and anti-earthward from a tail neutral line. Measurements of plasma flow in the outer magnetosphere, made during the past few years, have a very direct bearing on these essential features of the model and constitute tests which are perhaps more direct than those alluded to above. This paper treats those plasma flow measurements and associated dynamical processes in the outer magnetosphere.

2. DYNAMIC BEHAVIOR OF THE PLASMA SHEET DURING SUBSTORMS

The magnetotail plasma sheet becomes thin at an early stage of a substorm and then becomes thick again during a late stage (Hones et al., 1967). Observations made during a substorm on September 14, 1968, and treated in detail by Hones et al. (1971) nicely illustrate this behavior. The onset of the substorm's

expansive phase was clearly seen, in all-sky camera pictures taken
at College, Alaska, as a sudden appearance of a bright arc to the
south of the zenith (Figure 1) which occurred between 2330 and 2331
Alaska time (0930 - 0931 UT). (Henceforth, all times will be given
in UT unless noted otherwise.) The auroras advanced poleward,
reaching the zenith by $\sim$ 0934 and spreading far poleward of the
zenith by 0950. Between 1010 and 1020 auroras disappeared from
much of the camera's field of view but bright auroras appeared on
the northern horizon. Examination of all of the pictures from
College (taken at 30-second intervals) reveals that this poleward
shift of the auroral emission occurred rather suddenly at $\sim$ 1016 -
1017.

Figure 2 shows, at the bottom, the H-component of the magnetic
field at College. A 500-gamma negative bay began to develop rapidly
at $\sim$ 0930, reached its peak just before 1000 and was recovering
rapidly by $\sim$ 1015. For $\sim$ 30 minutes to an hour prior to 0930 the
magnetic field seemed to be decreasing gradually but erratically.

The counting rate of a Geiger counter, sensitive to electrons
(E_e > 40 keV), on the Vela 4A satellite is shown at the top of
Figure 2. Vela 4A was at $r \sim 18\ R_E$ in the late evening sector of
the magnetotail and situated $\sim$ 1 to 2 R_E from the estimated position
of the neutral sheet. From $\sim$ 0840 until 0930 the flux of the electrons decreased gradually. At 0930 the electron flux increased
suddenly but briefly and then the flux fell quite rapidly below the
background level of the counter. The flux remained below the background level for the next half hour until 1016 UT when it rose
suddenly and soon reached intensities $\sim$ 10 times greater than those
measured before the substorm.

The principal features of the substorm behavior of the plasma
sheet at the Vela orbit ($r \sim 18\ R_E$), as illustrated by this example
are:

a)   An immediate response of the plasma sheet to the expansive
       phase onset. In the September 14 substorm the immediate
       response was a brief increase of electron flux, quickly
       followed by a complete dropout of the flux. This response is
       fairly common but on other occasions the immediate response
       to the expansive phase is simply a rapid dropout of the flux.
       (Disappearance and reappearance of the flux of energetic
       electrons are usually good indications of the thinning and
       thickening of the plasma sheet (Hones et al., 1967) since
       the energetic electrons seem almost invariably to be confined
       to the region occupied by the lower energy protons and elec-
       trons which define the plasma sheet.)

b)   The plasma sheet remains thin for times ranging from tens of
     minutes to an hour or more while active auroras and the elec-
     trojet prevail at auroral latitudes, i.e., in the invariant
     latitude range $\Lambda \overset{\sim}{\sim} 65°$ to 70°.

c)   Plasma sheet thickening often coincides with a sudden shift
     of the auroras and electrojet to latitudes poleward of auroral
     latitudes, i.e., to $\Lambda \overset{\sim}{\sim} 75°$.

d)   The thickened plasma sheet contains hotter plasma and higher
     fluxes of energetic electrons than prevailed in the plasma
     sheet before thinning occurred.

Not every substorm expansive phase is preceded by such a pronounced
gradual thinning of the plasma sheet and gradual bay development
as was the September 14 event. Thus, consideration of these features as indicative of a substorm "growth phase" is subject to some
question.

   A study has been made of the distribution, at the Vela orbit,
of complete plasma sheet dropouts such as that of September 14, 1968.
Figure 3 shows that distribution, derived from several satellite
years of Vela data (Lui et al., 1975). Ignoring the points A
through H, which pertain to the unusual magnetospheric conditions,
one notes that the complete plasma dropouts are seen in a band
several earth radii thick that lies close to the tail's midplane
near local midnight and that lies well above the midplane in the
dawn and dusk portions of the tail. The distribution signifies
that the plasma sheet's half thickness does not often exceed $\sim 5\ R_E$

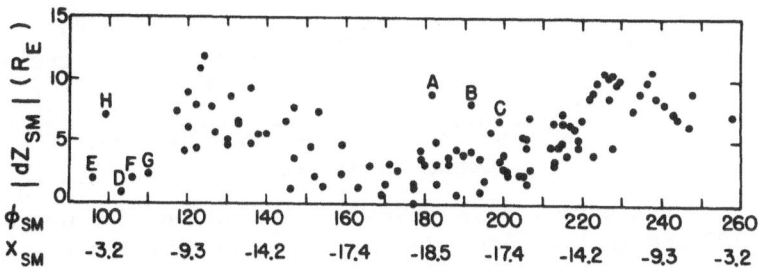

Figure 3 Distribution, at the Vela satellite orbit, of complete
plasma dropouts associated with substorms. The ordinate is the
estimated absolute distance of the satellite from the neutral sheet.
Φ_{SM} is the satellite's solar magnetospheric longitude and X_{SM} is the
solar magnetospheric X-component of its geocentric radial distance
from the earth.

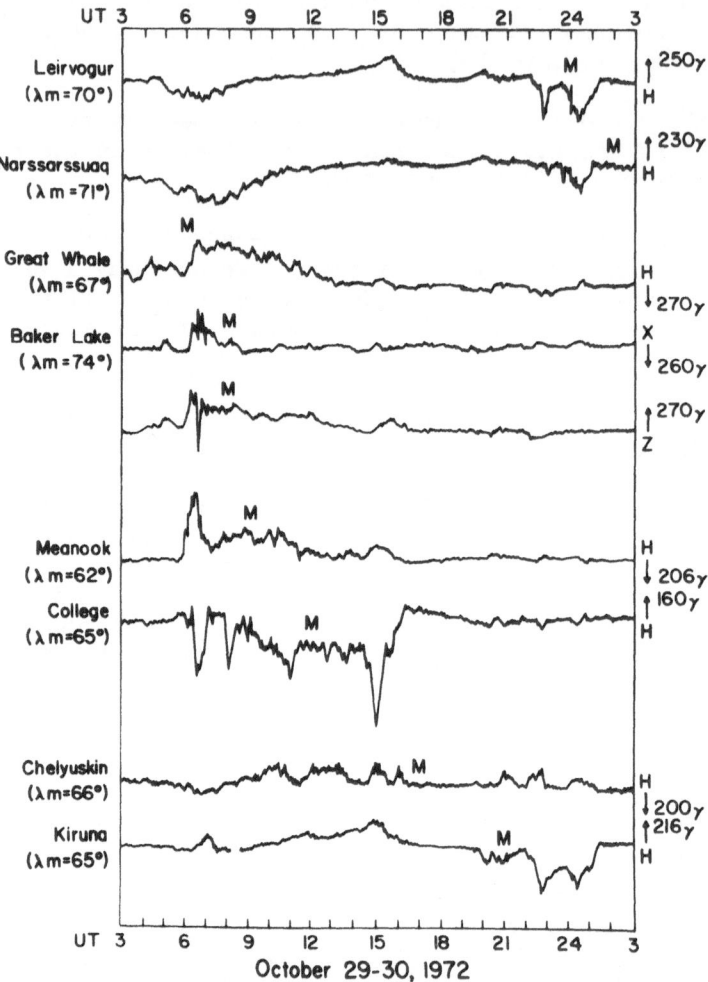

Figure 4 Magnetograms from auroral zone stations for October
29-30, 1972. Note that both the X and Z component records are
given for Baker Lake. M indicates the approximate universal
time of magnetic local midnight.

near local midnight and that it often thins to a half thickness
$\lesssim$ 1 R_E there. The half thickness of the dawn and dusk sections
often reaches $\sim$ 10 R_E and those regions seldom thin to a half
thickness $\lesssim$ 5 R_E.

It might be imagined that the deep dropouts of plasma in the
midnight sector of the tail imply a complete loss of the plasma
sheet there on those occasions. But this is not the case. Hones
et al. (1974) have shown, with Vela satellite measurements, that
a thin residual plasma sheet is found near the tail's midplane
throughout such events and that the plasma within it flows steadily
in the antisunward direction. This is found to be true also at
greater distances from the earth, as will next be illustrated with
data from the IMP 6 satellite. Magnetograms from several stations,
drawn in Figure 4, show that a fairly well-isolated substorm started
near 0600 on October 29, 1972. The negative bay at Meanook began at
0555, reached its peak at 0630 and began immediately to recover.
The sharp negative spike in the z-component of the Baker Lake record
shows that the electrojet surged poleward to the zenith of that low
polar cap station near 0630.

The vectors drawn in Figure 5 depict the magnitude and direction of the flow of plasma protons as measured by IMP 6, at r $\sim$
31 R_E, in the period encompassing the $\sim$ 0600 substorm. The vectors
are derived from measurements of the flux of protons in the energy
range 30 eV < E_p < 30 keV measured with the Los Alamos Scientific
Laboratory plasma probe on the satellite (Hones et al., 1975). The
measurements are made at many angles during the rotation of the
spacecraft, whose spin axis was perpendicular to the ecliptic plane.
The flow vectors are derived from measured azimuthal asymmetries
in the proton flux. From $\sim$ 0555 through $\sim$ 0630 the protons flowed
rapidly in a general anti-sunward direction. Measurements of the
plasma electrons with the same plasma probe showed that the plasma
density dropped dramatically between $\sim$ 0555 and $\sim$ 0615 and then
started to recover, reaching its original value by $\sim$ 0700. Thus,
the antisunward flow of plasma was a feature of the thinned plasma
sheet and both features prevailed from the substorm's onset until
the poleward surge of the electrojet about 35 minutes after onset.
The thickening of the plasma sheet was marked by a reversal of the
plasma flow to the sunward direction.

The anti-sunward flow of plasma in the thinned plasma sheet
during a substorm's expansive phase and the reversal to sunward
flow coincident with thickening of the plasma sheet as the electrojet
moves suddenly poleward later in a substorm, as illustrated in these
data from IMP 6, constitute another typical feature of the plasma
sheet's behavior during a substorm. This feature, along with features (a) - (d) listed earlier can be understood in terms of the substorm model in Figure 6. The plasma sheet may thin gradually for

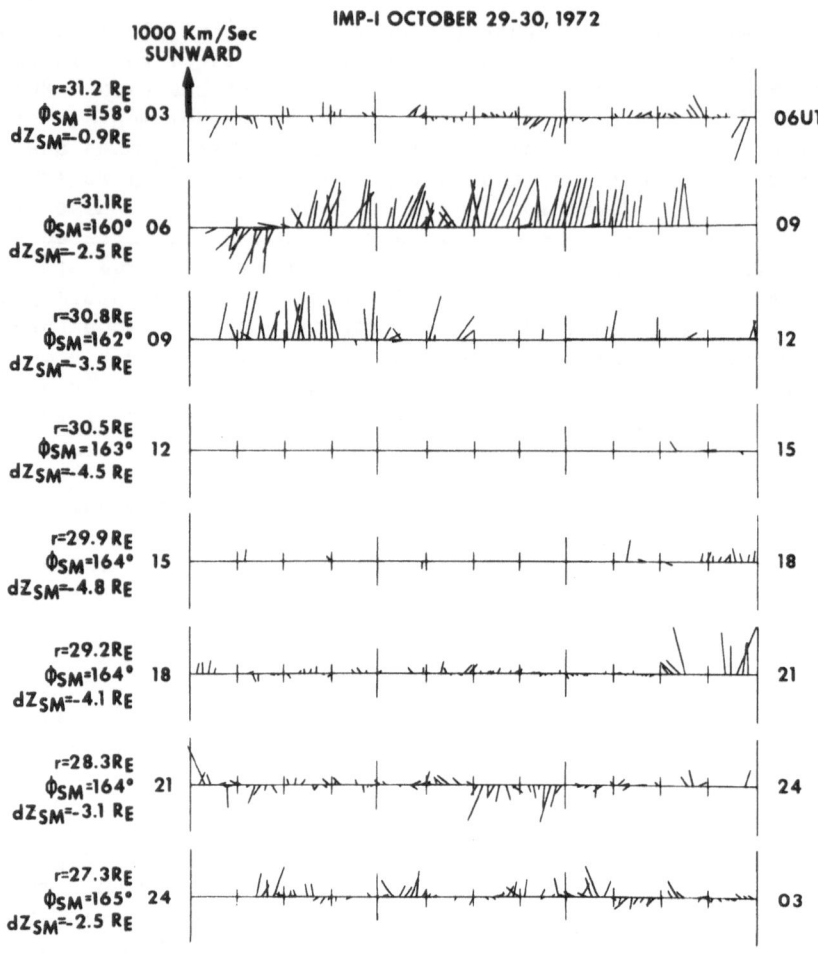

Figure 5   Proton flow vectors measured by IMP 6 in the plasma sheet on October 29-30, 1972. Each horizontal line represents three hours of data. The heavy vertical arrow at the left of the top line representes a sunward flow at 1000 km/sec. The location of the satellite and its position with respect to the neutral sheet (dZ_{SM}) are indicated at the left of each 3-hour line.

some time before the onset of the expansive phase (T = 0) as in
Figure 6 (a), but as we have indicated above, this is a questionable
issue and the arrows in the plasma sheet, together with the question
marks are meant to indicate that no characteristic flow pattern of
the plasma has been detected in this pre-expansive phase interval.
A magnetic neutral line forms across the tail, earthward of the Vela
orbit, at T = 0 (Figure 6 (b)), its formation possibly initiated by
a tearing mode instability (Schindler, 1974). This action initiates
very rapid thinning and rapid flow of plasma earthward and tailward,
as indicated. If a Vela satellite is in the plasma sheet at T = 0
it may see a brief tailward surge of plasma, then its disappearance,
as in the September 14 event (Figure 2) or, if the satellite is
appropriately situated with respect to the thin residual plasma
sheet it will see the continuing tailward flow of plasma that
prevails there throughout the expansive phase (T = 0 to T $\stackrel{\sim}{\sim}$ 60,
Figure 6 (c)), as in the October 29 event (Figure 5). When the
electrojet suddenly shifts poleward late in the substorm, the neutral
line moves rapidly farther out into the tail and a satellite, even
at a relatively large distance from the tail midplane, is enveloped
by the hot, earthward-flowing plasma (Figure 6 (d)).

The model of Figure 6, which specifically suggests that the
neutral line forms earthward of the Vela satellite orbit at r $\stackrel{\sim}{\sim}$
18 R_E (marked by V) represents occurrences in a moderate to strong
substorm. There are frequent instances in the Vela and IMP 6 data,
however, of the onset of rapid earthward flow of plasma near the
tail's midplane with little or no preceding tailward flow. Figure 7
shows an example of this, observed by IMP 6 at r $\stackrel{\sim}{\sim}$ 28 R_E. Rapid
plasma flow occurred for about one hour starting at $\sim$ 0130 on October 5, 1972. Initially, for $\sim$ 3 minutes the plasma flowed earthward
but it quickly reversed and flowed tailward for the remainder of
the hour. Examination of magnetic records from many stations in
the midnight sector of the earth revealed no substorm signatures.
At Leirvogur and Narssarssuaq, which were quite near local midnight,
only very weak micropulsations were observed to coincide with this
period of rapid plasma flow. We believe that the rapid flow observed
on this occasion may have been associated with a substorm whose
effects were confined to high latitudes at the earth (i.e., a substorm on the contracted auroral oval) and that the associated neutral
line formed initially quite near the IMP 6 satellite. Our interpretation of this event is supported by observations of other such
events (Hones et al., 1973).

3. ORIGIN OF THE PLASMA SHEET

The substorm model in Figure 6 has solar wind magnetic field
lines (and thus, solar wind plasma) reaching the plasma sheet by
merging of the solar wind field at the sunward magnetopause, sweeping

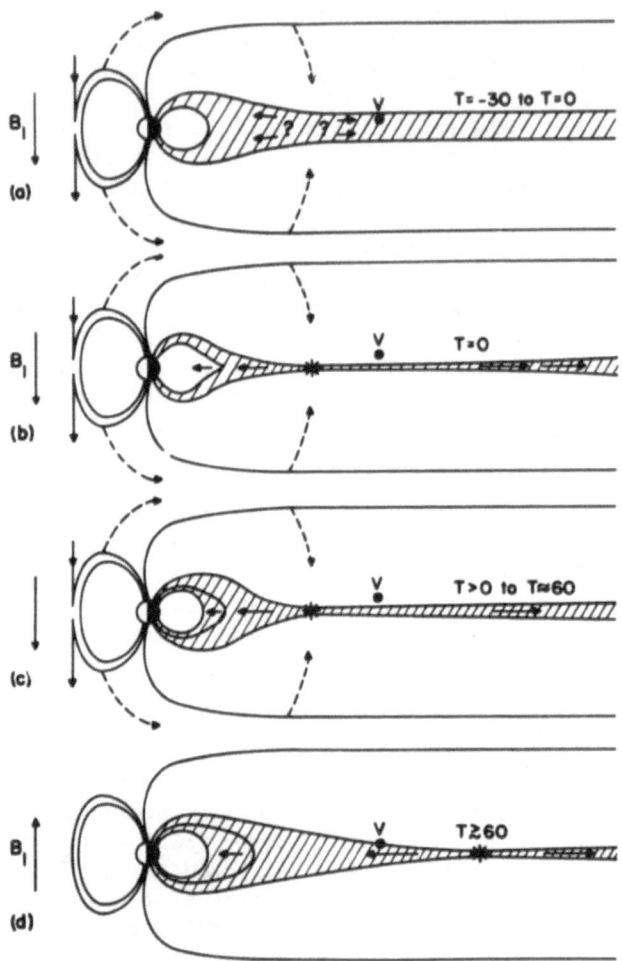

Figure 6   Schematic representation of substorm-related variations of the magnetotail plasma sheet. Dashed lines indicate suggested path of plasma convection into the tail lobes. Star represents magnetic neutral line. V indicates possible location of a Vela satellite. Solid arrows within plasma sheet indicate flow of plasma there.

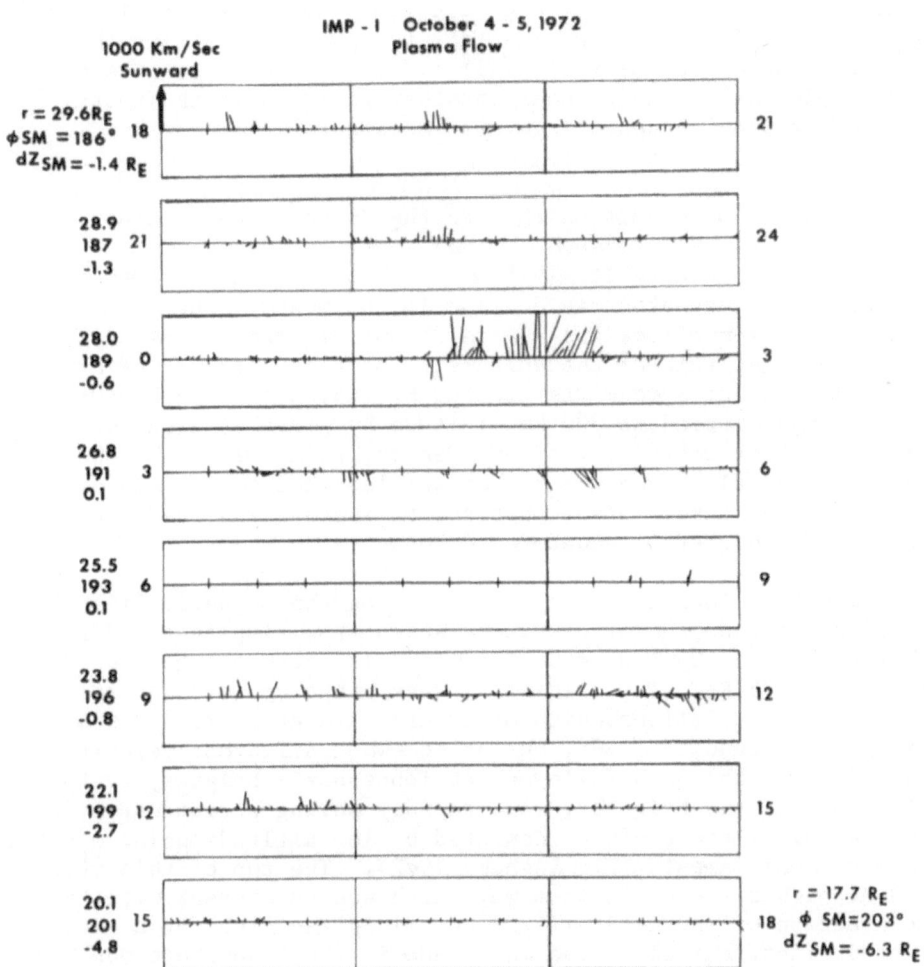

Figure 7 Vectors representing plasma flow measured by IMP 6 at a time when magnetic disturbance in the auroral zone was very weak (see text).

of plasma and the newly connected lines into the tail by the on-
flowing solar wind, convection of the plasma toward the tail's
midplane and, finally, reconnection of field lines across the midplane to energize and confine the plasma in the plasma sheet. This
course of entry of solar wind plasmas was depicted by Dungey (1961)
and by Levy et al. (1964). In the previous section we have described
evidence for reconnection of magnetic field lines across the tail's
midplane and injection of plasma into the plasma sheet (on the
earthward side of the neutral line). We shall next describe other
observations which tend to confirm other portions of the total
plasma entry process depicted by these authors.

A.   Convection of Plasma through the Magnetospheric Cleft

At 0032 on January 11, 1975 a streak of barium ions was projected upward along magnetic field lines in the magnetospheric cleft in
a joint operation of the Los Alamos Scientific Laboratory and the
Geophysical Institute of the University of Alaska (Jeffries et al.,
1975). The barium streak was created by explosion of a barium-lined
shaped charge carried to 500 km altitude by a rocket launched from
Cape Parry, Northwest Territories, Canada, which was at $\sim$ 1400
magnetic local time when the rocket was launched. A plasma probe
carried by the rocket showed that the barium was injected 1.6 degrees
poleward of the cleft's equatorward boundary.

The barium ions travelled upward along the magnetic field lines
at $\sim$ 14 km/sec and the streak was tracked from airplanes and from
the ground for $\sim$ 30 minutes during which it reached a geocentric
distance of $\sim$ 5 R_E. The original streak broke up into several
field-aligned striations which moved eastward at nearly constant
invariant latitude, $\Lambda \sim 78°$, but at somewhat different velocities.
Figure 8 shows the path followed, at ionospheric heights, by the
foot of one of the brighter, more rapidly moving striations. The
path lies along the cleft as depicted by low altitude polar orbiting
satellite measurements (Winningham, 1972). The top of this streak
moved at higher speed and in a way which was consistent with frozen-
in convection of the entire flux tube along the cleft by a southward-
directed electric field whose magnitude in the ionosphere was $\sim$ 24
mV/m (Stenbaek-Nielsen et al., 1975).

In the context of the present discussion, this barium streak
experiment illustrates the course that may be followed by solar
wind flux tubes shortly after they have merged with the earth's
field at the subsolar magnetopause. The z-component of the solar
wind magnetic field was negative (about -1.5 gammas) throughout the
experiment.

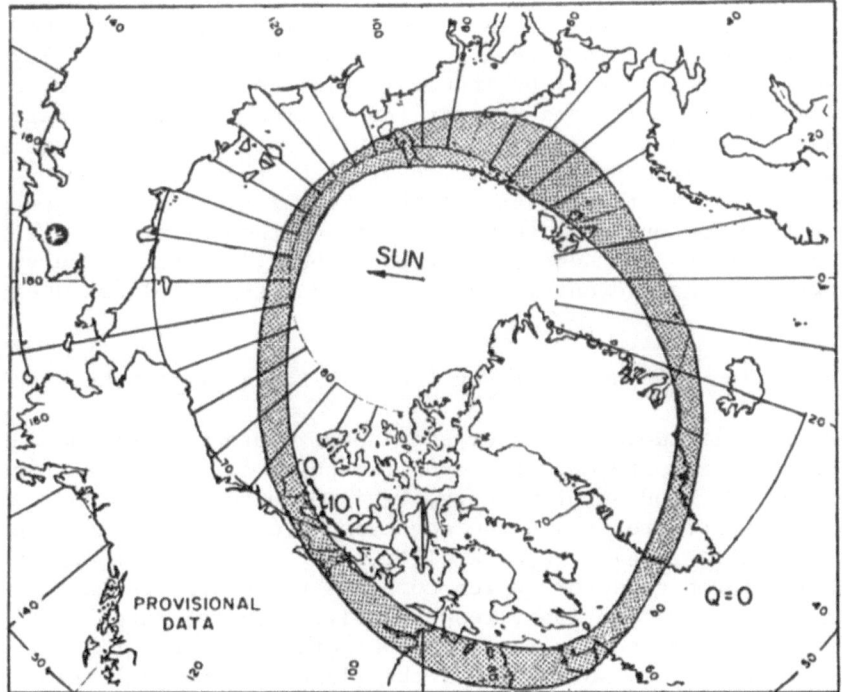

Figure 8 Track of the foot of the barium ion flux tube recorded
on January 11, 1975. Numbers along the track indicate the time,
in minutes, from the release of the barium ions at 500 km
altitude. The gray sector is the Feldstein oval for Q = 0.
(From Jeffries et al., 1975.)

B. Plasma Convection through the Magnetotail Surface Layer

The newly merged flux tubes are pulled back along the tail
where they form a region which has been observationally identified
as the magnetotail surface layer or boundary layer (Hones et al.,
1972). Some features of this surface layer are illustrated in
Figure 9. As the Vela 5A satellite passed from the magnetosheath
into the dusk side of the tail on June 9, 1969 it observed, for
∿ 1 to 2 hours, plasma flowing anti-sunward (as in the magnetosheath), but at ever decreasing velocity. The plasma density
also decreased rather monotonically. Such diminutions of plasma
flow velocity and density with decreasing distance from the tail's
axis are observed on many tail boundary crossings. Fig. 10 shows a
distribution of such observations derived from several months of
data from one Vela satellite. It is not known at present whether
the asymmetry of this distribution is a real feature of the surface
layer or simply reflects a bias in the sampling provided by the
data.

It is our view that the surface layer comprises erstwhile solar wind magnetic flux tubes whose feet are connected to the earth, probably through recent merging, and which are being layed down into the magnetotail by the on-flowing solar wind. The decreasing flow velocity and density result from partial drainage of plasma (the fastest-flowing component first) downstream as the flux tubes are incorporated into the tail by this convective process. Thus, in the context of the present discussion, passage of flux tubes through the surface layer and thence into the tail lobes constitutes the sequel to step A, above, in the course of solar wind plasma's eventual incorporation into the plasma sheet.

C.   Convection of Plasma through the Tail Lobes **toward** the Midplane

It is well known that energetic solar protons enter the earth's magnetosphere and are detected over the polar caps. In fact there have been many studies of magnetospheric structure based on distributions of solar protons in the magnetosphere (Morfill and Scholer, 1973; Paulikas, 1974). Vela satellites 5A, 5B, 6A and 6B carry energetic proton detectors which comprise three sensors making angles of 45°, 90° and 135°, respectively, with the satellites' spin vectors. These detectors, on the rotating satellites, provide three-dimensional measurements of the fluxes of energetic protons.

Figure 11 shows the intensity of 0.5 - 0.9 MeV solar protons in the north tail lobe as measured by a Vela satellite on April 16, 1970. A number of rather abrupt increases and decreases of the proton flux are seen. During such abrupt changes of intensity the solar proton flux is almost always found to be strongly spin modulated as is illustrated for the increase, R_3 (Figure 11) in Figure 12. The bottom three curves in Figure 12 are the fluxes, in highest time resolution, measured by the three sensors, N, O and S. It is noted that the three sensors see different time variations of the flux as the signals are modulated by the satellite's rotation.

Detailed analyses of these signals reveal that during the flux increase, R_3, and all other increases in Figure 10, the protons were found to be approaching the satellite from the dusk side of the tail. However, during all of the flux decreases, the protons approached the satellite from the dawn side of the tail. Consideration of the gyromotion of protons in the (sunward-directed) magnetic field in the tail's north lobe reveals that this association of flux directions with intensity changes is a consequence of a movement of rather sharply-bounded "slabs" of flux tubes, containing solar protons, downward towards the tail's midplane. Similar observations made at other times, have revealed flux tubes moving <u>upward</u> through the <u>southern</u> lobe of the tail.

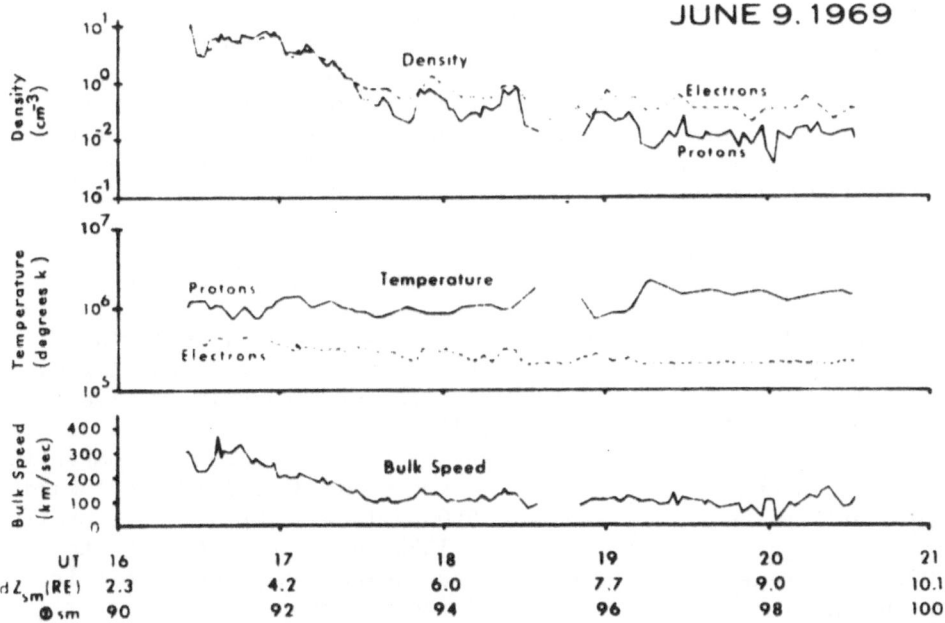

Figure 9   Some properties of the plasma in the magnetotail surface layer, measured by Vela 5A on June 9, 1969. Top: Number density of electrons and protons; Middle: Temperature of electrons and protons; Bottom: Bulk flow speed of protons. The direction of the flow was antisunward. The satellite's location is indicated at bottom.

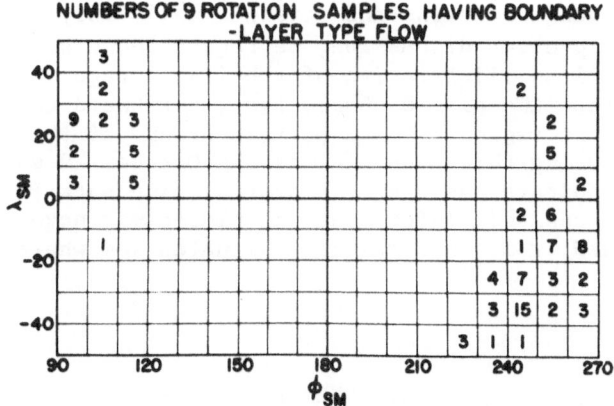

Figure 10   Distribution, at the Vela orbit, of surface-layer type flow observations during several months of operation of Vela satellite 4B. The grid marks 10-degree sectors of solar magnetospheric latitude and longitude within which the flow was observed. The numbers indicate the number of 9-minute data samples which showed such flow (see Hones et al., 1972).

The electric field required to drive the observed convection on April 16, 1970 was in the range $\sim$ 0.5 to 1.0 mV/m, yielding convective speeds of $\sim$ 50 to 100 km/sec. The interplanetary magnetic field was $\sim$ 20 gammas and pointed almost directly southward during the observations. In the context of this discussion, we regard the convection of plasma through the tail lobes, illustrated by these solar proton measurements, as the sequel to its entry through the surface layer in step B, above.

D. Plasma Convection Earthward through the Plasma Sheet

Projection of plasma earthward from a neutral line during a substorm was illustrated earlier (Figure 5). An even more impressive illustration of earthward projection of plasma is provided in data taken by IMP 6 on October 22, 1971 (Figure 13). The Explorer 41 satellite, in the solar wind, showed the interplanetary magnetic field, of intensity $\sim$ 20 gammas, to be oriented at large southward angles during the entire interval 0500 - 1900. IMP 6 was generally outside the plasma sheet prior to $\sim$ 1330 and again from 1500 to 1600. But whenever the satellite was inside the plasma sheet it sensed very rapid earthward flow of the plasma. IMP 6 was at $r \stackrel{\sim}{<}$ 22 to 27 R_E during the interval of interest and much of the time was less than 1 R_E from the neutral sheet. The satellite data show that plasma was flowing rapidly and continually earthward through a plasma sheet which, at least at times, was very thin, during this interval of continual southward interplanetary magnetic field. Figure 14 shows magnetic records from several auroral zone stations. The period was generally disturbed. However, the individual disturbances did not display temporal features of substorm magnetic bays but were longer lasting and more gradually varying. In the context of the present discussion, we regard these phenomena of October 22, 1971 as illustrative of the final stage of solar wind plasma entry into the plasma sheet, i.e., earthward convection from some distant region. Since no turn-around of the plasma flow was observed during this event (as in the October 29 substorm, Figure 5) to indicate passage of a neutral line outward past the satellite, one can only infer that the plasma may have been projected earthward from a neutral line situated somewhere beyond $r \stackrel{\sim}{<} 27$ R_E.

4. SUMMARY

Dimensional changes of the magnetotail plasma sheet associated with substorms and plasma flow within it at those times are consistent with a substorm model having the following features:

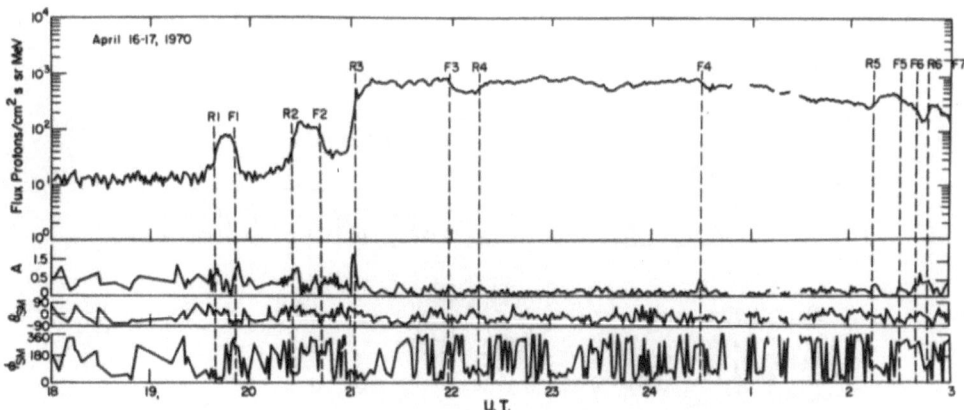

Figure 11 Flux of solar protons (0.5 MeV < Ep < 0.9 MeV) measured in the north magnetotail lobe by Vela 6A on April 16-17, 1970. (From Palmer et al., 1975.)

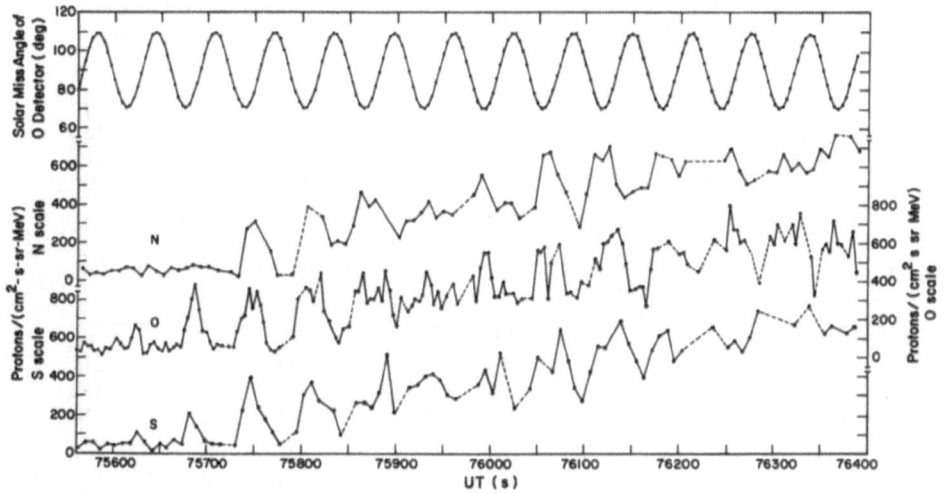

Figure 12 High time resolution record of the proton flux measured by the N, O and S sensors on Vela 6A during rise R_3 (Figure 11). The sinusoidal curve at top is indicative of the sensors' solar aspect angle.

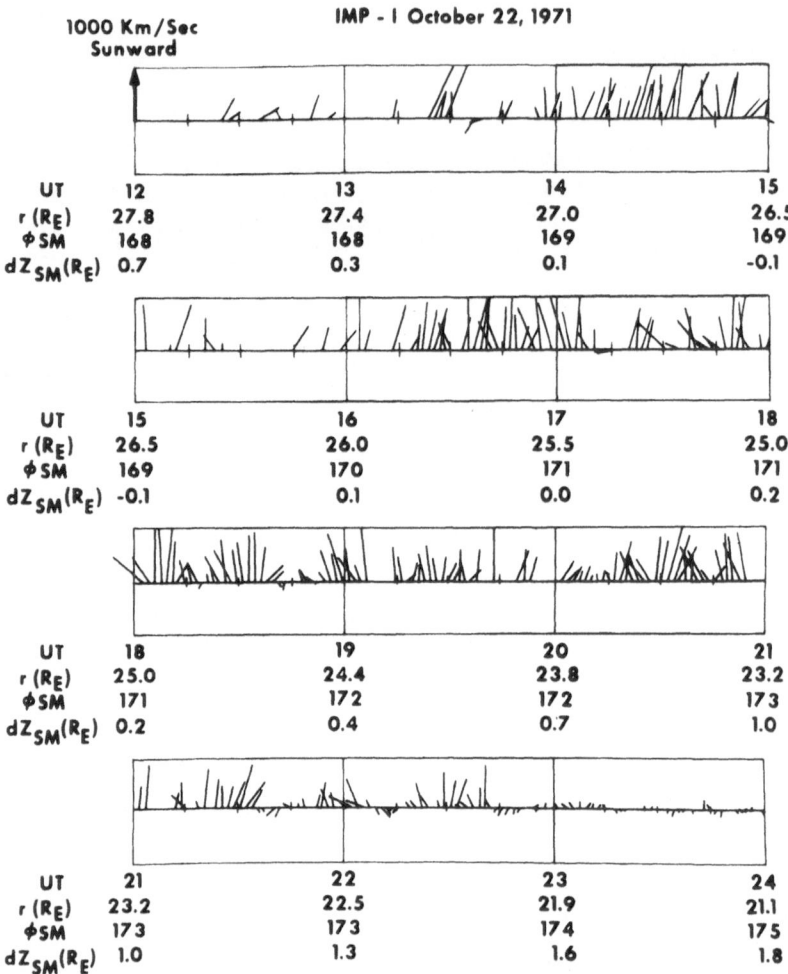

Figure 13   Vector representation of proton flow measured by IMP 6 in the plasma sheet on October 22, 1971. See caption of Figure 5 for details.

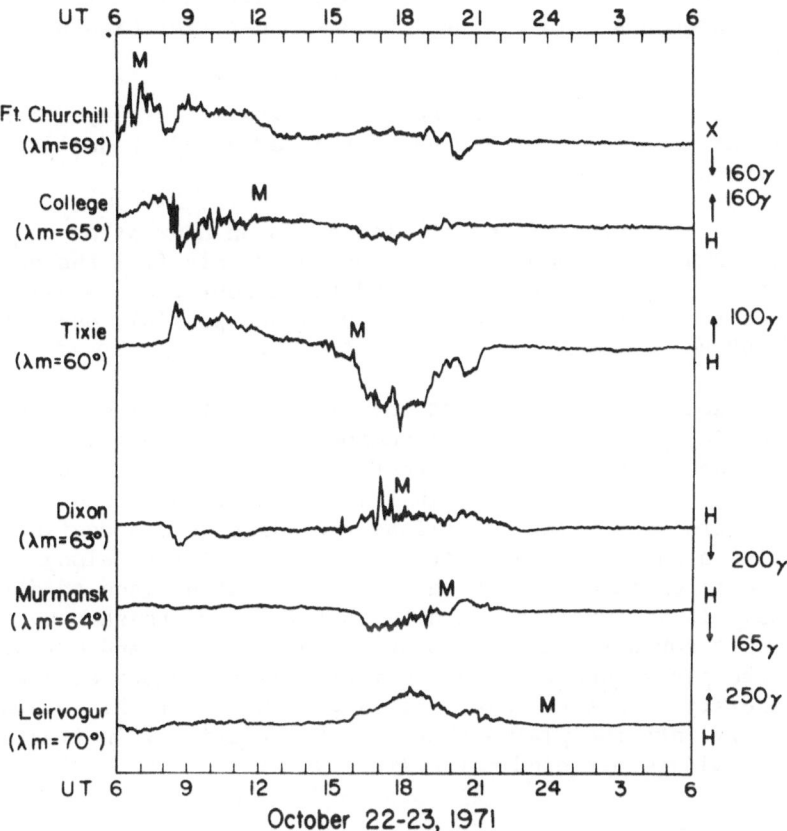

Figure 14 Magnetic records from auroral zone stations on October 22-23, 1971. M indicates universal time of magnetic local midnight.

a)   A magnetic neutral line forms across a substantial portion of
     the magnetotail at the onset of the substorm's expansive phase.
     The neutral line forms at geocentric distance < 18 R_E for
     moderate to intense substorms but at distances $\gtrsim$ 30 R_E for
     weak substorms. The neutral line clearly does <u>not</u> simply
     move to its near-earth position from some location farther
     out in the tail. Plasma flows rapidly earthward from the
     neutral line, once it is formed, causing substorm phenomena
     at the earth; it also flows rapidly tailward from the neutral
     line through a thin residual plasma sheet that exists through-
     out the substorm.

b)   Prior to the expansive phase onset there may or may not occur
     evident precursory phenomena at the earth and gradual thinning
     of the plasma sheet.

c)   Late in a substorm (tens of minutes to an hour after expansive
     phase onset) the neutral line moves suddenly from the near-
     earth location which it occupied throughout the expansive
     phase to a position beyond $r \stackrel{\sim}{\sim} 30$ R_E and possibly even beyond
     $r = 60$ R_E.

It is generally agreed that the plasma in the plasma sheet
originates in the solar wind. A commonly considered route of entry
of solar wind plasma into the magnetotail is by way of the magnetospheric cleft, convection through the tail lobes and thence projection earthward from a magnetotail neutral line to its confinement
in the plasma sheet. Several stages of plasma travel along this
route were illustrated here by actual observations made on different
occasions. It is important to note that three of these stages,
convection through the cleft, through the tail lobe and earthward
through the plasma sheet, were observed during periods of southward
interplanetary magnetic field. It is possible that this course of
plasma entry into the plasma sheet may be relatively ineffectual
when the field is directed northward.

ACKNOWLEDGEMENT

This work was performed under the auspices of the U.S. Energy
Research and Development Administration.

REFERENCES

Akasofu, S.-I., P. D. Perreault and F. Yasuhara, Auroral substorms
and the interplanetary magnetic field, J. Geophys. Res. 78,
1490, 1973.

Arnoldy, R. L., Signature in the interplanetary medium for substorms, J. Geophys. Res. 76, 5189, 1971.

Dungey, J. W., Interplanetary magnetic field and the auroral
zones, Phys. Rev. Letters 6, 41, 1961.

Fairfield, D. H. and L. J. Cahill, Jr., Transition region magnetic
field and polar magnetic disturbances, J. Geophys. Res. 71,
155, 1966.

Foster, J. C., D. H. Fairfield, K. W. Ogilvie and T. J. Rosenberg,
Relationship of interplanetary parameters and occurrence of
magnetospheric substorms, J. Geophys. Res. 76, 6971, 1971.

Hones, E. W., Jr., J. R. Asbridge, S. J. Bame and I. B. Strong,
Outward flow of plasma in the magnetotail following geomagnetic bays, J. Geophys. Res. 72, 5879, 1967.

Hones, E. W., Jr., R. H. Karas, L. J. Lanzerotti and S.-I. Akasofu,
Magnetospheric substorms on September 14, 1968, J. Geophys.
Res. 76, 6765, 1971.

Hones, E. W., Jr., J. R. Asbridge, S. J. Bame, M. D. Montgomery,
S. Singer and S.-I. Akasofu, Measurements of magnetotail
plasma flow made with Vela 4B, J. Geophys. Res. 77, 5503,
1972.

Hones, E. W., Jr., J. R. Asbridge, S. J. Bame and S. Singer,
Magnetotail plasma flow measured by Vela 4a, J. Geophys.
Res. 78, 5463, 1973.

Hones, E. W., Jr., A. T. Y. Lui, S. J. Bame and S. Singer, Prolonged tailward flow of plasma in the thinned plasma sheet
observed at r $\sim$ 18 R_E during substorms, J. Geophys. Res. 79,
1385, 1974.

Hones, E. W., Jr., S. J. Bame and J. R. Asbridge, Proton flow
measurements in the magnetotail plasma sheet made with IMP 6,
J. Geophys. Res., to be published, 1975.

Jeffries, R. A., W. H. Roach, E. W. Hones, Jr., E. M. Wescott,
H. C. Stenbaek-Nielsen, T. N. Davis and J. D. Winningham,
Two barium plasma injections into the northern magnetospheric
cleft, Geophys. Res. Letters, to be published, 1975.

Levy, R. H., H. E. Petschek and G. L. Siscoe, Aerodynamic aspects
of magnetospheric flow, AIAA Journal 2, 2065, 1964.

Lui, A. T. Y., E. W. Hones, Jr., D. Venkatesan, S.-I. Akasofu and
S. J. Bame, Distribution of plasma sheet thinnings at the Vela
orbit and some effects of geomagnetic activity and the IMF
upon their occurrence, J. Geophys. Res., to be published, 1975.

Morfill, G. and M. Scholer, Study of the magnetosphere using
energetic solar particles, Space Sci. Rev. 15, 267, 1973.

Palmer, I. D., P. R. Higbie and E. W. Hones, Jr., Gradients of
    solar protons in the high latitude magnetotail and the magne-
    tospheric electric field, J. Geophys. Res., to be published,
    1975.
Paulikas, G. A., Tracing of high-latitude magnetic field lines by
    solar particles, Rev. of Geophys. and Space Phys. 12, 117,
    1974.
Schindler, K., A theory of the substorm mechanism, J. Geophys.
    Res. 79, 2803, 1974.
Stenbaek-Nielsen, H. C., E. M. Wescott, T. N. Davis, R. A. Jeffries,
    W. H. Roach, E. W. Hones, Jr. and J. D. Winningham, Plasma
    convection and electric fields in the quiet time magnetospheric
    cleft from the Tordo Dos experiment, Paper presented at the
    56th annual meeting of the AGU, Washington, DC, June 1975.
Winningham, J. D., Characteristics of magnetosheath plasma obser-
    ved at low altitude in the dayside magnetospheric cusps, in
    Earth's Magnetospheric Processes, edited by B. M. McCormac,
    pp. 68-80, D. Reidel, Dordrecht, Netherlands, 1972.

HOT PLASMA DYNAMICS WITHIN GEOSTATIONARY ALTITUDES

Donald J. Williams

Space Environment Laboratory, ERL, NOAA

Boulder, Colorado

INTRODUCTION

We present observations and results pertaining to hot plasma
dynamics as observed in the earth's magnetosphere. Owing to the
fact that the earth's magnetosphere-ionosphere system is the only
naturally occurring plasma available to cost effective in-situ
study, these and similar results are not only valuable in their own
right, but can be used in possible direct tests of theories and
hypothesis concerned with the behavior of other remote naturally
occurring plasmas such as the solar corona, planetary magnetospheres
and ionospheres, pulsars, and the interstellar and intergalactic
media.

Since the present results stem from observations made by the
Explorer 45 satellite our discussion will generally be limited to
the region within geostationary altitude. We begin with a discussion
of protons during geomagnetic storms and present briefly considerations of energy densities, hot-cold plasma interactions, stable and
turbulent regions, the distribution function, charge exchange, SAR
ARC generation, and adiabatic and non-adiabatic effects. We then
present the behavior of the energetic electron population during
quiet and storm times along with a discussion of the mechanisms
responsible for the quiet time and post injection recovery behavior.
In all these discussions relevant wave-particle interactions are
considered and are seen to be an important factor in determining
hot plasma dynamics in the magnetosphere.

Explorer 45 (S^3) was launched on November 15, 1971 into an
elliptical orbit having an apogee of 5.24 earth radii, a perigee
of 220 km, a period of 7.82 hours and an inclination of 3.5°. An

electrostatic analyzer-channeltron instrument plus three separate
solid state detector systems measure proton intensities from 0.73
keV to 872 keV in 28 energy steps and electron intensities from
0.73 keV to 560 keV in 20 steps. (Williams et al, 1969; Longanecker
and Hoffman, 1973).

The satellite spin axis is maintained in the plane of the orbit
and pitch angle information is obtained by sectoring the satellite
spin (8.451 seconds) into 32 segments. All pitch angles presented
here are measured pitch angles using simultaneous data from the
onboard 3-axis fluxgate magnetometer.

Figure 1 shows the Explorer 45 orbit for December 17, 1971.
Apogee for the orbit shown is at $\sim$ 2100 UT and there is an $\sim$ 12°
per month precession towards dusk. Since much of the data to be
presented is from the December 17, 1971 geomagnetic storm, we show
in Figure 2 the D_{ST} values throughout this storm along with Explorer
45 orbit coverage.

PROTONS

Energetics

Strictly speaking this section should be termed "IONS" since
the Explorer 45 instruments were not able to uniquely identify ions
over the energy range of interest here (1-1000 keV). While hot

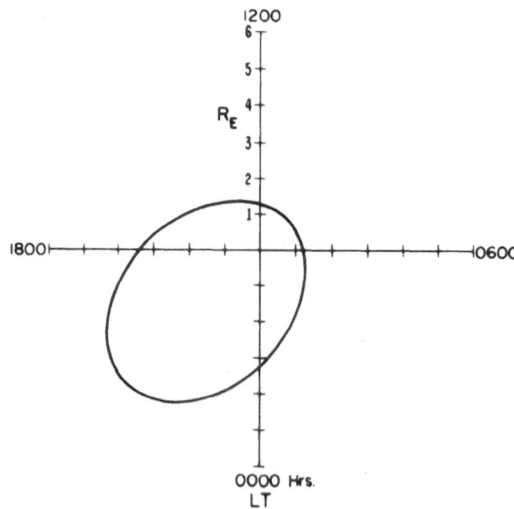

Figure 1. Explorer 45 Orbit, December 17, 1971, Shown in Polar
Coordinates of Local Time (Hours) and Altitude (Earth Radii).

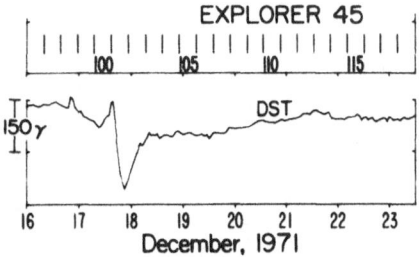

Figure 2. D_{ST} and Explorer 45 Orbit Coverage for December 16-23, 1971.

plasma composition studies at high altitudes have not yet been conducted, there exists speculation based on low altitude studies that other ions, mainly O_{16}^{n+}, may be significant contributors at times to the storm time ring current (Shelley et al, 1972; 1974a, b; Sharp et al, 1974).

As this is unresolved (Williams, 1974b), we shall continue to assume protons as the main contributor to the inner magnetospheric hot plasma in the results and analyses which follow.

Frank (1967a, b; 1970) has obtained the initial direct measurements of the hot ring current plasma in the magnetosphere. These early results, limited generally to a small pitch angle range at high altitudes, have been extended to all pitch angles by the Explorer 45 observations. We therefore shall consider the energetics of the proton component of the hot plasma as it develops during a geomagnetic storm using Explorer 45 results from the December 17, 1971 storm (Smith and Hoffman, 1973; Hoffman, 1973). Figure 3 shows energy density spectra for pre-storm conditions (orbit 97) and at the time of the formation of the symmetric ring current (orbit 102). Both spectra were taken where the energy densities were near maximum values (see Figure 4). The hot plasma responsible for the bulk of the storm time ring current is clearly seen. These protons (1-138 keV) account for > 90% of the particle energy densities shown during the storm and < 20% of the energy density for $L \lesssim 4$ during quiet times.

Figure 4 shows the equatorial energy density radial profiles for pre-storm quiet (orbit 97), main phase injection (orbit 101), and recovery phase (orbits 102 and 103). The development from an asymmetric to a symmetric hot plasma distribution is evident.

Figure 5 shows particle and field energy densities and their

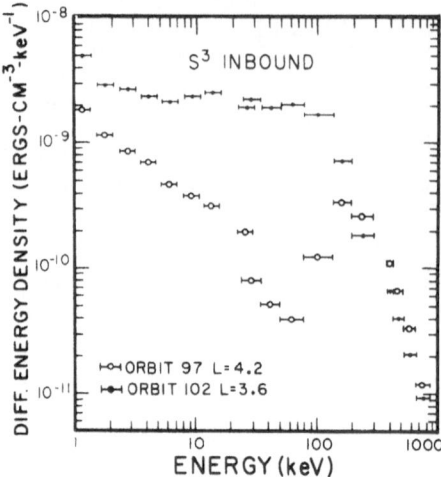

Figure 3.   Proton Energy Density Spectra as Measured by Explorer 45 (S³) During Pre-Storm Quiet (Orbit 97) and During Formation of Symmetric Ring Current (Orbit 102).

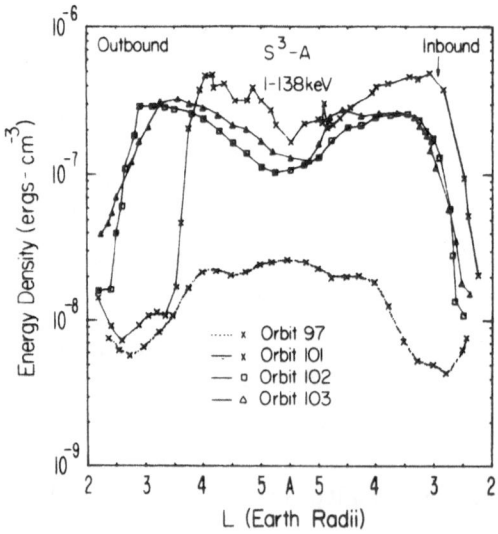

Figure 4.   Proton Energy Density Radial Profiles as Measured by Explorer 45 (S³-A) During Pre-Storm Quiet (Orbit 97) Main Phase Injection (Orbit 101), and Recovery Phase (Orbits 102 and 103).

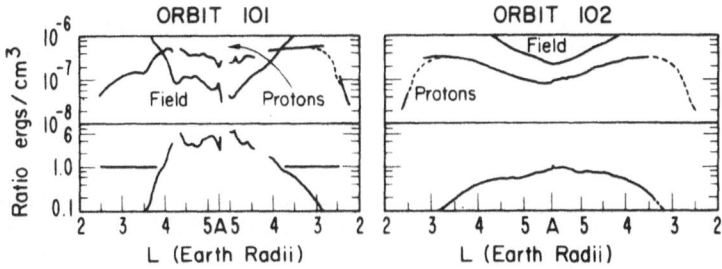

Figure 5. Proton and Magnetic Field Energy Densities and Their Ratio
(β) as Measured by Explorer 45 for Main Phase Injection (Orbit 101)
and the Beginning of Recovery Phase (Orbit 102).

ratio (β) for main phase injection (orbit 101) and recovery phase
(orbit 102). It is clear from Figures 4 and 5 that geomagnetic
storm main phase represents the injection of a high β (β > 1) plasma
to equatorial altitudes < 4 earth radii; that recovery phase
represents a moderate β (0.1 $\lesssim$ β $\lesssim$ 1) plasma at altitudes $\lesssim$ 5.5
earth radii; and that quiet times indicate a low β (β $\lesssim$ 0.1) plasma
at altitudes ≤ 5.5 earth radii.

In the following we show results from all three of the above
regimes. However we present analyses and explanations only for the
moderate and low β cases. We are almost entirely ignorant of which
high β processes are responsible for main phase hot plasma dynamics
and are just beginning studies towards this end.

Hot-Cold Plasma Interactions

Interactions between the magnetospheric hot and cold plasma
populations have been considered previously in order to explain a
variety of phenomena observed in the geophysical plasma (Kennel
and Petschek, 1966; Cornwall, 1966; Brice and Lucas, 1971; Cornwall
et al, 1970, 1971).

Of particular interest are the results of Cornwall et al (1970)
who predict the stimulation of ion-cyclotron waves as the hot ring
current plasma interacts with the cold plasmaspheric plasma. Wave
amplification is expected to be sufficient to drive the hot plasma
into strong pitch angle diffusion thereby resulting in an important
loss process for the hot particles. Such a process should be most

readily evident during the recovery phase of a geomagnetic storm since the overwhelming ring current source term, present during main phase, is absent.

A detailed analysis of Explorer 45 recovery phase data from the December 17, 1971 geomagnetic storm (Williams and Lyons, 1974a, b) yields basic agreement with this concept with the proviso that the resulting pitch angle diffusion is not strong (full loss cone) but moderate (fractionally full loss cone). Examples of the hot plasma distribution data sets used in this analysis are shown in Figure 6 in the form of an energy-spatial-pitch angle snapshot and in Figure 7 as phase space distribution functions, in the $v_{\parallel}$, $v_{\perp}$ plane.

The resonant energy equation describing the cyclotron resonance condition for protons and the amplified em waves is (Kennel and Petschek, 1966)

$$E_{\parallel,res} = \frac{B^2}{8\pi N} \ A^{-2} \ (1 + A)^{-1} \equiv \frac{B^2}{8\pi N} \ F(A)$$

where B is the local magnetic field magnitude and N is the total plasma density. A is a particle anisotropy factor obtained from a combination of integrals over the distribution function and its first derivative in velocity space. Williams and Lyons (1974a) estimate F(A) to be of order 1 within a factor of four.

From data such as shown in Figure 6, Williams and Lyons (1974a) are able to estimate $E_{\parallel,res}$. This estimate coupled with the measured value of B allows a normalized total plasma density to be plotted.

$$\frac{N}{F(A)} = \frac{B^2}{8\pi E_{\parallel,res}} \ (cm)^{-3}$$

The results of such an analysis are shown for three orbits during the December 17, 1971 storm in Figure 8. The results in Figure 8 are in good agreement with a simultaneous in-situ estimate of plasma density obtained from the saturation of an onboard DC electric field probe (Maynard and Cauffman, 1973; Morgan and Maynard, 1975). Furthermore, the apparent outward expansion with time shown in Figure 8 implies cold plasma refilling rates of $\sim 2(10)^8$ ions/cm^{2}sec at L = 3.5 to $\sim 2.5 (10)^7$ ions/cm^{2}sec at L = 4.5, in agreement with earlier observations by Chappell et al (1970) and Park (1970) and with theoretical expectations of Banks (1972).

From these studies Williams and Lyons (1974a, b) have concluded that:
1.   Above the plasmapause region, the hot ring current plasma is stable with negligible losses due to pitch angle diffusion. Thus, in the absence of cold plasma, a hot plasma can be stable in a trapping configuration on a scale size comparable to the earth's magnetosphere.

EXPLORER 45   ORBIT 103   INBOUND

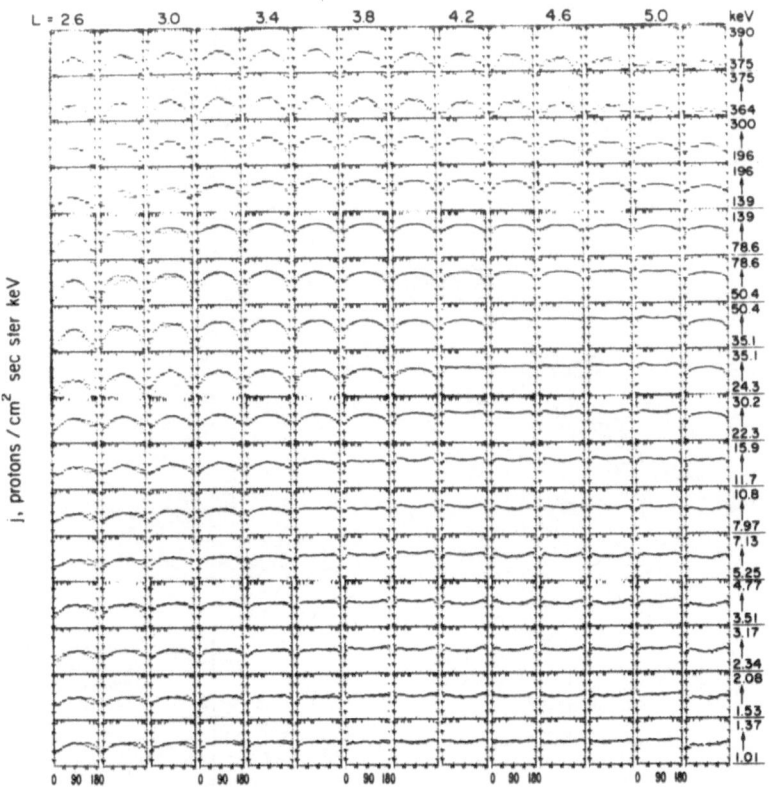

Measured Local Pitch Angle (Degrees)

Figure 6.   Proton ring current (hot plasma population) snapshot, December 18, 1971. Storm recovery phase. Note transformation of flat-top (concave-top) pitch angle distributions to rounded pitch angle distributions peaked at 90° with lower energies transforming at lower altitudes. When pitch angle scans in the flat-top distributions reach the loss cone region, intensity decreases are seen implying an empty loss cone. Each individual plot shows $\log_{10}$ differential flux vs measured local pitch angle for a specific energy and L value. Plots for sixteen energies covering 1-390 keV are stacked at a given L value and shown every 0.2 R_E. This is a subset of the full display covering 1-872 keV and every 0.1 R_E which was used for analysis. No data editing has been done. The region where the solid state detector (24.3-300 keV) often suffers a saturation problem (Williams et al, 1973) occurs for L $\lesssim$ 3.2. This saturation problem appears as unusual depressions in intensities for pitch angles 90° + 45°. Contamination of the channeltron instrument (1-30.3 keV) by reflected sunlight at near background count rates can be seen during the 90°-180° pitch angle sweep. This problem is related to spin axis orientation, is easily identified, and in the present data does not exist in the 0°-90° pitch angle sweep. Neither of the above effects has any influence on the present study. Data dropouts and telemetry noise effects can also easily be identified.

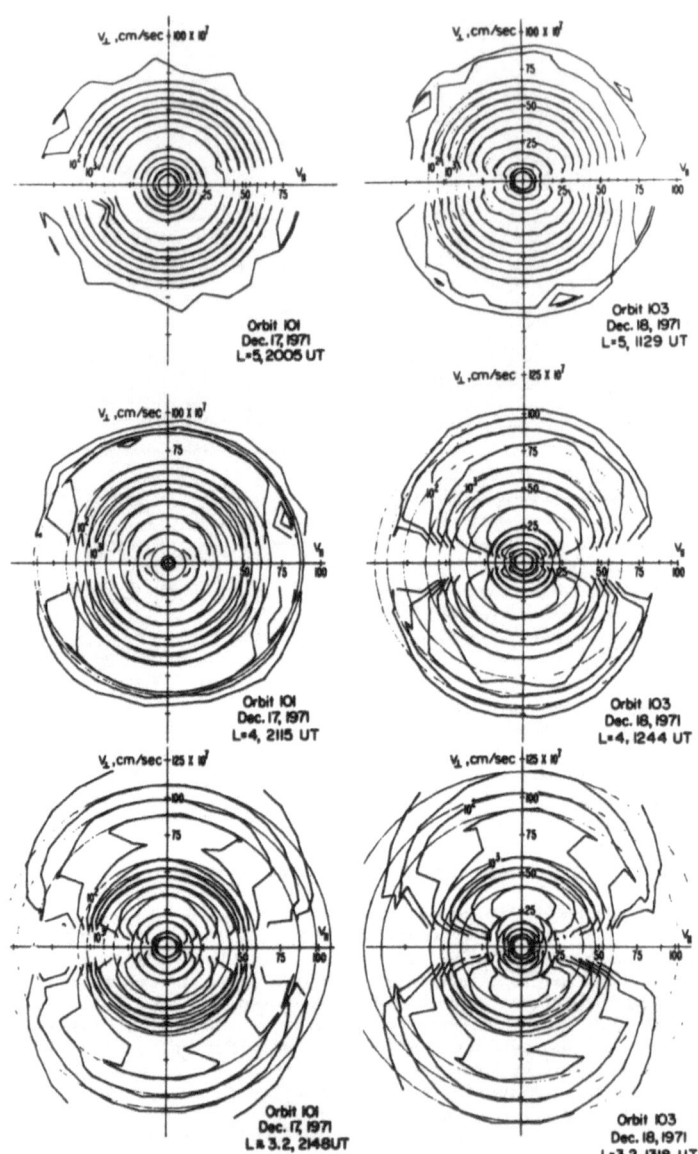

Figure 7. Contours of constant phase space density, f = J/E, plotted in $v_\perp$, $v_{||}$ plane. Circles centered at $v_\perp = v_{||} = 0$ (thin lines) added for reference. Contours generated every 0.5 units in $\log_{10} f$ by computer with no smoothing applied to the data. Contours generated for $v_\perp > 0$ and $v_\perp < 0$ separately and region of no available data separates the two sets of contours. Spatial evolution of distribution function shown during main phase (orbit 101) and recovery phase (orbit 103). Count rate statistics are the cause of jagged nature of several contours. Break in orbit 101, L = 5 contours is a telemetry problem.

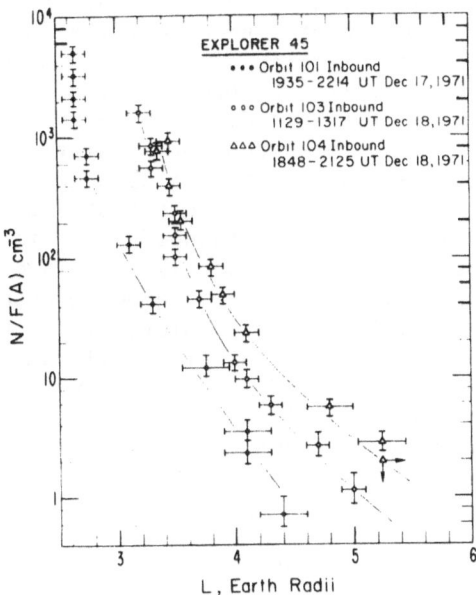

Figure 8.   Plot of N/F(A) vs altitude.   Resonant energy equation
used to obtain N/F(A) = $B^2/8\pi E_{\parallel,res}$.   Altitude and $E_{\parallel,res}$ determined
at point where flat-top distributions begin their transformation to
a rounded distribution (see for example Figure 6).   Analysis shown
for three orbits giving radial profiles and time history consistent
with the amplification of ion-cyclotron waves as the hot ring current
plasma interacts with the cold plasmaspheric plasma in the region of
the plasmapause.

        2.   The hot ring current plasma enters a moderate pitch angle
diffusion regime in the plasmapause region.   Therefore the addition
of cold plasma destabilizes the hot plasma.
        3.   The mechanism responsible for the destabilization of the
hot plasma is the amplification of ion-cyclotron waves due to the
interaction of the cold plasmaspheric plasma with the hot ring current plasma in a manner similar to that discussed by Cornwall et al
(1970).

        These results, obtained from observations made by Explorer 45,
are schematically summarized in Figure 9 where magnetospheric regions
of hot plasma stability, moderate turbulence, and strong turbulence
are shown during recovery phase.   Both an equatorial and a midnight
meridian projection are shown in Figure 9 in order to provide a guide
for correlating high and low altitude structures.   Williams and Lyons
(1974a) showed that a variety of low altitude observations (Amundsen
et al, 1972; Kleckner and Hoch, 1973; Bernstein et al, 1974; Mizera,

1974) were consistent with this picture and predicted certain low
altitude effects which have recently been verified (Hultqvist et al,
1974; Hultqvist, 1974). For example, low altitude observations showing a loss cone full of precipitating protons at geomagnetic latitudes above the plasmapause fit into Figure 9 by noting that such
observations would correspond to an equatorial altitude during recovery phase above that reached by Explorer 45. However, during main
phase the turbulent plasma sheet region is expected to penetrate to
lower equatorial altitudes due to an enhanced cross tail electric
field. During main phase of the December 17, 1971 storm, Explorer 45
did make observations consistent with its being immersed in a highly
turbulent region and thus, all regions depicted in Figure 9 were
probably sampled by Explorer 45 at that time (Williams and Lyons,
1974b). While Figure 9 may be valid schematically, note that the
relative sizes and locations of all regions depicted will vary considerably and will depend on past and existing magnetic and electric
field values and activity.

                          Charge Exchange Losses

     Charge exchange losses play an important role in the dissipation
of the hot ring current plasma (Swisher and Frank, 1968). Williams
(1974a, b) has qualitatively considered the relative effects of
charge exchange losses and ion cyclotron losses. It is apparent that
the hot plasma distribution function is dominated by ion cyclotron
resonant effects in the plasmapause region (Williams and Lyons, 1974a,
b). However above the plasmapause region in geomagnetic storm recovery phase, ion cyclotron effects are not seen in the hot ring current plasma. This hot plasma seems stable with little or no losses
occurring due to wave particle interactions. In this region of very
low cold plasma density, the specific effects to be expected on the
hot plasma distribution function will depend critically on the relative values of charged exchange lifetimes at a specific altitude and
plasmaspheric refilling times at that L shell. Well inside the plasmapause region, the remainder of the hot plasma distribution should
be dominated by charge exchange and Coulomb losses. A significant
hot plasma intensity exists inside the plasmapause region during recovery phase (Frank, 1971; Williams and Lyons, 1974a; Williams, 1974a)
presumably because the ion cyclotron resonance initiates only a moderate pitch angle diffusion mode and is relatively inefficient at re-
moving particles from equatorial pitch angles near $\alpha \sim \pi/2$ ($v_\parallel \sim 0$).
It is possible that the arrest of the ion cyclotron instability at a
moderate diffusion level may be due to absorption of wave energy by
cold electrons which in turn provide the heat source necessary for
stable auroral red arc stimulation (Cornwall et al 1971).

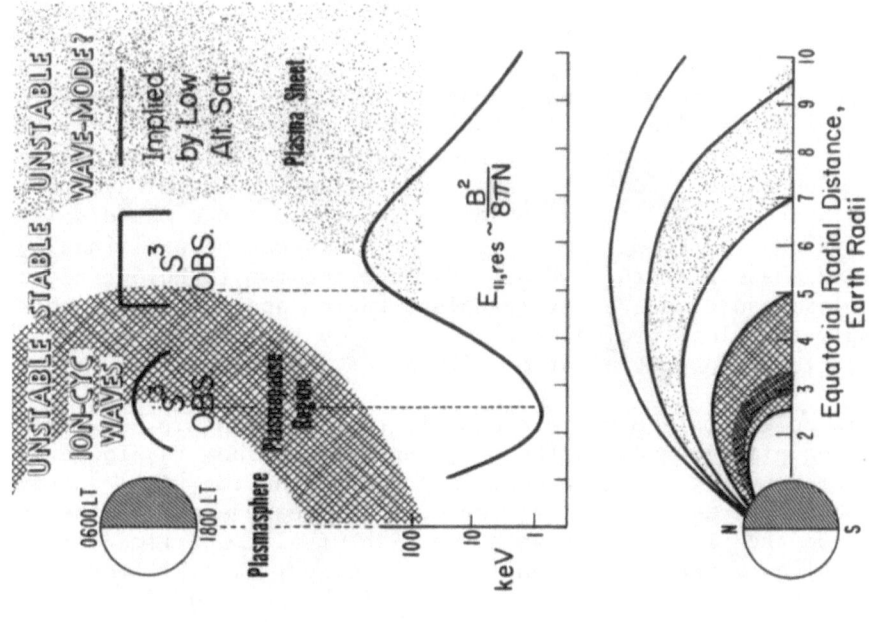

Figure 9. Schematic picture of geomagnetic storm recovery phase showing hot ring current plasma behavior in the geomagnetic field and its inter- action with the cold plasmaspheric plasma. The top panel shows an equatorial projection and the bottom panel shows a midnight meridian projection. Midnight meridian features are projected to low altitudes to indicate expected low altitude lat- itudinal structure. A semi-quantitative resonant energy versus altitude plot which uses Explorer 45 results is shown for reference. Moderate pitch angle diffusion occurs for the hot ring current plasma in the plasmapause region due to the amplification of ion-cyclotron waves. Each energy of the ring current plasma begins its interaction in the plasmapause region at the appropriate value of $B^2/8\pi N$. Heavier shading in the plasmapause region is used to indicate that energy flow to the ionosphere from the hot ring current plasma may be peaked at preferential locations and will depend on the plasmapause shape and hot plasma energy spectrum. More than one peak in the energy flow is possible. Above the plasmapause region the hot ring current plasma is stably trapped with negligible or no losses due to pitch angle scattering. Above this region, and above Ex- plorer 45 (s³) apogee during recovery phase, there is the plasma sheet region exhibiting strong turbulence and a full loss cone as implied by low altitude measurements.

SAR ARCS

It seems clear that both the plasmapause region and the existence of the hot ring current plasma are involved in the formation
of stable auroral red (SAR) arcs (see review by Hoch, 1973). For
example SAR arcs occur just after geomagnetic storm main phase and
are located in the plasmapause region where the ring current overlaps
the plasmapause, i.e., in the "unstable-ion cyclotron wave" region
of Figure 9. Furthermore, Kleckner and Hoch (1973) have reported
that, when observations were available, a SAR arc was always detected
when a hydrogen (H) arc was observed and the SAR arc was displayed
1-2 L values equatorward of the H arc. This can be explained by the
recovery phase picture of Figure 9 since the two turbulent regions
(the plasmapause moderately turbulent region and the plasma sheet
strongly turbulent region) are separated by a stable region exhibiting little or no precipitation of the hot ring current plasma.

In an attempt to better quantify the relationship between SAR
arcs, the plasmapause and the ring current, we show in Figure 10
(Williams, Lyons and Hernandez, in preparation) the total energy
available in the ring current and our best measurement of the energy
lost from the ring current as a function of altitude immediately following ring current injection in the December 17, 1971 storm. It can
be seen that sufficient energy is lost from the ring current to establish and maintain a SAR arc although the actual mechanism is not yet
confirmed (Cole, 1965; Cornwall et al, 1971). We have no convenient
way of estimating an energy loss rate from the ring current since we
only have single satellite observations. However by noting that our
measurements were made $\sim 10^4$ seconds after the beginning of storm
main phase, we can obtain a lower limit to this loss rate by dividing
the ΔE_s values in Figure 10 by 10^4 seconds. Available data (Rees and
Akasofu, 1963; Cornwall et al, 1971) indicate that this lower limit
loss rate (~ 0.3 ergs/cm^{2}sec at L ~ 2.8) may be sufficient to establish a SAR arc consistent with the December 17, 1971 storm.

Thus from a consideration of both spatial location and energetics, it appears that SAR arcs are established by the energy lost from
the proton ring current as it impacts the cold plasmaspheric plasma.
Detailed correlations and calculations are now underway in an attempt
to identify the responsible mechanism(s) (Williams, Lyons, and Hernandez, in preparation).

Adiabatic and Non-Adiabatic Responses

We show in Figures 11 and 12 (Lyons and Williams, in preparation
1975) the behavior of the hot plasma distribution function during the
December 17, 1971 geomagnetic storm. In Figure 11 we show the distribution function $2m_p f = j/E$ as a function of time at several altitudes for specific values of both energy and first adiabatic

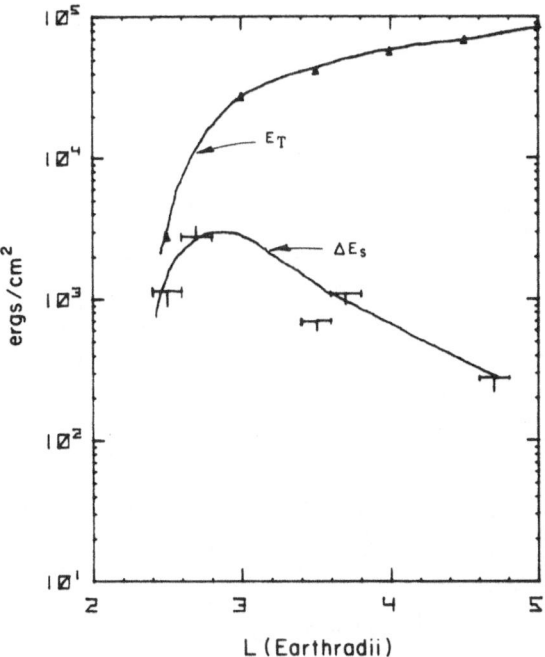

Figure 10. The total energy (E_T) available in the proton ring current above 1 cm^2 in the ionosphere and an estimate of the total energy lost from equatorial regions (ΔE_s) to each cm^2 in the ionosphere as a function of L. Data are from Explorer 45 Orbit 101 at the end of main phase (injection) for the December 17, 1971 geomagnetic storm.

invariant μ. The non-adiabatic storm response is clearly seen throughout the distribution function. It is of interest to note that apparently all the variations at high energies (high μ) are due to an adiabatic response to the changing B field magnitude. To further look at the distribution functions through the storm, we show in Figure 12 a plot of 2m$_p$ f versus μ (shown as an equivalent corrected energy based on the changing B field) for several altitudes during the storm. Non-adiabatic increases are clearly seen with the increase for orbit 101 being especially marked. Figures 11 and 12 are shown for nearly equatorially-mirroring particles. We plan to extend this analysis into three dimensions (to non-equatorially-mirroring particles) in the future. Long term time histories of the distribution function such as this will allow an accurate separation of adiabatic and non-adiabatic effects occurring in the hot plasma in the magnetosphere.

PROTON DISTRIBUTION FUNCTION vs TIME

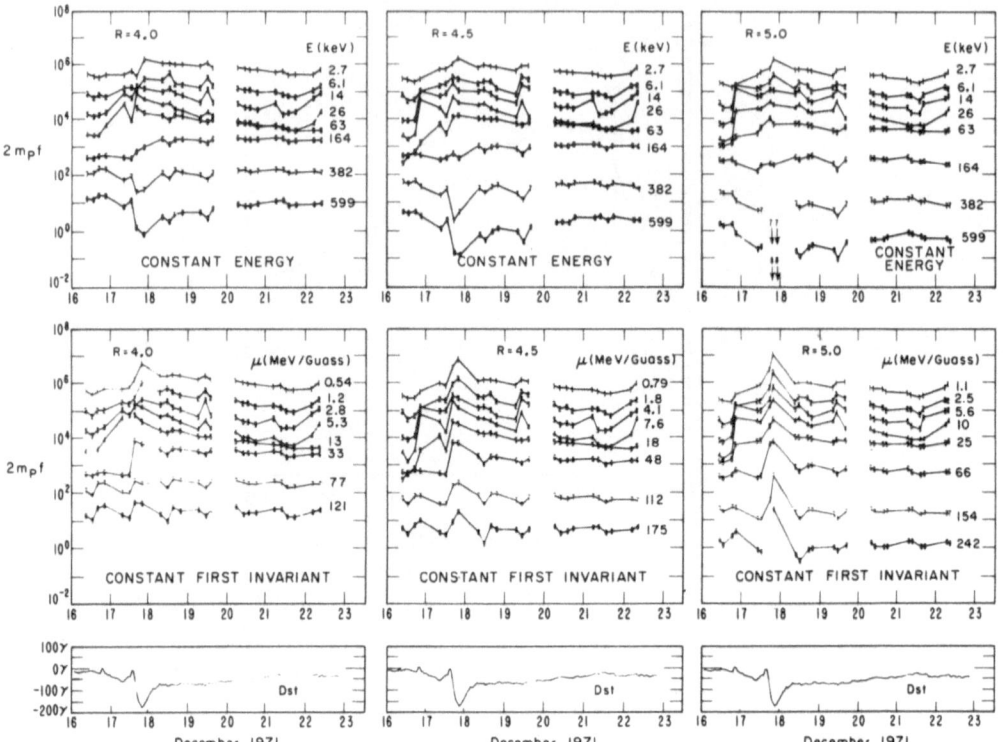

Figure 11. Distribution function, multiplied by two times the proton mass, for equatorially mirroring protons as a function of time
throughout the period of the December 17, 1971 storm. Curves at
constant energy E and constant first adiabatic invariant μ are
shown in the upper and lower panels, respectively. D_{ST} is also
shown. The values of μ, indicated in each of the lower panels,
have been selected so that the first point of each curve (from
orbit 97 outbound) is for the same energy as for the corresponding
curve in the upper panels. No data points suffering from the saturation problem with the solid state detector (Williams et al, 1973)
have been included. Radial distances R = 4.0, 4.5 and 5.0 R_E are
shown. The four points at R = 5 R_E with downward directed arrows
indicate upper limits, since the flux decreased below the detectable
limit. These points have been treated as missing data when con-
structing the constant first invariant curves. However, due to
the decrease in B, this results in only one missing point in the
constant first invariant curves.

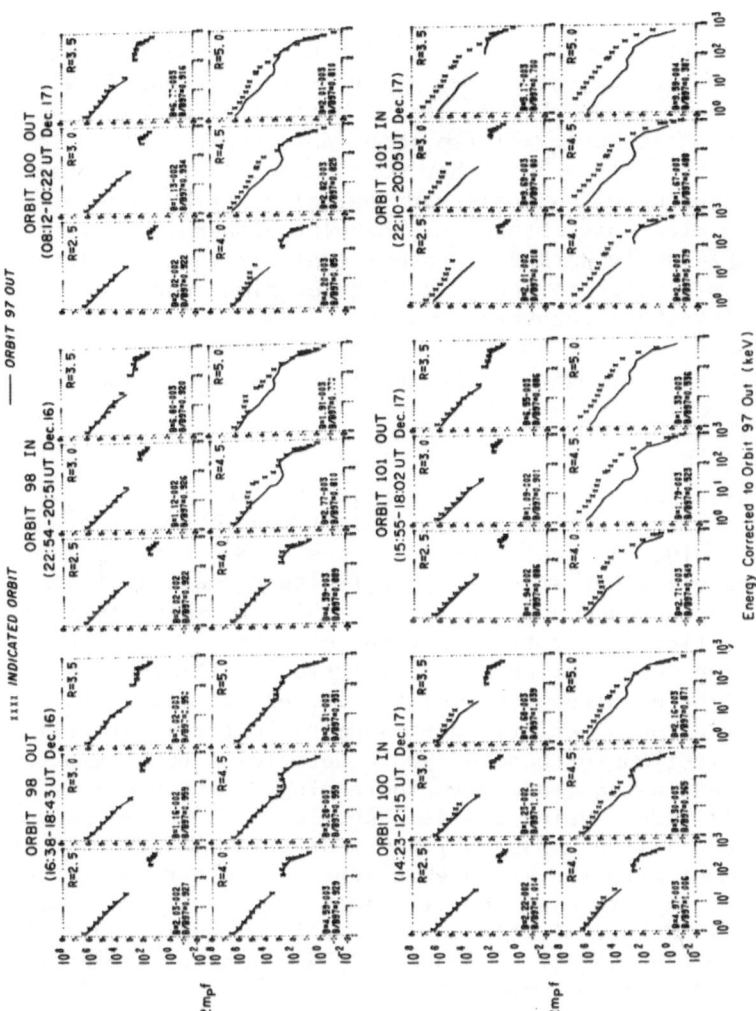

Figure 12. Distribution function, multiplied by two times the proton mass, for equatorially mirroring protons as a function of the first adiabatic invariant μ for orbits preceding and during the December 17, 1971 storm main phase. The horizontal axes are labeled in units of energy corrected to orbit 97 outbound (E_{97}). Thus the appropriate values for μ are equal to E_{97}/B_{97}, where B_{97} = .0219, .0121, .00739, .00494, .00342, .00248 Gauss for R = 2.5, 3.0...5.0 R_E, respectively. In each panel, the solid line gives the spectrum from orbit 97 outbound, and the X's give the observations from the indicated orbit.

ELECTRONS

In this section, we present the behavior of energetic (35-560 keV) electrons as observed throughout the plasmasphere ($L \lesssim 5.2\ R_E$). As with protons we are just beginning studies of the injection (main) phase and remain ignorant of injection mechanism specifics. Consequently although we show injection data, we attempt explanations only of the pre-storm quiet and post-injection recovery behavior.

In recent reports, Lyons et al (1972) and Lyons and Thorne (1973) have placed the subject of radiation belt electrons in a firm quantitative context. By combining the effects of pitch angle scattering with the plasmaspheric whistler mode wave band, Coulomb interactions and cross L diffusion, they have been able to satisfactorily explain the energy, pitch angle, and spatial distributions of 20 keV-2 Mev trapped electrons throughout the plasmasphere. We very briefly review this early work and present Explorer 45 observations which extend these results to the general quiet-time condition as well as to the post-storm recovery phase.

Figure 13 shows the bounce averaged diffusion coefficient

$$D_\alpha = D_{o,\alpha} + \sum_{n \neq o} D_n$$

obtained by Lyons et al (1972) for three electron energies. Note that the bounce averaged diffusion coefficient consists of two parts--that containing all cyclotron resonance harmonic terms ($n \geq 1$) which are efficient at scattering finite $P_\parallel$ particles down the field line and the Landau resonance ($n = 0$) term which is effective at moving particles away from near the geomagnetic equator. The result of using such a pitch angle diffusion coefficient to determine the equilibrium pitch angle distributions for energetic electrons is that a characteristic bump appears in the equatorial pitch angle distribution surrounding 90° equatorial pitch angle. This bump decreases in magnitude with increasing electron energy and increasing altitude. Using a plasmaspheric whistler mode wave band modeled after existing observations Lyons et al (1972) obtained reasonable agreement with the measured pitch angle distributions of West et al (1973) and with electron lifetimes throughout the plasmasphere. A further advance of this understanding of radiation belt electrons occurred when Lyons and Thorne (1973) combined the whistler mode turbulent loss with Coulomb losses and cross L radial diffusion effects. Figure 14 shows relevant time scales for the various processes and can be used to indicate regions of preferential inward diffusion or loss for the indicated values of the magnetic moment, μ. Combining these processes equilibrium radiation belt intensity profiles were obtained through the plasmasphere assuming the existence of a source on the high altitude side of the plasmapause. Comparison with quiet time solar minimum observations of Pfitzer et al

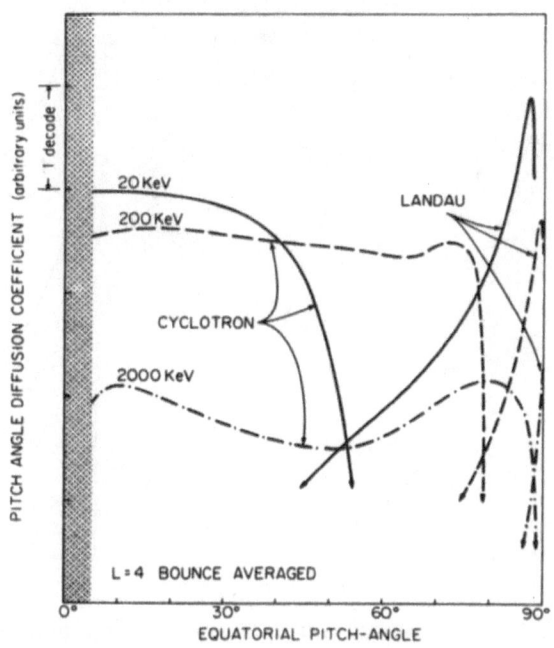

Figure 13. Bounce-orbit averaged cyclotron and Landau resonant pitch-angle diffusion coefficients as a function of equatorial pitch angle at L = 4 for 20, 200, and 2000 keV electrons.

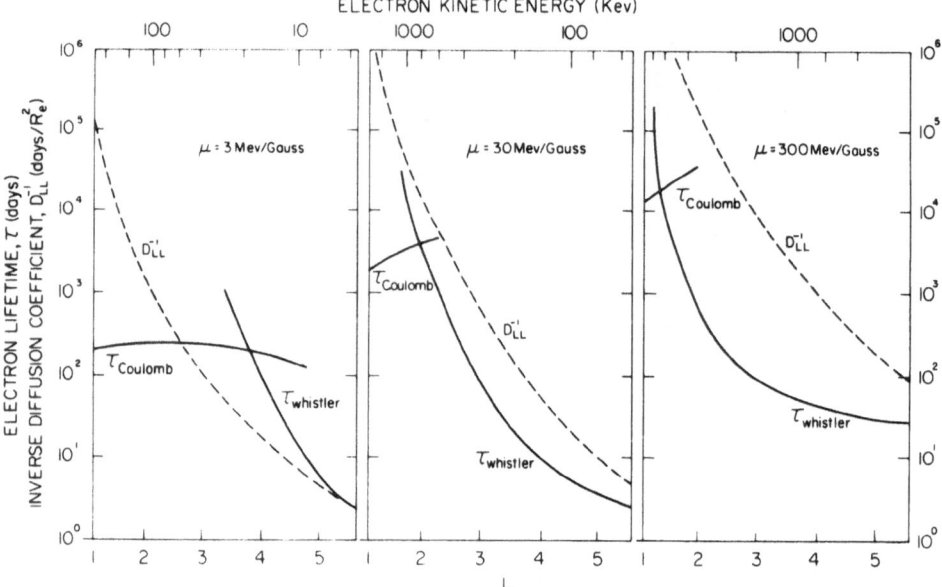

Figure 14. Lifetimes due to Coulomb scattering and whistler mode turbulent loss (B_w = 10 mγ) are plotted against L for three values of the electron first adiabatic invariant. Regions of preferential inward diffusion or loss can be determined by comparison with the radial diffusion rate D_{LL}^{-1} (E = 0.10 mv/m).

(1966) yielded good agreement. To extend these results to a general
quiet time condition and to the post-injection recovery phase, we
briefly summarize recent work of Lyons and Williams (1975a, b).

Quiet Time

Using Explorer 45 data to obtain a measured source spectrum
(1-560 keV) at the edge of the plasmapause, Lyons and Williams
(1975a) calculated the radial profiles and pitch angle distributions
expected for energetic (35-560 keV) electrons throughout the plasmasphere (1.8-5.2 R_E) during geomagnetic quiet time conditions. This
source spectrum coupled with the same basic plasmaspheric hiss and
radial diffusion models employed by Lyons and Thorne (1973) reproduced the observed energy and spatial structure reasonably well with
the flux magnitudes correct to within an order of magnitude--an accuracy consistent with the uncertainties associated with present
knowledge of plasmaspheric hiss and radial diffusion parameters.

Figure 15 shows predicted and observed pitch angle distributions
for 35-560 keV electrons during a quiet time orbit. Note that not
only is the expected structure of Lyons et al (1972) around 90° pitch
angle seen, but the expected energy and spatial dependence of this
structure is also observed. Although theory overestimates the magnitude of the 90° structure, this difference might be removed if the
actual wave spectrum were known since increasing the spread of wave
energy with frequency reduces the size of the bumps (Lyons et al,
(1972).

Figures 16 and 17 demonstrate the stability of the quiet time
electron distribution for the magnetically quiet period extending
from December 9, 1971 to December 16, 1971. Lyons and Williams
(1975a) inspected three additional geomagnetically quiet periods
from January through June 1972 and found that the electron distribution varied from the December 15, 1971 reference orbit (orbit 94)
no more than did the data in Figures 16 and 17.

The quiet-time two zone structure is evident in all the profiles
of Figure 16 with the flux decreases in the slot region becoming more
significant with increasing energy. The radial profiles exhibit
little change throughout the slot region and inner zone (L $\lesssim$ 4).
Outer zone (L $\gtrsim$ 4) fluxes are more variable and are associated with
slight changes in the level of magnetic activity signaling changes
in the position of the plasmapause and in the electron source function (Lyons and Williams, 1975a).

This long term stability of the quiet-time energetic electron
distribution plus the agreement with the calculations of Lyons and
Thorne (1973) leads to the conclusion that the major source and loss
processes controlling the equilibrium radiation belt electron

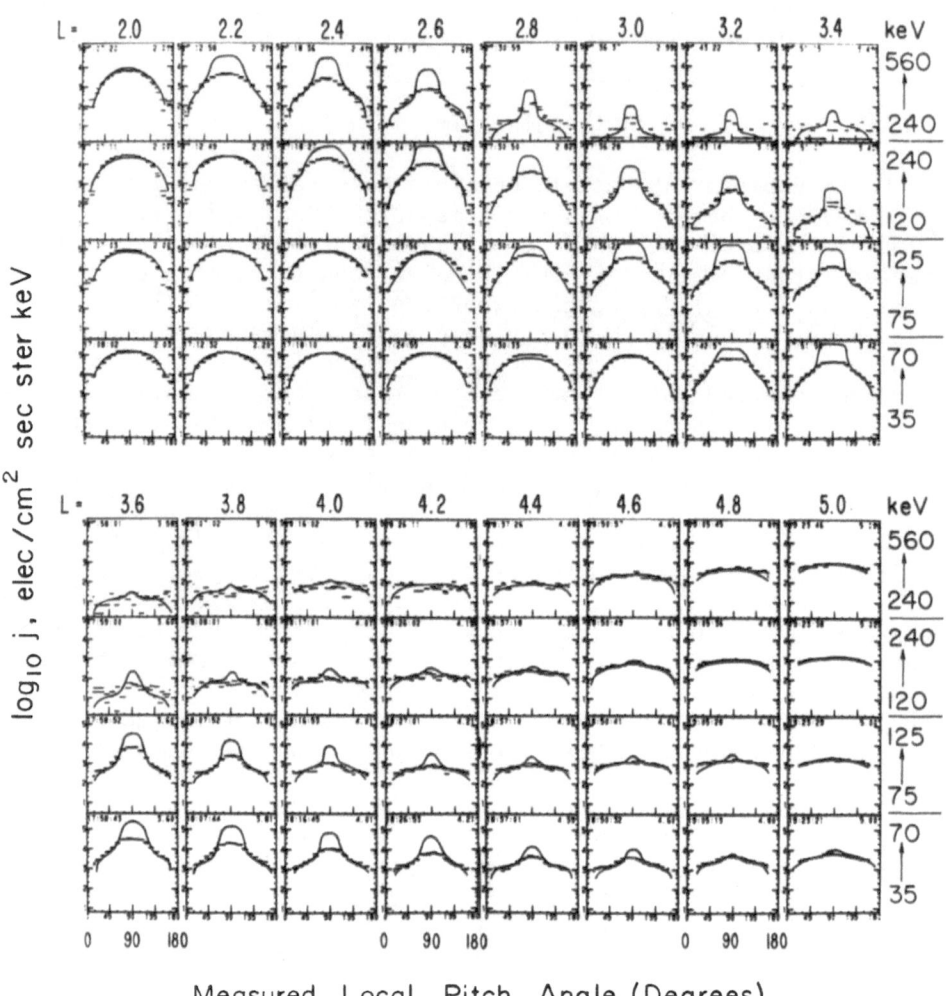

Measured Local Pitch Angle (Degrees)

Figure 15.   Equatorial pitch angle distributions observed on December 15, 1971 (orbit 95) and those predicted by Lyons et al (1972).
Distributions are shown from L = 2 to L = 5 for four energies. The
dashes give the measured electron flux with their horizontal extent
indicating the pitch angle scan for each measurement. The solid
curves are the theoretically predicted pitch angle distributions
for the geometric mean energy of each energy interval. The vertical
positionings of the theoretical distributions are arbitrary on a
logarithmic scale and have thus been adjusted to best illustrate
the comparison with the observations.

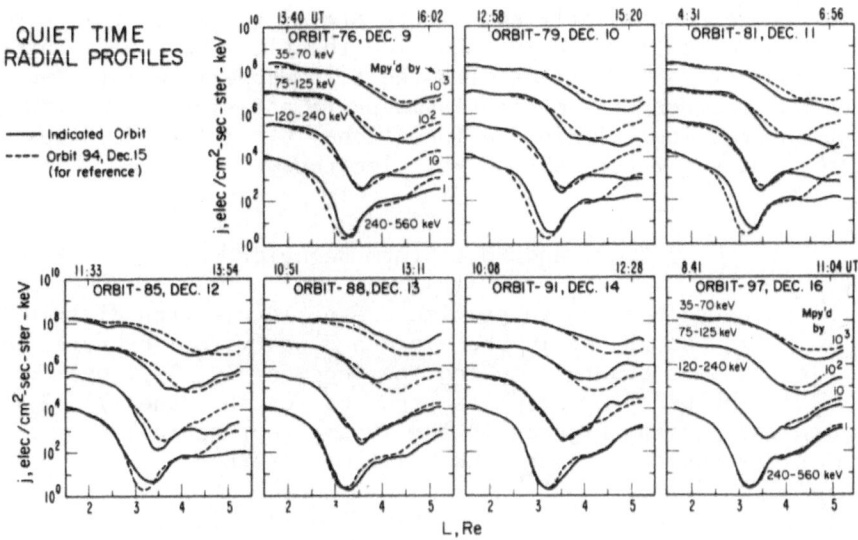

Figure 16. Radial profiles of the perpendicular (90° local pitch angle) electron flux obtained near the geomagnetic equator, approximately once per day, for the quiet period of December 9-16, 1971. Solid curves give the profiles from the orbit indicated in each panel, while the dashed curves (shown for reference) give the profiles from orbit 94 on December 15, 1971. To clearly display the data, the 120-240 keV, 75-125 keV, and 35-70 keV fluxes have been multiplied by 10^1, 10^2, and 10^3, respectively.

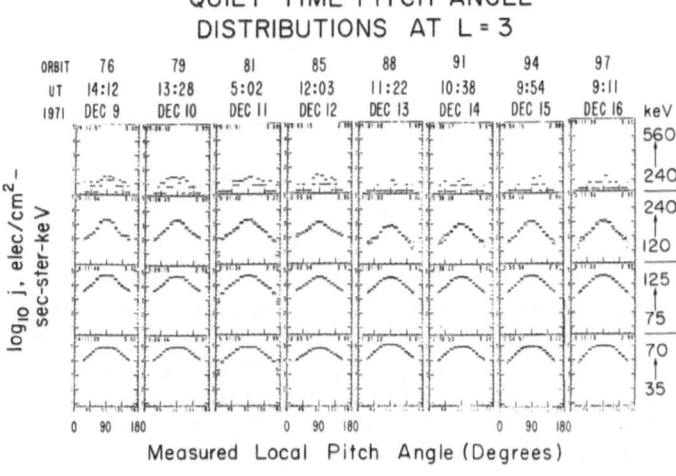

Figure 17. Pitch angle distributions at L = 3, approximately once a day, for the period of December 9-16, 1971.

structure have been identified. These are 1) radial diffusion of
an electron source located just outside the plasmapause, 2) resonant
interactions of the electrons with the plasmaspheric whistler mode
wave band throughout the outer plasmasphere and slot region and 3)
Coulomb interactions within the inner zone, $L \lesssim 3$ for 35-70 keV
electrons and $L \lesssim 2$ for 240-560 keV electrons.

Storm and Post-Storm Behavior

Having described the quiet time energetic electron distribution
we show in Figures 18 and 19 the storm and post-storm behavior of
these electrons as observed during the December 17, 1971 geomagnetic
storm. Similar behavior also was reported for the June 17, 1972
storm indicating that Figures 18 and 19 are representative of a
general storm response (Lyons and Williams, 1975b).

It is evident that storm injection greatly distorts the quiet

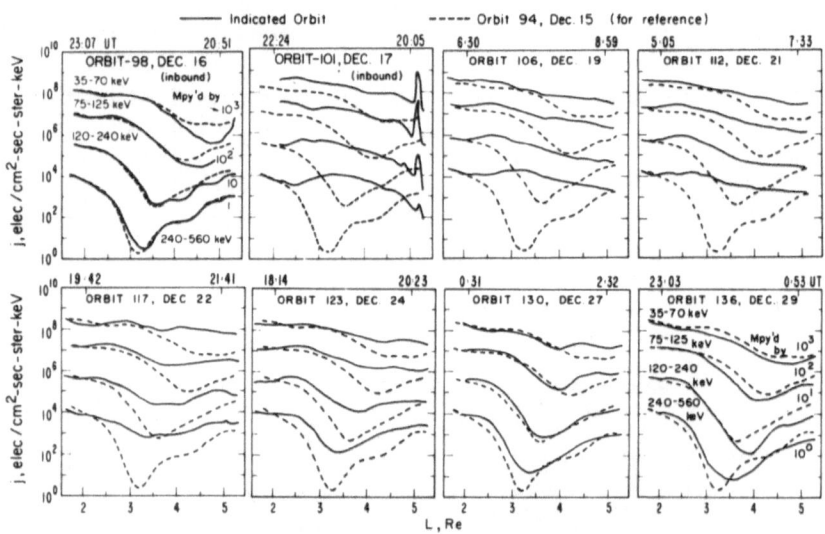

Figure 18. Observed radial profiles of the perpendicular (90°
local pitch angle) electron flux obtained near the geomagnetic
equator for the periods preceding, during and following the storm
of December 17, 1971. Solid curves give the profiles from the
orbit indicated in each panel, and the dashed curves give the pre-
storm profiles from orbit 94 outbound (December 15, 1971) for comparison. To clearly display the data, the 120-240 keV, 75-125 keV,
and 35-70 keV fluxes have been multiplied by 10^1, 10^2, and 10^3, respectively. Note the relaxation of the post-storm profiles to
their pre-storm shapes and intensities.

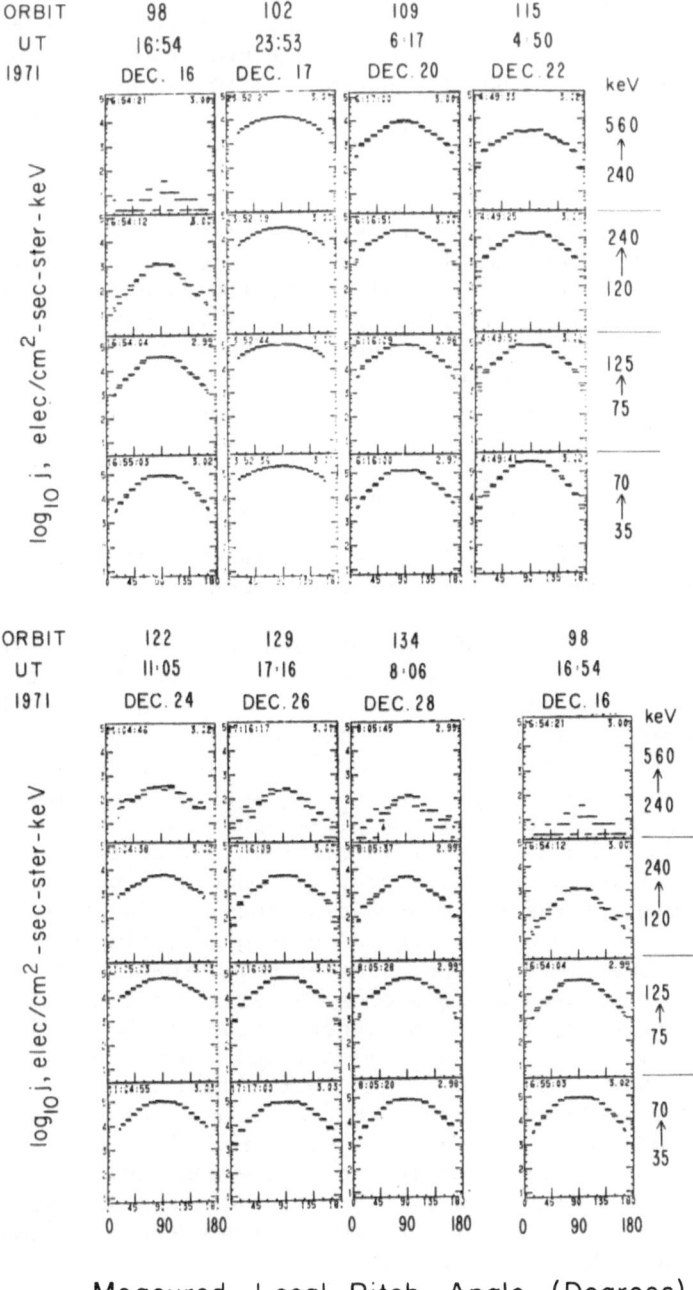

Measured Local Pitch Angle (Degrees)

Figure 19. Observed pitch angle distributions at L = 3 for the periods preceding, during and following the storm of December 17, 1971. Pre-storm distributions are repeated at end of figure to indicate return to pre-storm shapes and intensities.

time distributions, both in the pitch angle distributions and radial profiles. However during the recovery phase following injection, pitch angle distributions return to their pre-storm shape over a period of several days and then decay back to the quiet time flux levels. The return to quiet time levels is seen in Figure 19 where the radial profiles steadily evolve to the quiet time pre-storm reference orbit.

It is concluded by Lyons and Williams (1975b) that since the electrons return to and maintain their pre-storm structure, the source and loss processes responsible for the quiet time equilibrium structure of radiation belt electrons are also responsible for the post-storm recovery of the electron distribution. Furthermore, the relative strengths of the source and loss processes apparently are the same for the overall recovery phase as they are before the storm.

SUMMARY

We have presented a variety of results obtained primarily from the Explorer 45 satellite concerning the behavior and interactions of the hot magnetospheric plasma. These results are aiding in improving our understanding of hot-cold plasma interactions in the magnetosphere, SAR arc formation, the correlation of low altitude structures with high altitude phenomena, separation of adiabatic and non-adiabatic effects, and the development and behavior of the overall hot plasma distribution function. The situation for the pre-storm equilibrium and post-storm recovery phase energetic electron distributions throughout the plasmasphere seems to be well in hand. While not emphasized, it is clear in all of these studies that wave particle interactions play an important role in determining the behavior of the hot plasma distribution in the magnetosphere. While our understanding has been greatly improved, there is much that is as yet unknown. For example, the results presented herein are for the general case of a magnetized plasma with $\beta < 1$. We have just begun to attack the general case of the establishment and behavior of a high β ($\beta > 1$) magnetized plasma. From the brief discussion on SAR arcs, it is clear that single satellite measurements will not yield an accurate enough estimate of the energy lost from the hot plasma to the ionosphere for definitive tests of specific models. A strong requirement for this case and many other magnetospheric-ionospheric research problems is a minimum of dual satellite observations, allowing a first approximation towards a separation of spatial and temporal effects.

We look forward to both the application of newly existing data to these problems as well as appropriate data from future magnetospheric research programs which will be required to answer many of the questions remaining concerning the behavior of naturally-occurring plasmas.

ACKNOWLEDGMENTS

It is a pleasure to acknowledge the work of Dr. L. R. Lyons, a co-researcher, in much of the work presented in this paper. I also wish to acknowledge the Explorer 45 experiment principals: R. Hoffman, T. Fritz, L. Cahill, N. Maynard and D. Gurnett, without whose cooperation these studies would not have been possible.

Figures 1 and 2 are after Williams and Lyons (1974a). Figure 3 is from Smith and Hoffman (1973). Figure 4 is after Smith and Hoffman (1973). Figure 5 is after Hoffman (1973). Figures 6 and 7 are from Williams and Lyons (1974a). Figure 8 is from Williams and Lyons (1974b). Figure 13 is from Lyons et al (1972). Figure 14 is from Lyons and Thorne (1973). Figures 15, 16 and 17 are from Lyons and Williams (1975a). Figures 18 and 19 are after Lyons and Williams (1975b).

These studies were partially funded under NASA Contract No. S-50028.

REFERENCES

Amundsen, R., F. Söraas, H. R. Lindalen, and K. Aarsnes, Pitch angle distributions of 100 to 300 keV protons measured by the ESRO IB satellite, J. Geophys. Res., 77, 556, 1972.

Banks, P. M., Behavior of thermal plasma in the magnetosphere and topside ionosphere, in Critical Problems of Magnetospheric Physics, Proc. of the Joint COSPAR/IAGA/URSI Symposium, Madrid, Spain, May 11-13, 1972, ed. E. R. Dyer, Pub. IUCSTP Secretariat November, 1972.

Bernstein, W., B. Hultqvist, and H. Borg, Some implications of low altitude observations of isotropic precipitation of ring current protons beyond the plasmapause, Planet. Space Sci., 22, 767-776, 1974.

Brice, N., and C. Lucas, Influence of magnetospheric convection and polar wind on loss of electrons from the outer radiation belt, Geophys. Res., 76, 900, 1971.

Chappell, C. R., K. K. Harris, and G. W. Sharp, The morphology of the bulge region of the plasmasphere, J. Geophys. Res., 75, 3848, 1970.

Cole, K. D., Stable auroral red arcs, sinks for energy of D_{st} main phase, J. Geophys. Res., 70, 1689, 1965.

Cornwall, J. M., Micropulsations and the outer radiation zone, J. Geophys. Res., 71, 2185, 1966.

Cornwall, J. M., F. V. Coroniti, and R. M. Thorne, Turbulent loss of ring current protons, J. Geophys. Res., 75, 4699, 1970.

Cornwall, J. M., F. V. Coroniti, and R. M. Thorne, Unified theory of SAR arc formation at the plasmapause, J. Geophys. Res., 76, 4428, 1971.

Frank, L. A., Several observations of low-energy protons and elec-
    trons in the earth's magnetosphere with OGO 3, J. Geophys. Res.,
    72, 1905, 1967a.
Frank, L. A., On the extraterrestrial ring current during geomag-
    netic storms, J. Geophys. Res., 72, 3753, 1967b.
Frank, L. A., Direct detection of asymmetric increases of extra-
    terrestrial ring current proton intensities in the outer
    radiation zone, J. Geophys. Res., 75, 1263, 1970.
Frank, L. A., Relationship of the plasma sheet, ring current, trap-
    ping boundary, and plasmapause near the magnetic equator at
    local midnight, J. Geophys. Res., 76, 2265, 1971.
Hoch, R. J., Stable auroral red arcs, Rev. of Geophys. and Space
    Phys., 11, 935, 1973.
Hoffman, R. A., Particle and field observations from Explorer 45
    during the December 1971 magnetic storm period, J. Geophys.
    Res., 78, 4771, 1973.
Hultqvist, B., The ring current and particle precipitation near the
    plasmasphere, Kiruna Geophysical Institute Preprint 74:304,
    September, 1974.
Hultqvist, B., W. Riedler, and H. Borg, Ring current protons in the
    upper ionosphere within the plasmasphere, Submitted for pub.
    to Planet. Space Sci., 1974.
Kennel, C. F., and H. E. Petschek, Limit of stably trapped particles,
    J. Geophys. Res., 71, 1, 1966.
Kleckner, F. W., and R. J. Hoch, Simultaneous occurrences of hydro-
    gen arcs and mid-latitude auroral red arcs, J. Geophys. Res.,
    78, 1187, 1973.
Longanecker, G. W., and R. A. Hoffman, S^3-A spacecraft and experi-
    ment description, J. Geophys. Res., 78, 4711, 1973.
Lyons, L. R., R. M. Thorne, and C. F. Kennel, Pitch-angle diffusion
    of radiation belt electrons within the plasmasphere, J.
    Geophys. Res., 77, 3455, 1972.
Lyons, L. R., and R. M. Thorne, Equilibrium structure of radiation
    belt electrons, J. Geophys. Res., 78, 2142, 1973.
Lyons, L. R., and D. J. Williams, The quiet time structure of ener-
    getic (35-560 keV) radiation belt electrons, J. Geophys. Res.,
    80, 1975a (in press).
Lyons, L. R., and D. J. Williams, The storm and post-storm evolution
    of energetic (35-560 keV) radiation belt electron distributions,
    Space Environment Laboratory Preprint, February, 1975, sub-
    mitted to J. Geophys. Res., 1975b.
Maynard, N. C., and D. P. Cauffman, Double floating probe measure-
    ments on S^3-A, J. Geophys. Res., 78, 4745, 1973.
Mizera, P. F., Observations of precipitating protons with ring cur-
    rent energies, J. Geophys. Res., 79, 581, 1974.
Morgan, M. G., and N. C. Maynard, Comparisons of whistler-derived
    equatorial densities with plasmapause locations determined by
    Explorer 45, Dartmouth College-NASA Goddard Space Flight Center
    Preprint, February, 1975.

Park, C. G., Whistler observations of the interchange of ionization between the ionosphere and protonsphere, J. Geophys. Res., 75, 4249, 1970.

Pfitzer, K., S. Kane, and J. R. Winckler, The spectra and intensity of electrons in the radiation belts, Space Res., 6, 702, 1966.

Rees, M. H., and S. Akasofu, On the association between subvisual red arcs and the D_{st}(H) decrease, Planet. Space Sci., 11, 105, 1963.

Sharp, R. D., R. G. Johnson, E. G. Shelley, and K. K. Harris, Energetic O^+ ions in the magnetosphere, J. Geophys. Res., 79, 1844, 1974.

Shelley, E. G., R. G. Johnson, and R. D. Sharp, Satellite observations of energetic heavy ions during a geomagnetic storm, J. Geophys. Res., 77, 6104, 1972.

Shelley, E. G., R. G. Johnson, and R. D. Sharp, Morphology of energetic O^+ in the magnetosphere, in Magnetospheric Physics, edited by B. M. McCormac, D. Reidel Pub. Co., Dordrecht, Holland, (in press) 1974a.

Shelley, E. G., R. D. Sharp, and R. G. Johnson, The ionosphere as the source of ring current particles, EOS, 55, 1015, 1974b.

Smith, P. H., and R. A. Hoffman, Ring current particle distributions during the magnetic storms of December 16-18, 1971, J. Geophys. Res., 78, 4731, 1973.

Swisher, R. L., and L. A. Frank, Lifetimes for low-energy protons in the outer radiation zone, J. Geophys. Res., 73, 5665, 1968.

West, H. I., Jr., R. M. Buck, and J. R. Walton, Electron pitch angle distributions throughout the magnetosphere as observed on OGO 5, J. Geophys. Res., 78, 1064, 1973.

Williams, D. J., R. A. Hoffman and G. W. Longanecker, The small scientific satellite (S^3) program and its first payload, IEEE Trans. on Nuclear Sci., NS-16, #1, 322, 1969.

Williams, D. J., Hot-cold plasma interactions in the earth's magnetosphere, Invited paper presented at the Neil Brice Memorial Symp. on the Magnetospheres of Earth and Jupiter, Frascati, Italy, May 28-June 1, 1974a. To be published in the Symposium Proceedings.

Williams, D. J., Magnetospheric proton dynamics, EOS, 55, 1021, 1974b.

Williams, D. J., and L. R. Lyons, The proton ring current and its interaction with the plasmapause: Storm recovery phase, J. Geophys. Res., 79, 4195, 1974a.

Williams, D. J., and L. R. Lyons, Further aspects of the proton ring current interaction with the plasmapause: Main and recovery phases, J. Geophys. Res., 79, 4791, 1974b.

ACCELERATION PROCESSES IN THE PLASMA SHEET

J.W. Dungey

Imperial College
Physics Department
London SW7

ABSTRACT

The environment of the plasma sheet is described and the possible importance of noise and parallel electric fields near the outer boundary noted, but attention is concentrated on the neutral sheet. A significant feature in the plasma sheet is that the parallel pressure must exceed the perpendicular pressure.

The motion of particles in the neutral sheet is reviewed for models of increasing complexity. Cowley's modification of Alfvén's model raises the question of space charge in the middle of the sheet and this provides a natural explanation for field aligned currents, so it is fortunate that observations of these yield information about the neutral sheet. Recent observations suggest that polar wind plasma entering the neutral sheet from the lobes contributes only a small proportion of the plasma sheet population and recent work emphasises plasma entering from the magnetosheath. Acceleration in the neutral sheet leading to double streaming of electrons is likely to cause instability and hence anomalous resistivity, so that the system is explosive and may explain the sudden onset of substorms.

Away from the neutral line the normal component of magnetic field is important and drifts are discussed, but it is hard to decide whether a typical particle has come through the neutral sheet. Systematic variations in the particle population with position in the plasma sheet should be studied.

INTRODUCTION

The title of this paper overlaps several other titles of this
symposium and reading the minds of the program committee I shall
devote most of my time and space to the neutral sheet; but this
introduction will attempt the dual purpose of sketching the whole
area and providing an environment for the neutral sheet.   The plasma
sheet is defined by a population of electrons and protons whose
pressure is comparable to the magnetic pressure and whose characteristic energy is of the same order as that of auroral primaries,
namely a few keV.   There are sharp outer boundaries between the
plasma sheet and the tail lobes, where the density is the lowest
yet found by spacecraft.   Gurnett's (1974) continuum radiation indicates values which are sometimes less than 10^{-2} electrons cm^{-3}.
If the plasma in the lobes is polar wind, it has a low temperature
and its speed, though supersonic, is not large.   In some theoretical
models the plasma sheet also has a sharp boundary with the magnetosheath, but this is less striking in the plasma measurements.   Theoretical discussion of this requires interpretation of the plasma
sheet in terms of field topology; it is widely accepted that the
plasma sheet is on "closed" field lines, while the lines in the
lobes are "open" (at one end).   Topologically it could then be
argued that everywhere there are lobelike lines between the plasma
sheet and the magnetosheath, but at the sides of the sheet the
layer of lobelike lines might be very thin and not of practical importance.

It is obviously important to relate the features observed on
distant satellites to the aurora, "inverted V's" and other phenomena
at ionospheric altitudes, and Frank's group have a convincing picture
of this, in which the discrete aurora are near the magnetic surface,
which further out is the outer boundary of the plasma sheet.   Field
aligned currents are the common feature: they are found in association with both discrete auroras and the sheet boundary.   The association of upward currents with auroras immediately suggests parallel electric fields and hence anomalous resistivity.   Both these
topics are the subjects of other speakers, but I wish to stress
two aspects of the observations, which have simple but important
consequences.   First, the observation of field alignment of the
electron flux puts a constraint on pitch angle diffusion due to
plasma turbulence and the narrowness of the energy spectrum puts
a similar constraint on energy diffusion.   Second, instruments
like the Lepedea in principle measure the distribution function f,
which is used by theorists, and it is useful to compare the observed values with the value of f for some Maxwellian plasma.   The
values observed by Craven and Frank (1974) range up to 10^{8} electrons
$(cm^{2}\text{-sec-sr-eV})^{-1}$ at energies around 2keV and then $f \sim 10^{-26}$ c.g.s..
It may be noted that, while satellites might be expected to miss the
highest fluxes because they are so localised, sounding rockets in
discrete auroras (Arnoldy et al. 1974) did not see such high values.
Now for comparison consider a Maxwellian plasma at a temperature of
1eV as might occur in the polar wind.   For density n, the maximum

value of f is ~ 10^{-24} n c.g.s. and for densities near the earth
this is much larger than the above value obtained from electron
flux.   The maximum value of f can be reduced by diffusion, or in
a very similar way by the limited resolution of the instrument, but
there is an observed limit on the amount of diffusion as pointed
out above.   These constraints should set a lower boundary for any
region of parallel $\underset{\sim}{E}$.

Another feature with significant consequences is the extension
of the neutral sheet towards the earth.   We now need some coordinate
directions and as usual take x towards the sun and z northwards.
Hedgecock and Thomas (1975) on HEOS I have seen very sudden reversals
in B_x as close as 10 earth radii, where B_z is at least 10γ.   A
simple conclusion follows from consideration of stress.   Considering
the magnetic stress alone the reversal of the off-diagonal component
$B_x B_z$ over a very short distance in the z-direction implies a force
density in the x-direction which could not conceivably be balanced
by any contribution involving the much larger scale of variation in
the x-direction.   It is then necessary that the particle contribution to the stress tensor should almost exactly cancel the off-
diagonal component or in other words the total stress on either
side of the current sheet must be isotropic.   If the particle stress
expressed as $p_{\shortparallel}$ and $p_{\perp}$ includes contributions due to flow (which
are unlikely to be large) then $p_{\shortparallel} - p_{\perp} = B^2/4\pi$ and there should be
a strong preference for small pitch angles at least for those par-
ticles most responsible for the stress.   Hones (1971) has found
such an anisotropy for electrons, but the statistics of his instrument are inadequate for protons and even for electrons he can only
set a lower limit to the anisotropy.   The argument in terms of
stress is hard to evade and the origin of the anisotropy should be
sought in accelerating mechanisms.   The possibility of acceleration
in the outer current-bearing layer of the plasma sheet by parallel
E or turbulence should be mentioned again in this connection,
though further discussion of the mechanisms will be left to others.
It should be noted here that the cross-tail electric field, which
is generally believed to exist over the plasma sheet as well, must
cause particles from the low density population in the lobes to
drift inwards across the sheet boundary, probably crossing the current layer in about a minute. This may be long enough for acceleration, but note that protons of say 2 keV energy can travel only a
few earth radii in a minute and it seems that protons in this layer
must come from the lobes and have a low density.   As the plasma
drifts further into the plasma sheet it is compressed correspondingly to the contraction of the field lines, but this can still hardly
provide the necessary stress.   The minimum density of 2 keV particles required to isotropise the stress in a 20γ field is about
$1\ cm^{-3}$.   Furthermore the particles in the plasma sheet are believed
to come mainly from the solar wind rather than the polar wind, because measurements of helium ions show many more doubly ionised
than singly ionised (Reasoner 1973).   Another important consequence
of the presumed anisotropy of particle pressure in the plasma sheet

is that it implies a limit on any turbulence which would cause
pitch angle scattering.

In relation to the general question of the origin of plasma in
the plasma sheet there is quantitative uncertainty about the relative
importance of the inward drift due to the cross-tail electric field
and the cross-tail drift due to nonuniformity of the magnetic field.
This is equivalent to the question of the magnitude of the time
averaged cross-tail potential in relation to the characteristic
particle energy and it seems likely that neither drift is negligible.
Then much of the plasma has entered through the neutral sheet which
occupies the body of this paper.

SIMPLIFIED MODELS OF THE NEUTRAL SHEET

Over the past decade the theoretical modelling of the neutral
sheet has developed into quite an extensive subject. There is no
acute controversy, but progress requires considerable simplifying
approximations and different models neglect different features
thereby of course emphasising the others. The most basic simplification is made by reducing the number of dimensions to two and
it seems plausible at first that the essentials may survive in a
model which neglects any variation in the y-direction. We shall
see that this approximation is unsatisfactory, however, and indeed
auroral observations, particularly the Harang discontinuity, require
consideration of variations in the y-direction. Other studies
neglect variation in the x-direction, but only claim to model a
part of the sheet and should prove useful in finally assembling a
three-dimensional model. The sheet has been broken down into different regions, where different approximations are appropriate,
and current work concerns these separate regions. I shall endeavour
to describe them with decreasing simplicity. Two assumptions can
be made throughout: B_x is an odd function of z and $B_y = 0$. Even
these are not justified by observation, but they are common to all
theoretical work of which I am aware. It should be remarked that
the "neutral sheet" is the plane z = 0, but that strict neutrality,
meaning $|B| = 0$, occurs only where $B_z = 0$. Usually B_z is taken
to be a function of x and there are one or more neutral lines.
One other universal assumption is that E_y has the same sign as the
current in the sheet.

The objective is self-consistency with distribution functions
for electrons and protons satisfying Liouville's theorem and also
giving charge and current densities which agree with the sources
required by the field. Most of the work therefore concerns the
trajectories of the particles, which are needed in order to apply
Liouville's theorem. All the models are static, so that an
electrostatic potential ϕ exists, such that $\underset{\sim}{E} = -\underset{\sim}{\nabla}\phi$ and the particle
energy $\frac{1}{2} mv^2 + q$ is a constant of the motion. For sufficiently

simple models integrals of x- and y- canonical momentum are useful
and it is convenient to note them here.   Omitting only B_y the
general momentum equations are

$$m\left(v_x\right) = q\{\int B_z \, dy + \int E_x \, dt\} \tag{1}$$

$$m\left(v_y\right) = q\{\int B_x \, dz - \int B_z \, dx + \int E_y \, dt\} \tag{2}$$

the integrals following the trajectory.   Use of these equations
is seldom advantageous unless the contributions from the electric
field can be neglected.   Schindler and Soop (1968) have given a
complete treatment for self-consistent fields in which all quantities
are independent of y and the distribution functions depend only on
the energy and the y-component of canonical momentum.   They obtain
elegant results and can treat the tearing mode instability as well
as the steady state.   Other aspects must be understood, however,
before applying their results.

While (1), (2) and the conservation of energy determine the
trajectories, it is useful to consider them pictorially in the
simplest case in which all field components other than B_x are
neglected and some are shown in Fig. 1.   Particles which do not
cross the plane z = 0 have a gradient drift which is in the opposite
direction to the current given by curl $\underset{\sim}{B}$ often allowing for the sign
of the particle.   Particles which cross z = 0 have a z-motion which
oscillates symmetrically about z = 0 and a motion in the y-direction
which can be rapid and has a preference for the direction of the

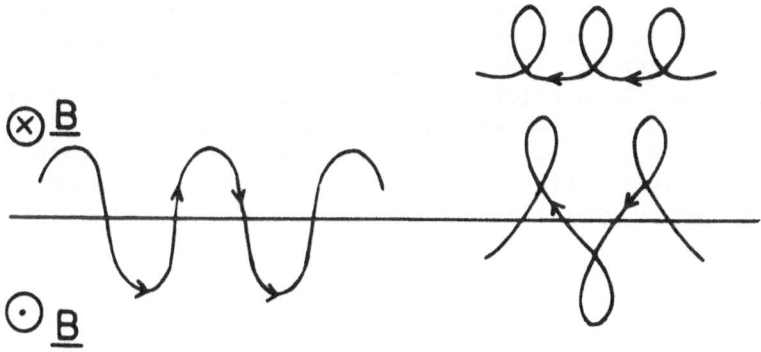

Fig. 1   Trajectories

current, but can go either way.   Consider next the effect of E_y in the same direction as the current which is duskwards.   Particles not crossing z = 0 now drift towards z = 0 until they do cross it. If E_y is weak the trajectories crossing z = 0 will appear very similar to those in Fig. 1, but evidently the particle's energy will change corresponding to its displacement in y.   Particles moving "the wrong way" lose energy until eventually their y-motion is reversed and those moving "the right way" gain energy indefinitely. If E_z is omitted the equation for the z-motion is

$$m \, d \, v_z/dt = -q \, v_y \, B_x$$

and for small amplitudes of z-oscillation B_x may be taken to be proportional to z.   The frequency of oscillation is

$$(\frac{q \, v_y}{m} \frac{dB_x}{dz})^{\frac{1}{2}}$$

which is real for particles going "the right way" and, if v_y increases slowly due to E_y, the WKB approximation is valid, or equivalently there is an adiabatic invariant, so that the amplitude of the oscillation is proportional to $v_y^{\frac{1}{4}}$.   It is the indefinite gain in energy from E_y which frustrates the combination of simplicity with reality.   The includsion of B_z is one way to remove the difficulty, but this will be deferred to section 4.   Now following Alfvén (1968) we will make the model finite in the y-direction keeping $B_z = 0$, so that there is no x-dependence.   The finite width of the model makes it more realistic and at first sight does not appear to add much complication.

Modifications of Alfvén's model will occupy the next section, but his original simple treatment will be described now.   He used a short cut which yielded a formula for the cross-tail potential without requiring any knowledge of the structure of the neutral sheet and this short cut is still extremely useful in more complicated modifications.   He made the assumptions that all the current was carried by particles from the lobes, that all the electrons went to the dawn side and all the protons to the dusk side.   The total current is related to B_x by Ampère's law and the rate at which the particles drift into the neutral sheet is proportional to the cross-tail potential and these relations determine Alfvén's potential

$$\Phi_A = B_x^2/4\pi \, ne$$

where B_x and n refer to the lobes.   For $B_x = 15\gamma$ $\Phi_A \sim n^{-1}$ kV, so that values of n now expected, say $\sim 10^{-2}$ cm^{-3}, give potentials in agreement with estimates obtained indirectly from various observations and with the measured potential difference across the polar

ionosphere, say tens of kilovolts. These high values may be confined to expansion phases, however, and it will be suggested in
the next section that the actual cross-tail potential is substantially less that ϕ_A under quiet conditions, but is increased when
noise causes anomalous resistivity so as to again agree with estimates from observation.

DEVELOPMENTS BASED ON ALFVÉN'S MODEL

The simplification $B_z = 0$ will be retained throughout this
section and it should be reemphasised that we are dealing with an
idealised model which at best can apply to a very small region near
the neutral line. The motivation for such studies is partly the
importance of the neutral line, but they are also aimed at the
study of microinstabilities which can cause anomalous resistivity
and it seems best to begin such studies on relatively simple
models of the unperturbed state. These idealised models are also
found to suggest an explanation for the field-aligned currents
found on the outer boundary of the plasma sheet.

Cowley (1971) retained Alfvén's assumptions, but drew attention
to the occurrence of space charge near the neutral sheet. A major
cause of space charge is that, whereas the speed of drift in the
electric field is the same for electrons and protons, protons move
much more slowly after crossing $z = 0$ and therefore spend longer
in the neutral sheet. Cowley investigated the self-consistent
field problem and found that the plasma approximation remains valid
and found a solution with positive space charge in the sheet. All
the equipotentials go nearly to the dusk edge of the sheet as in
Fig. 2, so that most of the protons have only a short way to go to
the dusk edge where they emerge, and this eliminates the original
cause of the space charge. Fig. 2 includes an even thinner sheet

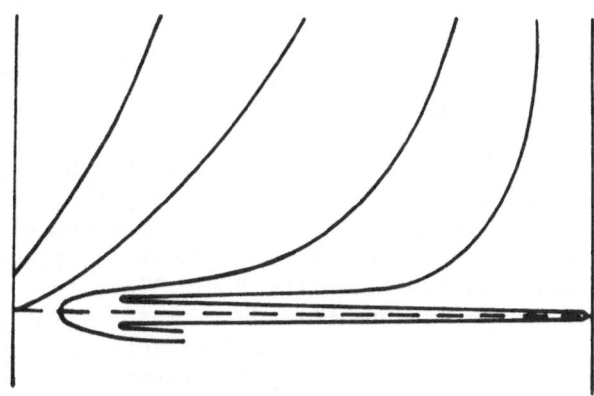

Fig. 2   Equipotentials

of reduced potential right in the middle, which is expected as a
consequence of the electron beam travelling to the dawn end.   The
problem of determining the electric field outside the sheet is
similar to that of a flow with rectangular walls with a sink in
one corner and is well-behaved.

Since Cowley's picture shows nearly all the particles from
the lobes going near the dusk edge, it is necessary to consider
the importance of particles entering the edges of the sheet from
the magnetosheath, which were previously neglected, probably because
of the thinness of the sheet.   It is instructive to consider the
magnetosheath extending in between the lobes of the tail and in
a steady state this magnetosheath plasma should supply a quantity
of current corresponding to pressure balance, so that we should ask
what is the pressure of this plasma in relation to the magnetic
pressure in the lobes.   The observation that the lobes do not
separate suggests that the plasma pressure is less than the magnetic pressure, and visualising the whole magnetosphere as an obstacle one would also guess that the plasma pressure is reduced,
but possibly by only a few per cent.   No better estimates have
been made and we will now assume that particles from the magnetosheath are important, carrying most of the current in the
neutral sheet, but that particles from the lobes are still required to carry some small fraction of that current.   Alfvén's
argument still survives and the cross-tail potential is simply that
fraction of Φ_A, but the structure of the neutral sheet must be
modified, and as yet only a few preliminary points can be made.

If the magnetosheath were able to provide pressure balance, so
that one could model the neutral sheet as a layer of magnetosheath
between the lobes, this would be equivalent to a pair of Rosenbluth
sheaths back to back.   There would be a potential difference between the sheath and the lobes due to the greater penetration of
protons.   Its magnitude would be only of the order of the proton
energy, but its sign would be negative in the middle, which is the
opposite of Cowley's.   A reexamination is necessary and has not
been completed.   The solution might not even be unique, but one
possibility seems to give a positive charge to the sheet as a whole
with a thin central layer of negative charge as would be the case
in Fig. 2.   It should now be pointed out that any potential difference between the middle and outside of the sheet provides a natural explanation of the field aligned currents on the boundary of
the plasma sheet.   In Cowley's case the magnitude of the potential
is a good fraction of the cross-tail potential so that E_z is large.
The voltage would be fed along the field lines to a thin layer in
the ionosphere and drive currents determined by the sum of the
ionospheric resistance and any anomalous resistance on the field
lines.   It may be noted that this is opposite to the situation
inside the magnetosphere, where quasiequilibrium of the trapped
particles determines what currents flow to the ionosphere and

ionospheric resistivity determines the electric field.   At the
neutral sheet closing of the field aligned currents requires some
j_z, but this is spread over a very large area, and is unimportant.
Consequently the neutral sheet acts as a voltage stabiliser, but
somewhere in the plasma sheet there must be a transition region
which is more complicated.   Now spacecraft may never go very near
the neutral line, but they commonly cross the outer surface of the
plasma sheet, so that observation of field-aligned currents near
this surface is an importance source of information about the neutral
sheet, and it is fortunate that so simple a phenomenon results.
Nevertheless there is still uncertainty whether the currents are
on the first open field lines, as would result by the above argument,
or the last closed field lines, which have yet to be discussed,
and most likely there is a combination.

The question of stability in the neutral sheet is expected to
be essential to the understanding of the sudden onset of the
expansion phase of a substorm and attempts have been made on the
stability problem in models of the type just described.   The first
observations of noise were reported by Scarf et al. (1974), but in
this sole event B_z was quite large and the authors point out that
the noise observed can be interpreted in the same way as noise
observed within the magnetosphere.   Unfortunately there are long
odds against a spacecraft crossing the neutral sheet near the neutral
line around the time of an expansion phase onset.   However it is
important to study the stability of our models, as soon as they are
sufficiently precise to be used as the unperturbed state in stability analysis. Simplification is possible, if, in the middle of
the sheet, E_y is negligible and also variations with y, so that the
normal modes vary like e^{iky}.   This is the case mentioned in section
2, where equation (2) is useful and Schindler and Soop (1968) have
given a general treatment of the self-consistent problem, but the
above discussion of how the particles enter this region gives a semi-
quantitative model for the distribution functions.   If there is a
cross-tail potential concentrated at the dusk end as in the Cowley
model this acts like an electron gun and some kind of double stream
instability is expected, though application of uniform plasma theory
can hardly be assumed for such an extremely nonuniform system.

Bowers (1973a) used the Vlasov formulation for the stability
problem just described and later (1973b) included electrons from the
magnetosheath.   The exact integrals are complicated by the closely
spaced resonances occurring when the average frequency seen by a
particle is a multiple of the frequency of its oscillation in z,
though the coefficients for the higher harmonics may well be small.
Consequently sweeping approximations were called for and Bowers
considered short wavelength electrostatic waves.   His solutions
resemble ion acoustic waves with phase varying more rapidly with
z than with y.   Electromagnetic waves should also be considered,
but Bowers' formulation demonstrates the feasibility of a general
computer search.   It is probable that any waves which grow are

evanescent in the lobes as discussed by Gjøen (1971), both because
otherwise energy would be lost by radiation and because observation
shows that the lobes are remarkably quiet.

With the above formulation the wave is static in a frame moving
in the y-direction at speed ω/k and it is possible to visualise the
direction of quasilinear diffusion in phase space.   For particles
which cross $z = 0$, the energy in this wave frame is a constant of
the motion and the diffusion is subject to this constraint.

In this respect the system is similar to the double stream
instability in a uniform plasma and instability is expected from
the accelerated beam of electrons coming from the dusk magnetosheath.
The instability leads to anomalous resistivity, hence to increased
cross-tail potential and greater acceleration of the beam and therefore selfexacerbation, which is the rather unusual situation required
for explosive onset. However, much further work is needed on the
nonlinear development and there seems to be a problem concerning
the location of the anomalous resistivity.   If waves are amplified
as they travel dawnwards, the intensity of turbulence should increase
towards the dawn edge.   However changes in E_y in general cause
changes in space charge and this may propagate by means of surface
waves, which have yet to be investigated.

MODELS WITH FINITE B_z AND GENERAL DISCUSSION

Speiser (1965) described trajectories in a model with a weak
but finite B_z for particles arriving in the sheet with low energies.
For the type of trajectory he found v_y is positive while the particle
is in the sheet and there is an oscillation in z of small amplitude
which behaves just as in the case with $B_z = 0$.   The motion in the
plane of the sheet is barely affected by the oscillation in z and is
like the motion in crossed fields E_y and B_z, but the particle leaves
the sheet when v_y becomes small as it must within a time $\pi m/e\, B_z$.
The amplitude of v_z decreases with v_y, so that both are small compared to v_x when the particle leaves the sheet and consequently outside the sheet the particle has a small pitch angle.   The argument
is reversible in time and it is now seen that particles with small
pitch angles outside the sheet correspond to Speiser orbits in the
sheet and, it may be noted, have large pitch angles at $z = 0$, though
they perform only half a gyration.   The discussion of p" in the
introduction shows that such particles are important.

Eastwood (1972) obtained self-consistent models corresponding
to Speiser orbits with incoming cold streams representing the polar
wind. He neglected any variation in x and then, provided the orbits are smaller than the width of the tail, the problem is reduced
to one dimension. Unfortunately the variation in x must be important anywhere near the neutral line, but Eastwood's results are use-

ful for future model building. He showed that the current sheet can
be extremely thin, that there may be a positive potential in the middle and that electrons trapped in such a potential can thicken the
sheet. He also found that iteration for the current density profile
converged rapidly, because the current computed from the trajectories
was not very sensitive to the magnetic field model, and this is en-
couraging for future computations of two dimensional models. Little
further progress has been made, but it is clear that drifts must be
studied and these provide a view of the complete problem.

A view of the drift motion is provided by the x-component of
canonical momentum (2), if E_x is omitted. For Speiser orbits the
particle has a small pitch angle and therefore mirrors far from the
neutral sheet before returning to perform another half gyration.
Now v_x suffers a reversal both in the mirroring and the half-gyration
and (2) relates the change to the displacement in y. The value of
B in the mirror region must be large and unless this region happens
to be near geomagnetic latitude 60° B_z will also be much larger than
its value in the neutral sheet. Then the drift comes mainly from
the half-gyration in the neutral sheet and the resultant drift speed
is in fact just given by the formula for a dipole field with the
field strength replaced by the actual value of B_z in the sheet. For
any other orbits the pitch angle outside the sheet is not small, so
that the orbit cannot extend over a large range of x and it may be
assumed that B_z varies linearly with x. Then equation (2) is again
useful and shows that the drift is like a gradient drift in just B_z
and proportional to $\bar{v}_x^2$. Thus orbits with small v_x have particularly
slow drifts, but otherwise the drift velocities are not very different from the familiar values. In fact typical observed values of
B_z are not very different from the dipole values and we cannot really
improve on the dipole formula, which, for an energy of 2 keV gives
~ $(L/10)^2$ earth radii per hour.

We should finally ask what has been said here about accelerating
mechanisms. We have seen that noise may be important in the thin
current layers, but let us ignore noise and assume a steady state.
Then the change in energy of a particle results simply from going to
somewhere where the potential is different. It is reasonable that
this should be the important accelerating mechanism and now we should
consider its consequences. The problem is seen to be the origin of
particles in the plasma sheet and is very close to another paper on
entry of particles into the magnetosphere. The problem should be
posed by considering a point in the middle of the plasma sheet and
following back all trajectories through that point to see where they
come from. Particles which come from the magnetosheath on the
"right" side (dusk for electrons) have gained energy and the drift
is indeed the "right" way. If all trajectories above a certain
energy were like this the distribution function above this energy
would be determined by that in the magnetosheath and that for elec-

trons would be almost independent of pitch angle. As pointed out
by Hill (1974) the flow in the magnetosheath is important for the
proton velocity distribution and might be expected to lead to an
asymmetry between the dawn and dusk sides of the plasma sheet. Alternatively most trajectories may come from the neutral sheet and
it is then necessary to ask where they came from before that. Some
will still come from the magnetosheath, but it is likely that some
with large values of v_z in the neutral sheet will come from the
lobes. For particles coming from the lobes the distribution function is small except in very small regions in velocity space corresponding to small initial energies and the lobe density is so
small that the contribution of these particles is probably unimportant. This may offer an explanation of anisotropy in the plasma
sheet, which in a much simplified form would say that the Speiser
orbits have magnetosheath distribution functions while the others
have lobe distribution functions.

It must be admitted that these conclusions are vague and unsatisfactory. Modelling continues and much remains to be done.
It is very clear that we think of some typical velocity distribution
for the whole plasma sheet, but theory predicts significant differences depending on position and it is time to reexamine the data.
In the nearest part of the plasma sheet some particles could drift
right round the earth, if they had small enough pitch angle and
high enough energy. Shell splitting then offers a completely different explanation of anisotropy and some difference should be
noticed in particle spectra. Differences should also be sought in
the proton distribution near the dawn side and in the electron distribution near the dusk side.

ACKNOWLEDGEMENT

This paper has been influenced by conversations with numerous
colleagues, but was finished too late for anyone to read before it
was committed to type.

REFERENCES

Alfvén, H., 1968, J. Geophys. Res. 73, 4379.

Arnoldy, R.L., P.B. Lewis and P.O. Isaacson, 1974, J. Geophys. Res. 79, 4208.

Bowers, E.C., 1973a, Astrophys. Space Sci. 21, 399.
    1973b, Astrophys. Space Sci. 24, 349.

Cowley, S.W.H., 1971, Cosmic Electrodyn. 2, 90.

Craven, J.D. and L.A. Frank, 1974, U. of Iowa 74-30.

Eastwood, J.W., 1972, Plan. Space Sci. 20, 1555.

Gjøen, E., 1971, Plan. Space Sci. 19, 635.

Gurnett, D.A., 1974, U. of Iowa, 74-39.

Hedgecock, P.C. and B.T. Thomas. 1975, in Formisano and Kennel (eds.), Neil Brice Memorial Symposium, Frascati.

Hill, T.W., 1974, Rev. Geophysics and Space Physics, 12, 379.

Hones, E.W., J.R. Asbridge, S.J. Bame and S. Singer, 1971, J. Geophys. Res. 76, 63.

Reasoner, D.L., 1973, Rev. Geophysics and Space Physics 11, 169.

Scarf, F.L., L.A. Frank, K.L. Ackerson and R.P. Lepping 1974, Geophys. Res. Letters 1, 189.

Schindler, K. and M. Soop 1968, Phys. Fluids 11, 1192.

Speiser, T.W. 1965, J. Geophys. Res. 70, 4219.

VLF ELECTROSTATIC WAVES IN THE MAGNETOSPHERE

M. ASHOUR-ABDALLA

C.F. KENNEL[*]

CENTRE DE RECHERCHES EN PHYSIQUE DE L'ENVIRONNEMENT
3, avenue de la République, Issy-les-Moulineaux, France

ABSTRACT

The stability of electrostatic waves above the electron cyclotron frequency in an electron plasma consisting of a cold component
and a hot component with a weak "loss-cone" perpendicular velocity
distribution is investigated. A simple marginal stability algorithm
permits a rapid estimate of the unstable frequency band as a function of two ratios, the hot to cold density, and the cold upper hybrid to electron cyclotron frequency. The growth rate peaks near but
below the cold upper hybrid frequency. However, the group velocity
parallel to the magnetic field can have zeroes near the odd half-
harmonics of the cyclotron frequency, suggesting that the instability is non convective under certain conditions, possibly accounting
for observations of narrow band emissions near 3/2, and sometimes
near 5/2 of the electron cyclotron frequency in the magnetosphere.

[*]Also at Centre de Physique Theorique, Ecole Polytechnique, Palaiseau
On leave from Department of Physics and Institut of Geophysics and
Planetary Physics UCLA, Los Angeles 90024, USA

1. INTRODUCTION

Magnetospheric research has identified three problems where
fundamental plasma processes affect the macroscopic state of the magnetosphere - magnetic field line reconnection at the nose and tail
of the magnetosphere, anomalous resistance in auroral arcs and perhaps elsewhere, and pitch angle scattering on auroral field lines
and in the radiation belts. In these three areas, intelligent interpretation of satellite results requires theories in which plasma
physics is placed in a geophysical context. In two of the three,
reconnection and anomalous resistance, many interesting plasma physical ideas have been discussed but not integrated into a theory capable of making geophysical predictions. With pitch angle scattering,
the situation is somewhat better, though as we will argue shortly,
not as good as is commonly thought, and certainly not as good as it
ought to be. We do understand how electromagnetic ion cyclotron and
whistler instabilities scatter *high energy* ions and electrons, respectively, from the radiation belts ; there is even a reasonable
convergence of theory and observation. We lack geophysical understanding of electrostatic instabilities, which scatter ions and
electrons in the *main* part of their velocity distributions. We have
not even conclusively identified the relevant ion electrostatic instabilities, experimentally or theoretically : a gross impediment to
understanding magnetic storms. For electrostatic waves affecting
electrons, we are somewhat better off. OGO-5, and subsequently IMP-6,
have identified a class of electrostatic waves, typically with frequencies near but above the electron cyclotron frequency Ω (the so-
called "3/2" waves), which are found sufficiently often in the region
$4 < L < 14$ that they could be responsible for the diffuse auroral
1-10 Kev electron precipitation. (Kennel et al, 1970 ; Scarf et al,
1973 ; Gurnett and Shaw, 1973). In addition, a series of theoretical
papers stimulated by the observations has identified the basic plasma
physics of the instability (Fredericks, 1971 ; Young et al, 1973 ;

Karpman et al, 1975a - 1975b ; Ashour-Abdalla et al, 1975 ; Ashour-Abdalla, 1975), and a promising start, albeit heuristic, has been made on the problem of non-linear saturation and pitch angle diffusion (Lyons, 1974). Nonetheless, we shall argue here that our current theoretical understanding is not yet sufficiently realistic in geophysical terms to have a significant impact on the experimentalist.

It is not the purpose of this paper to dwell on the relative successes of whistler and ion cyclotron turbulence theory, but it is worthwhile pointing out that they stem partly from a series of historical accidents. On the experimental side, whistlers had been observed on the ground before the satellite era, and so the first generation of satellite experiments were equipped to measure them. Moreover, the first experiments measured only the high energy particles which, as it happens, resonate with electromagnetic waves. On the theoretical side, there existed from 1961 (Sagdeev and Shafranov, 1961) an instability theory so simple that it could be applied unambiguously to the magnetosphere. Luck was with us even at the non-linear level, since the quasi-linear diffusion of energetic particles reduces to almost pure pitch angle diffusion (Kennel and Petschek, 1966) rather than mixed energy and pitch angle diffusion. By contrast, the measurements of low energy particles and the electrostatic waves resonating with them had to wait until the second generation of satellite experiments, and the theory, while no more difficult in principle, is much more complicated in practice.

Let us compare some features of whistler and electrostatic theories to illuminate where the electrostatic complexities lie. First of all, for whistlers, basically only one resonant interaction, at the first harmonic of the electron cyclotron frequency, is important, whereas for electrostatic waves, at least two are important. For example, for a wave with frequency $3/2 \; \Omega$, at least the $n = 1$ and $n = 2$ cyclotron harmonic resonances must be included in the theory. And if the wave number parallel to the magnetic field, $K_{//}$,

is sufficiently large - as it turns out to be - more than two resonances play a role. To determine the stability of whistlers, the
theorist has only to model the tail of the electron distribution,
whereas for electrostatic waves, he must model the *entire* distribution. His electrostatic results are therefore much more model-depen-
dent. Since he must also use the computer, it is more difficult to
arrive at general conclusions even granted his model. The fact that
the cold plasma determines the whistler phase and group velocity
and the resonant particle energy also simplifies the task of the
whistler theorist. As we shall see, the phase and group velocities
of electrostatic waves are determined by a mixture of hot and cold
plasma effects ; they too depend upon the choice of the model distribution. Finally, while whistlers are always most unstable when
$K_\perp = 0$ and $K_{//} \neq 0$, where $K_\perp$ is the wave number component perpendicular to the magnetic field, electrostatic waves have both $K_{//}$ and
$K_\perp$ non zero, and the most unstable $K_{//}/K_\perp$ depends upon the model
electron distribution function.

In this paper, we rephrase the theory of electrostatic waves
in terms hopefully useful for interpretation of satellite experiments. In Chapter II, we review theoretical work to date, and in
Chapter III, we present a simple algorithm for determining which
waves can be unstable. In Chapter IV, we discuss selected solutions
for unstable frequencies and growth rates. Here we conclude that, at
least for the simple model distribution function we will choose, the
only instability is of the upper hybrid wave, based on cold electron
density. In Chapter V, we add one factor of realism hitherto over-
looked: since the waves propagate the instability is weakly convective. In several particular cases, we show that the hot plasma could
conceivably induce a nonconvective instability at $3\ \Omega/2$ and $5\ \Omega/2$,
since the parallel group velocity has zeroes in those frequency ranges.

II. A BRIEF HISTORY OF MAGNETOSPHERIC ELECTROSTATIC WAVE THEORIES

The basic plasma physics of electrostatic cyclotron instabilities was set out by Dory et al. (1965), who solved the dispersion relation for electrostatic waves with propagation vectors at an arbitrary angle to a uniform magnetic field using a "loss-cone" electron distribution function, one with a "hole" at small velocities $v_\perp$ perpendicular to the magnetic field. Such an electron distribution is subject to a variety of instabilities. Spurred on by the OGO-5 observations, Fredericks (1971) solved the same general dispersion relation for the simplest possible such distribution, a ring distribution of the form $F = (2\pi)^{-1} \delta(v_\perp - U) \delta(V_{/\!/})$. He found that instability above the electron cyclotron frequency was possible, and that in general an intermediate range of $K_{/\!/}/K_\perp$, neither large nor small, would be the most unstable. Young et al. (1973), critizised the ring distribution as too severe to be realistic for the magnetosphere. In fact, they found that more gentle distributions would be *stable* without the admixture of a cold electron component with a density n_c the order of, or less than, the hot density n_H. They thus established the basic model used in all subsequent investigations - a mixture of a hot component with a loss cone distribution and a cold component. They investigated briefly what happens when the cold component is given a small non-zero temperature T_c, and argued that finite T_c could restrict the range of unstable frequencies, thereby accounting for the fact that the emissions near 3/2 Ω are typically observed to be narrow band.

Young et al (1973), and Ashour-Abdalla and Cowley (1974) argued that a pure pitch angle anisotropy A, which can destabilize whistlers, cannot be responsible for electrostatic instabilities above the electron cyclotron frequency, for two reasons. First, for an instability with frequency ω in the range $n\Omega < \omega < (n + 1) \Omega$, a large anisotropy is required, A > n. Second, instability is possible only in the range

$(n + 1/2) \, \Omega < \omega < (n + 1) \, \Omega$, and waves seem to be observed as much below the half-harmonic frequencies as above.

A group in the Soviet Union, under the leadership of V. Karpman, performed the first extensive numerical calculations of wave dispersion relations and growth rates, first without a cold component and then adopting the Young et al (1973) model (Karpman et al, 1974a - 1974b). From their work emerged a realization of the importance of the cold upper hybrid mode. They varied the strength of the positive slope of the distribution, the cold density and temperature, and the magnetic field. They found that many normal modes could exist for $\omega > \Omega$, especially, it seems, when the positive slope is strong and $T_c \neq 0$. By adding complexity to the plasma distribution function, one adds new possible normal modes of oscillation.

Ashour-Abdalla et al (1975), returned to a simpler model of the electron distribution, with $T_c = 0$ and with a gentle slope of the hot electron distribution. They invented a graphical algorithm which gives the range of unstable frequencies and wave numbers as a function of the measurable parameters n_c/n_H and the magnetic field strength. In Chapter III, we exploit their insight, extending considerably the range of parameters they studied.

III. A MARGINAL STABILITY ALGORITHM

We start with the general dispersion relation of Harris (1959) :

$$D(\omega^{\star}, \, K_{\perp}, \, K_{/\!/}) \equiv 1 + \sum_{j} \frac{\omega_{pj}^2}{K^2} \sum_{n} \int \frac{G_{nj} \, (v_{/\!/}) \, dv_{/\!/}}{\omega^{\star} - K_{/\!/} v_{/\!/} - n\Omega_j} = 0 \qquad (3.1)$$

where $\sum\limits_{j}$ denotes a sum over species, or in our case hot and cold components, ω_{pj} is the plasma frequency of that component, $\omega^{\star}$ is the complex frequency and G_{nj} is given by :

$$G_{nj}(v_{/\!/}) = 2\pi \int_0^\infty v_\perp d\,v_\perp J_n^2 \left(\frac{K_\perp v_\perp}{\Omega_j}\right) \left[\frac{n\Omega_j}{v_\perp} \frac{\partial F_j}{\partial v_\perp} + K_{/\!/} \frac{\partial F_j}{\partial v_{/\!/}}\right] \qquad (3.2)$$

and $F_j(v_\perp, v_{/\!/})$ is the distribution function. When $|\gamma/\omega| \ll 1$, (3.1) can be expanded as follows :

$$D(\omega^\star, K_\perp, K_{/\!/}) \simeq D(\omega, K_\perp, K_{/\!/}) + i\,\gamma \frac{\partial}{\partial\omega} D(\omega, K_\perp, K_{/\!/}) \qquad (3.3)$$

Separating into real and imaginary parts, we have, to lowest order :

$$\text{Re } D(\omega, K_\perp, K_{/\!/}) = 0$$

$$\gamma = -\frac{\text{Im } D(\omega, K_\perp, K_{/\!/})}{\partial D/\partial\omega} \qquad (3.4)$$

To provide the positive slope in the hot distribution, Ashour-Abdalla et al (1975) chose the substracted Maxwellian below :

$$f_H = \frac{n_H}{\pi^{3/2}\,\alpha_\perp^2\,\alpha_{/\!/}\,(1-\beta)}\, e^{-v_{/\!/}^2/\alpha_{/\!/}^2} \left[e^{-v_\perp^2/\alpha_\perp^2} - e^{-v_\perp^2/\beta\alpha_\perp^2}\right] \qquad (3.5)$$

with $0 < \beta < 1$. f_H is normalized to unity. $\alpha_\perp$ and $\alpha_{/\!/}$ are thermal speeds perpendicular and parallel to the magnetic field. When $v_\perp$ is of order $\sqrt{\beta}\alpha_\perp$, $\partial f_H/\partial v_\perp > 0$, $\partial f_H/\partial v_\perp < 0$ if $v_\perp \gtrsim \alpha$.

When $\beta \to 1$ and $(1-\beta)\,v_\perp^2/\alpha_\perp^2 \ll 1$:

$$f_H \to \frac{1}{\pi^{3/2}\,\alpha_\perp^2\,\alpha_{/\!/}} \left(\frac{v_\perp}{\alpha_\perp}\right)^2 e^{-v_{/\!/}^2/\alpha_{/\!/}^2} e^{-v_\perp^2/\alpha_\perp^2} \qquad (3.6)$$

When $\beta \to 1$, the substracted Maxwellian reduces to the weakest Dory-Guest-Harris loss-cone distribution $(j = 1)$.
We will restrict our attention to this weak slope limit, as this seems most realistic for the magnetosphere.

If T_c is exactly zero, cold electrons make no contribution to Im D, and the marginal stability condition is independent of n_H as well :

$$\sum_n \left(\frac{\alpha_\perp^2}{\alpha_{//}^2} \; C_n \; (\lambda) \; z_n + \frac{n \; D_n \; (\lambda)}{K_{//}\rho_{//}} \right) e^{-z_n^2} = 0 \qquad (3.7)$$

where $\lambda = \frac{1}{2} K_\perp^2 \rho_\perp^2 \equiv \frac{1}{2} \frac{K_\perp^2 \alpha_\perp^2}{\Omega^2}$, $z_n = \frac{x - n}{K_{//}\rho_{//}}$, $X = \frac{\omega}{\Omega}$

$\rho_{//} = \alpha_{//}/\Omega$, and C_n and D_n are functions coming from the integrations over $v_\perp$ in (3.2)

$$C_n \; (\lambda) = \frac{1}{1-\beta} \left\{ e^{-\lambda} \; I_n(\lambda) - \beta e^{-\beta\lambda} \; I_n(\beta\lambda) \right\} \qquad (3.8a)$$

$$D_n \; (\lambda) = \frac{1}{1-\beta} \left\{ e^{-\lambda} \; I_n(\lambda) - e^{-\beta\lambda} \; I_n(\beta\lambda) \right\} \qquad (3.8b)$$

where $I_n \; (\lambda)$ is the modified Bessel function of order n. Even if $T_c \neq 0$, equation (3.7) is approximately satisfied for ω sufficiently different from $n\Omega$ so that cold electron cyclotron damping is negligible.

Assuming $T_c = 0$ and substituting the subtracted Maxwellian into the real part of the dispersion relation, we find

$$\text{Re D} = 1 - \frac{\omega_{pc}^2}{\Omega^2} \left(\frac{K_\perp^2/K^2}{X^2-1} + \frac{K_{//}^2/K^2}{X^2} \right) - \frac{\omega_{pH}^2}{\Omega^2} \; \text{Re H}_H \qquad (3.9)$$

where Re H_H is

$$\text{Re H}_H = \frac{K_\perp^2}{2K^2\lambda} \sum_n \left(\frac{\alpha_\perp^2}{\alpha_{//}^2} \; C_n \; \text{Re Z}' \; (z_n) - \frac{2n}{K_{//}\rho_{//}} \; D_n \; \text{Re Z} \; (z_n) \right) \qquad (3.10)$$

and $Z(z_n)$ is the plasma dispersion function. Since, for the unstable modes of interest, $K_{//}/K_\perp < 0.3$, we can neglect the term involving $K_{//}$ on the left hand side of equation (3.9). Then equation (3.9) can be simplified to :

$$\frac{n_c}{n_H} - \frac{\Omega^2}{\omega^2_{pH}} (X^2-1) = - (X^2-1) \text{ Re } H_H \equiv \hat{K}(X^2-1) \tag{3.11}$$

Ashour-Abdalla et al's (1975) graphical marginal stability algorithm consists of choosing, say $K_{//}$, solving equation (3.7) for $X(\lambda)$, inserting $X(\lambda)$ into Re H_H and computing K as a function of X^2-1 with $K_{//}$ as a parameter. Then, on the plot of $\hat{K}$ (X^2-1), we choose the ordinate at $(X^2-1) = 0$ to be n_c/n_H and draw a straight line with slope $-\Omega^2/\omega^2_{pH}$. Where this straight line intersects $\hat{K}(X^2-1 ; K_{//})$, we have determined a marginally stable solution.

Note that the intersection of the straight line with the $\hat{K} = 0$ axis is given by the cold upper hybrid frequency ω_{UH} :

$$X^2 - 1 = \frac{\omega^2_{pc}}{\Omega^2} \quad \text{or} \quad \omega^2 = \omega^2_{pc} + \Omega^2 \equiv \omega^2_{UH} \tag{3.12}$$

Thus, the marginally stable solutions can also be determined by choosing n_c/n_H and $\omega^2_{UH}/\Omega^2 - 1$ and drawing a straight line between the two.

Figure 1 shows a plot of $\hat{K}$ versus $X^2-1 = \omega^2/\Omega^2-1$ for $\alpha_\perp/\alpha_{//} = 1$ (or $T_\perp/T_{//} = 1$) and $\beta = 0.9$ for $K_{//}\rho_{//}$ in the range $0.2 < K_{//}\rho_{//} < 2.5$. While they are not shown here, the $\hat{K}$ functions for $K_{//}\rho_{//} < 0.2$ lie close to the $K_{//}\rho_{//} = 0.2$ curve everywhere. The frequencies $X = 3/2$ and $X = 5/2$ are marked on the $(\omega/\Omega)^2 - 1$ axis by small crosses. Since the typically marginally stable $K_\perp\rho_\perp \simeq 3$, the neglect of $K^2_{//}/K^2 X^2$ in equation (3.9) becomes suspect for the larger values of $K_{//}\rho_{//}$. From the point of view of growth rate however, the interesting $K_{//}\rho_{//}$ are less than unity.

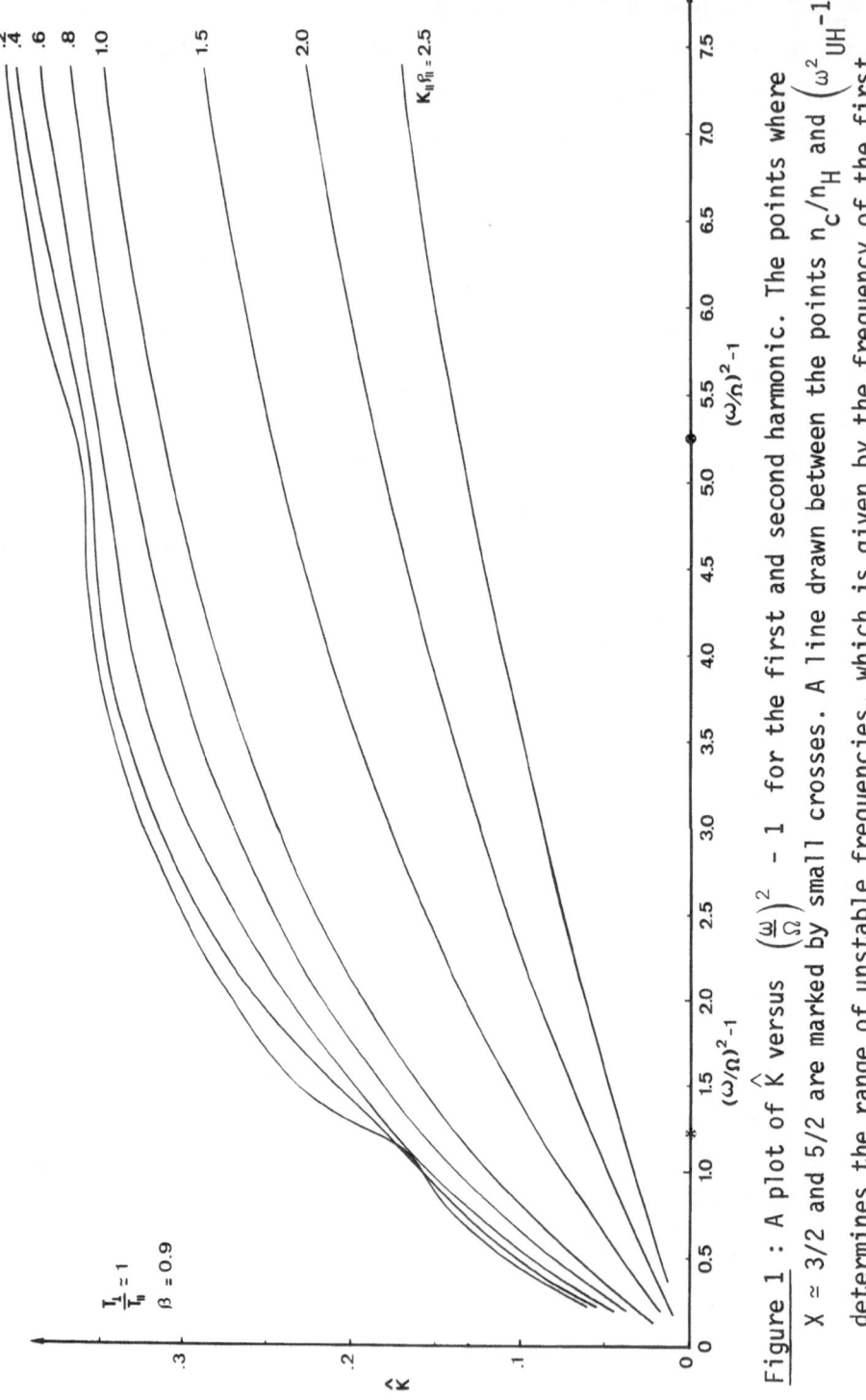

Figure 1 : A plot of $\hat{K}$ versus $\left(\frac{\omega}{\Omega}\right)^2 - 1$ for the first and second harmonic. The points where $\overline{X} \simeq 3/2$ and $5/2$ are marked by small crosses. A line drawn between the points n_c/n_H and $\left(\omega^2_{UH} - 1\right)$ determines the range of unstable frequencies, which is given by the frequency of the first intersection with the $\hat{K}$ function and ω_{UH}.

Now, let us apply our algorithm. We pick an n_c/n_H and $\frac{\omega^2_{UH}}{\Omega^2} - 1$ and draw a straight line between the two. The lowest unstable frequency, Xmin, will be given by the first intersection of that straight line with one of the $\hat{K}$ functions. Furthermore, we show in section 4 that the largest unstable frequency, Xmax, is very near X_{UH}.

Using the graphical algorithm, we can easily deduce many of Young et al's (1973) conclusions on how n_c and n_H determine the range of unstable frequencies. For example, keeping n_H fixed and increasing n_c increases *both* Xmin and Xmax. For a given n_H there is a critical n_c for which Xmin < 1.75. Thus since X < 1.75 is most commonly observed, n_c must typically be below this critical value. Note that increasing n_c further eventually moves Xmin above 2. While, it is possible to have waves near X = 5/2 unstable, with those at X = 3/2 stable, according to Fredericks (1974) this happens rarely if at all.

If we keep n_H fixed, but let $n_c \rightarrow 0$, the entire range of unstable frequencies shrinks to zero, another way of saying that gentle loss cone distributions are stable in the absence of cold electrons.

Keeping n_c fixed and increasing n_H decreases Xmin ; decreasing n_H increases Xmin leaving Xmax $\simeq X_{UH}$ fixed. Thus, when $n_H/n_c \ll 1$ only frequencies near the upper hybrid frequency can be unstable. On the other hand, when n_c/n_H is less than about 0.1, frequencies very near the first electron cyclotron harmonic can be unstable. Since X > 1.25 is typically observed, Young et al (1973) concluded that, for a given n_H, n_c usually exceeds a certain lower bound. In order that the most commonly observed frequencies lie in the range 1.25 < X < 1.75, n_c must fall between two extreme values which they estimated roughly to be $10^{-2} < n_c < 1$ electron/cm³

As a further demonstration, if one is needed, that the loss-cone property, and not a pitch angle anisotropy, is reponsible for instability, we plot in figures 2a and 2b $\hat{K}(X^2 - 1)$ for the cases $T_\perp/T_\parallel = 1/2$ and 1/10, and β = 0.9. $\hat{K}$ has roughly the same magnitude as when $T_\perp/T_\parallel = 1$, and so all the statements made with reference to figure 1 are true also for figures 2a and 2b. Figures 2a and 2b suggest two general conclusions. First, electrostatic waves can be unstable when whistlers are not, and that as far as marginal stability is concerned, their properties seem relatively insensitive to variations in $T_\perp/T_{//}$ when $T_\perp/T_{//} < 1$. Remember, however, that a loss-cone distribution can masquerade as a pitch angle anisotropy to anyone measuring the pitch angle distribution at constant energy. It is easy to see that the effective anisotropy A_{eff} is given by :

$$A_{eff} = \frac{T_\perp}{T_{/}} (1 + \beta) - 1$$

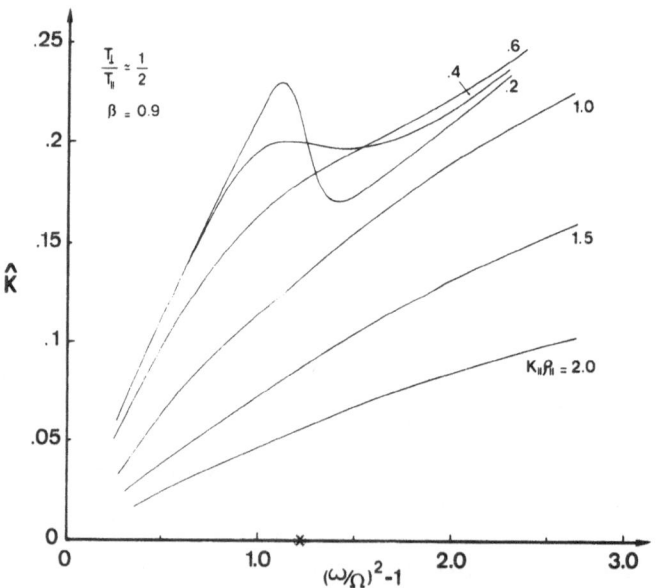

Figure 2a : A plot of $\hat{K}$ versus $\left(\left(\frac{\omega}{\Omega} \right)^2 - 1 \right)$ for the first harmonic, for $T_{//}/T_\perp = 2$.

and so for figure 2a, A_{eff} = - 0.05 and for 2b, A_{eff} = - 0.81.

Finally, we emphasize again that the unstable frequency range is determined only by n_c/n_H. Since cold electrons originate in the ionosphere, and hot electrons in the plasma sheet, and since there can be exchange of hot and cold electrons between the magnetosphere and ionosphere, we would expect a considerable range of n_c/n_H and X_{UH}, and therefore Xmin and Xmax, to exist at different times. And nothing in the structure of $\hat{K}$ emphasizes the regions near X = 3/2 and 5/2. Thus, why the waves are observed as *narrow band* emissions near half-harmonic frequencies appear to be a puzzle.

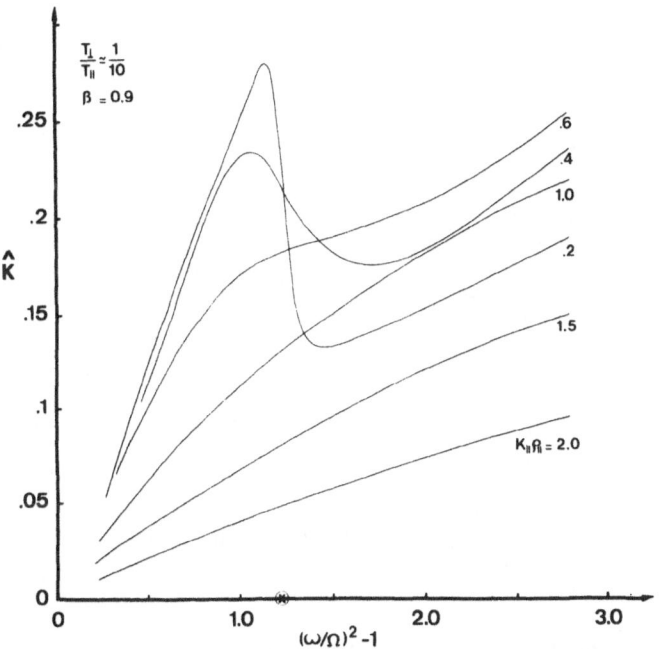

Figure 2b : A plot of $\hat{K}$ versus $\left(\left(\frac{\omega}{\Omega}\right) - 1\right)$ for the first harmonic, for $T_{//}/T_{\perp}$ = 10.

IV. PARTICULAR SOLUTIONS OF THE DISPERSION RELATION

In figure 3a, we plot $X = \omega/\Omega$ against $K_\perp \rho_\perp$ for $\beta = 0.9$, $T_\perp/T_{//} \simeq 1$, and one parallel wavelength $K_{//}\rho_{//} = 0.4$, for several different ω_{UH}/Ω ranging from 1.8 to 3, keeping n_H fixed. Where the wave is damped, ω is plotted as a solid line ; where it is unstable it is plotted as a dashed line. In figure 3b, the growth γ/Ω is plotted against $X = \omega/\Omega$ for the same parameters. From 3a, we conclude that the frequency approaches the cold upper hybrid frequency as $K_\perp \rho_\perp$ increases.

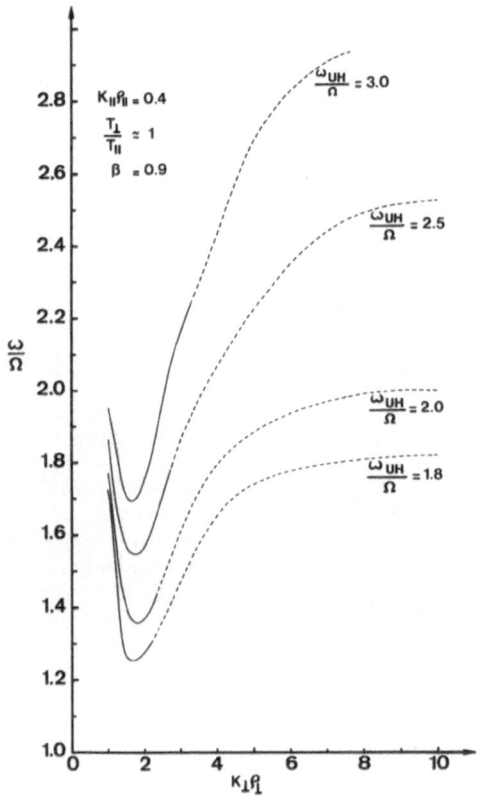

<u>Figure 3a</u> : The real part of the frequency versus $K_\perp \rho_\perp$ for different cold upper hybrid frequencies. The solid lines correspond to damping while the dotted lines show the unstable region.

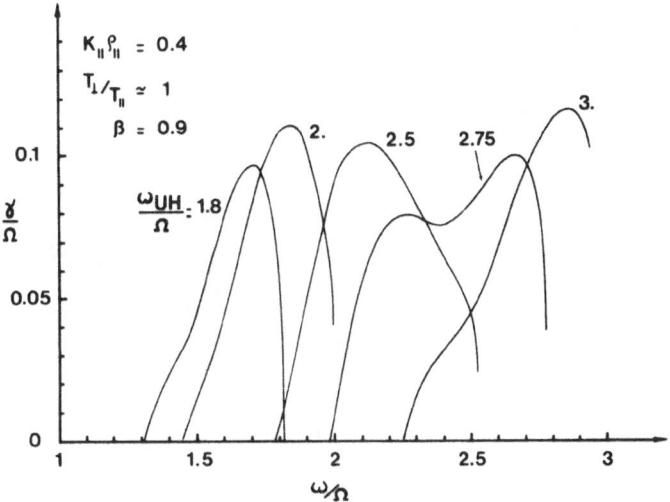

Figure 3b : The corresponding growth rates for Figure 3a. The frequency corresponding to peak growth rate follows ω_{UH}.

Secondly, the unstable frequencies are always below the upper hybrid frequency. No other modes were found in the range of parameters shown. The maximum growth rates in figure 3b are near but below the cold upper hybrid frequency, and, because we did not vary n_H, have the same order of magnitude for all ω_{UH}. The growth rate goes to zero at ω_{UH} and the Xmin correspond to what can be deduced from figure 1.

In figure 4a, we plot $X = \omega/\Omega$ versus $K_\perp \rho_\perp$ for the standard values $T_\perp/T_{//} \simeq 1$, $\beta = 0.9$, keeping n_c and n_h fixed, with $\omega_{UH} = 1.8$. We choose a range of $K_{//}$, $0.1 < K_{//}\rho_{//} \le 2$. All the curves for $0.4 \le K_{//}\rho_{//} \le 2$ are sufficiently close together that we shaded this region in. Our calculations indicate that, the curves for $K_{//}\rho_{//} = 0.1$ and 0.2 finally do approach the upper hybrid frequency as $K_\perp \rho_\perp \to \infty$. In figure 4b, we plot the growth rate γ/Ω versus $X = \omega/\Omega$ for the same parameters as in figure 4a.

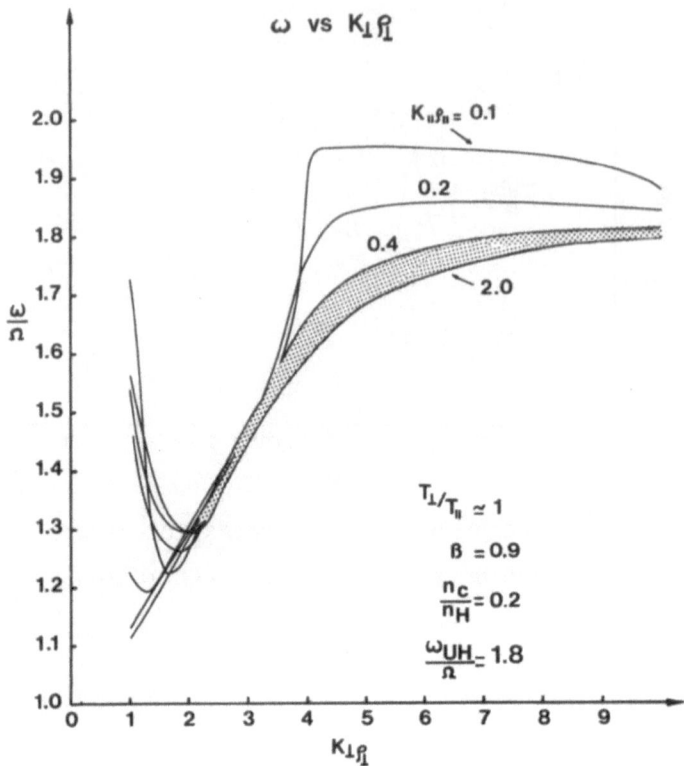

Figure 4a : The real part of the frequency versus $K_\perp \rho_\perp$ for different values of $K_{//}\rho_{//}$. The frequency is nearly independent of
   $K_{//}\rho_{//}$ in the shaded region.

Though not shown here, the peak growth rate for $K_{//}\rho_{//} = 0.1$ is less
than that for $K_{//}\rho_{//} = 0.2$ by a factor 2, and so figure 4b contains
the wave of maximum growth rate. Taken together, figures 4a and 4b
support, for one hybrid frequency and many $K_{//}\rho_{//}$, the conclusions
drawn from figures 3a and 3b for one $K_{//}\rho_{//}$ and many ω_{UH} : the growth
rate is zero as $K_\perp \to \infty$ and $\omega \to \omega_{UH}$, and the peak growth rates are
near but below ω_{UH} for all $K_{//}\rho_{//}$.

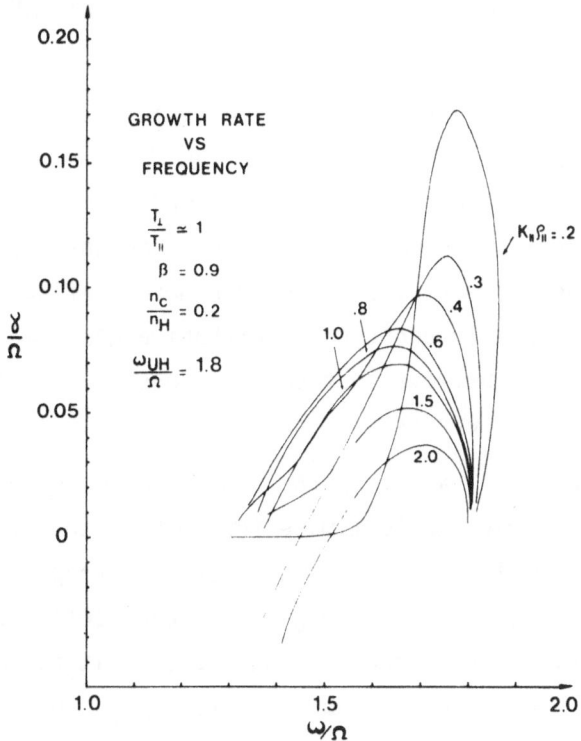

<u>Figure 4b</u> : The growth rate versus frequency for different values
of $K_{//}\rho_{//}$. The peak growth for $K_{//}\rho_{//}$ = 0.1 (not shown) is less
than for $K_{//}\rho_{//}$ = 0.2.

In fact, the largest growth rate of all, for $K_{//}\rho_{//}$ = 0.2, occurs
virtually at the upper hybrid frequency. We conclude that for the
parameters studied, we are dealing with an instability of the cold
upper hybrid mode. Once again, no structure that picks out a narrow
region around odd half-harmonic frequencies is evident.

V. PROPAGATION CHARACTERISTICS OF UNSTABLE WAVES

A proper consideration of electrostatic instabilities in the magnetosphere would necessarily feature the fact that they propagate in a spatially inhomogeneous medium . Instead of computing the temporal growth rate γ, we should compute the imaginary part of the wave vector. Since computing Im K is much more difficult than computing γ, people usually don't do it without a good reason. However, we illustrate why we believe that computations of γ alone might lead to serious conceptual errors.

When $|\gamma/\omega| \ll 1$, as it is for all the cases studied here, a reasonable but not rigorous estimate for Im K is to divide the growth rate by the group velocity. Here, we study the group velocity. We see that from figure 4a for $\omega < \omega_{UH}$ the perpendicular group velocity is roughly the electron thermal speed, and that it approaches 0, gradually for $K_{//}\rho_{//} > 0.4$ and more abruptly for $K_{//}\rho_{//} < 0.4$, near the hybrid frequency. But the perpendicular group velocity may not matter all that much in the magnetosphere. We expect the scale lengths of the particle distributions to be very long in at least the azimuthal direction. Thus, waves which propagate azimuthally can be effectively unstable, in the sense that if their parallel group velocity didn't remove them from the equatorial plane region, they would have a sufficient perpendicular growth length to reach non-linear saturation. Therefore, the magnitude of $\partial\omega/\partial K_{//}$ could determine the relative convective growth rates.

The most interesting feature of figure 4a is that the parallel group velocities ought to be *very small*. Precisely how small is shown in figure 5, where $\partial\omega/\partial K_{//}$, normalized to the electron thermal speed a , is plotted for the parameters in figures 4a and 4b for the range $0.4 < K_{//}\rho_{//} < 1.0$. We note two features. First $\partial\omega/\partial K_{//}$ tends to zero at the upper hybrid frequency, as it should. The dispersion relation for the cold plasma alone gives zero group velocity as $K_\perp \to \infty$.

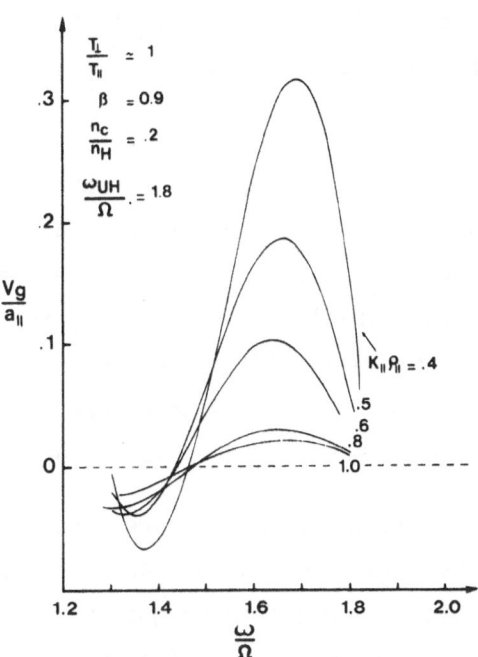

Figure 5 : The parallel group velocity normalized to the parallel electron thermal speed versus frequency for different values of $K_{//}\rho_{//}$. Note the zero crossings of $\partial\omega/\partial K_{//}$ near but slightly below $\omega/\Omega = 3/2$.

Secondly, and more interesting, all the $\partial\omega/\partial K_{//}$ have a zero crossing
near $X \simeq 3/2$. Thus, we reach the following tentative conclusions
for this case :

(1) There is a possible non-convective instability near the
     cold upper hybrid frequency. However, Young et al (1973)
     and Karpman et al (1974a - 1974b) have argued that the
     $K_{\perp} \to \infty$ region is sensitive to T_c. Choosing a small but
     non zero T_c is stabilizing. This point requires further
     study.

(2) There is a possible effectively non-convective instabi-
     lity in a narrow range of frequencies near $X \simeq 3/2$. Thus,
     even though there are peaks near the upper hybrid fre-
     quency, the parallel group velocity there is very large,
     and the minimum amplification length could occur near
     $X \simeq 3/2$. We offer this as an alternate explanation for
     the observed narrow band emissions near $X \simeq 3/2$.

(3) The zero crossing of $\partial\omega/\partial K_{//}$ are a hot plasma effect,
     since the cold plasma alone can have zero group velocity
     only at the upper hybrid frequency. Thus, in principle
     there is a minimum n_H/n_c which induces $\partial\omega/\partial K_{//} = 0$ and
     therefore, presumably, an effectively non-convective
     instability.

The Im $K_{//}$ gotten by dividing γ by $\partial\omega/\partial K_{//}$ are so large that they
cannot be trusted. However, the small group velocities mean that very
much weaker conditions than one would expect on the basis of considerations of γ alone could still lead to appreciable convective amplification. Within the present computational scheme, we cannot reduce
the slope in the loss cone distribution, but we can reduce its density
n_H. From figure 1, decreasing n_H means that we anticipate a narrow
band instability around ω_{UH}. In figure 6, we have chosen $\omega_{UH} = 1.8$,

$T_\perp/T_{/\!/} \simeq 1$, $\beta = 0.9$, again but we have taken n_c/n_H significantly larger than visualized by Young et al (1973), $n_c/n_H = 2$. $\partial\omega/\partial K_{/\!/}$ is plotted as a dotted line with its scale to the left, and γ/Ω as a solid line with its scale to the right. Im $K_{/\!/}$ is the order of 0.1. Presumably n_H an order of magnitude smaller would also produce a significant Im $K_{/\!/}$.

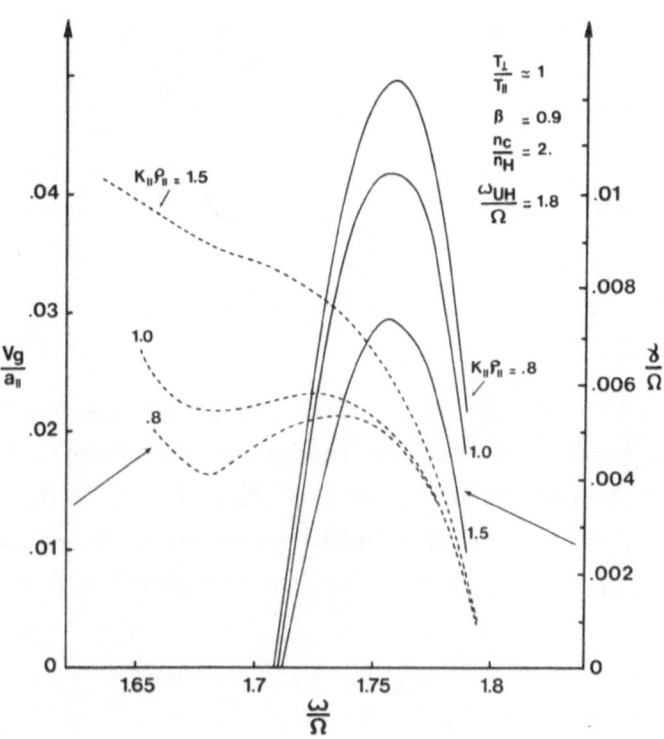

Figure 6 : The parallel group velocity and the growth rate versus frequency. Here n_H was chosen very small.

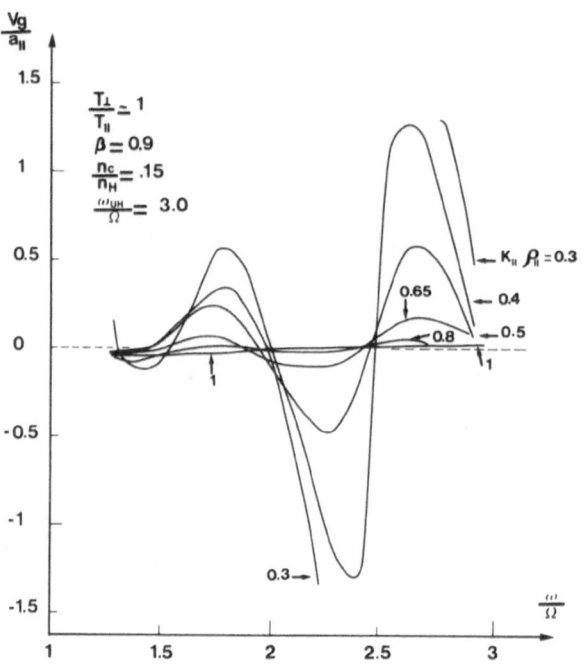

Figure 7 : The parallel group velocity versus frequency for modes
   which are temporally unstable throughout the range $1.4 \leqslant \omega/\Omega < 3$.
   The parallel group velocity is zero at $\omega/\Omega = 2$, (which should
   be damped if $T_c = 0$) at the cold upper hybrid resonance, near
   $\omega/\Omega \simeq 5/2$, and in a range of frequencies near $\omega/\Omega \simeq 3/2$.

Fredericks (1974) suggested that a major default of all electrostatic wave theories to date has been their failure to explain
the *simultaneous* occurence of narrow band emissions near X = 3/2
and 5/2. To see if two bands might be due to convective effects, we
chose parameters using figure 1, such that Xmin < 3/2 and ω_{UH}/Ω > 5/2
and then plotted $\partial\omega/\partial K_{//}$ against X = ω/Ω for 0.3 < $K_{//}\rho_{//}$ < 1.0, a range
of $K_{//}$ that is unstable everywhere between 3/2 < X < 3 (figure 7).
There are 4 zero crossings, two relatively uninteresting, and two
interesting. $\partial\omega/\partial K_{//}$ goes to zero at the hybrid frequency, and near
X = 2. Presumably, when $T_c \neq 0$, any non-convective instability near
X = 2 would be destroyed by cold electron-cyclotron damping. There
is an interesting zero crossing at X ≈ 5/2 and while all the zeroes
of $\partial\omega/\partial K_{//}$ for X < 2 do not occur at nearly the same frequency (as
they do near X = 5/2) it is significant that there exist zeroes of
$\partial\omega/\partial K_{//}$ for X < 2. Thus, the possibility of two narrow regions of
effectively non-convective instability cannot be excluded.

SUMMARY AND DISCUSSION

We may summarize our considerations as follows :

(1)   The hot electrons must have a loss cone distribution
        ($\partial f/\partial v_\perp$ > 0) to be unstable (Dory et al, 1965 ; Fredricks,
        1971). Pitch angle anisotropy alone is an unlikely cause of
        the observed electrostatic waves (Young et al, 1973 ; Ashour-
        Abdalla and Cowley, 1974).

(2)   Weak loss cone distributions are stable without cold electrons
        (Young et al, 1973).

(3)   When T_c = 0, the ratios n_c/n_H and ω_{UH}/Ω determine the range
        of unstable frequencies, in a manner that may be deduced
        from figures 1 and 2.

(4)   When T_c = 0, the unstable wave is basically the cold upper
        hybrid mode, whose dispersion relation is modified by the hot
        electrons. Its growth rate peaks near the cold upper hybrid
        frequency.

(5)   Within the present model, none of the calculations of the
        growth rate indicate that narrow band instabilities near odd
        half harmonics of the cyclotron frequency should be expected.

(6)   However, in a spatially inhomogeneous plasma we must also com-
        pute the group velocity to estimate the amplification length.

(7)   Both components of the group velocity $\partial\omega/\partial K_\perp$ and $\partial\omega/\partial K_{/\!/}$, tend
        to zero near the cold upper hybrid frequency.

(8)   In several particular cases, the parallel group velocity
        $\partial\omega/\partial K_{/\!/}$ has zeroes near $3\Omega/2$ and $5\Omega/2$, suggesting that the
        observed narrow band emissions are due to the fact that the
        instability is *effectively* non-convective there.

(9)   The zeroes in $\partial\omega/\partial K_{/\!/}$ are induced by the hot plasma, since the
        cold electrons alone produce a mode which has zero group velo-
        city only at the cold upper hybrid frequency.

        Clearly, much more theoretical work remains ahead of us. For
example, until the dependence of $\partial\omega/\partial K_{/\!/}$ on plasma parameters is sys-
tematically investigated, we will not know how general the existence
of zeroes of $\partial\omega/\partial K_{/\!/}$ is. A more serious objection is that even our
"weak" positive slope hot electron distribution leads to unrealisti-
cally short parallel amplification lengths. Even weaker slopes in
the electron velocity distribution might also be convectively unstable. Finally, until we have a good solution for the quasi-linear
diffusion equation to insert into the growth rate calculations, we
cannot represent accurately a self-consistent steady state. Any-
how, the theoretical advances made in the past five years are sufficient to convince us that new experimental investigations are needed.

First of all, a new survey of the phenomenology of electrostatic waves is important. How do the emissions depend on n_c and n_H ? on T_c ? Do 5/2 emissions ever occur without 3/2 ? What fraction of the time do 5/2 emissions occur ? And so on. The list is long. An attempt should be made to measure ω_{UH}, and therefore n_c, by plasma sounding techniques, or by searching for weak emissions at ω_{UH} and the cutoff above ω_{UH} using broad band detectors with a wide dynamic amplitude range. The theory demands measurements of $\partial f/\partial v_\perp$ at constant $v_{//}$ for a variety of $v_{//}$; the electron data reduction should carefully distinguish between true pitch angle anisotropy and a loss cone distribution.

ACKNOWLEDGEMENTS

We acknowledge with pleasure many useful discussions with Randy Burton, Roger Gendrin and René Pellat. Special thanks are due to C. Davoust for his help in the numerical computations. We are deeply grateful for the help of M. Lefloch in the curve plotting.

The work of M. Ashour-Abdalla was supported by CNES (French National Center for Space Research) and that of C. Kennel by the Alfred P. Sloan Foundation and by NSF, Grant 34148X.

REFERENCES

ASHOUR-ABDALLA, M., Instability theory of electrostatic half-harmonic electron gyrofrequency waves in the magnetosphere : a review, *Ann. Geophys.*, 1975 (in Press).

ASHOUR-ABDALLA, M., and S.W.H. COWLEY, Wave particle interactions near the geostationary orbit in *"Magnetospheric Physics"* edited by B.M. Mc Cormac, P. 270, D. Reidel Publishing Company, Dordrecht, Holland, 1974.

ASHOUR-ABDALLA, M., G. CHANTEUR and R. PELLAT, A contribution to the theory of the electrostatic half-harmonic electron gyrofrequency waves in the magnetosphere, *J. Geophys. Res.*, 1975 (in Press).

DORY, R.A., G.E. GUEST and E.G. HARRIS, Unstable electrostatic plasma wave propagation perpendicular to a magnetic field, *Phys. Rev. Lett.*, 14, 131, 1965.

FREDRICKS, R.W., Plasma instability at $(n+1/2)$ f_c and its relationship to some satellite observations, *J. Geophys. Res.*, 76, 5344, 1971.

FREDRICKS, R.W. and F.L. SCARF, Recent studies of magnetospheric electric field emissions about the electron gyrofrequency, *J. Geophys. Res.*, 78, 310, 1973.

FREDRICKS, R.W., Wave particle interactions in the outer magnetosphere : a review, presented at the Neil Brice Memorial Symposium, Frascati, 1974.

GURNETT, D.A. and R.R. SHAW, Electromagnetic radiation trapped in the magnetosphere above the plasma frequency, *J. Geophys. Res.*, 78, 8136, 1973.

GURNETT, D.A. and L.A. FRANK, Thermal and suprathermal plasma densities in the outer magnetosphere, *J. Geophys. Res.*, 79, 2355, 1974.

HARRIS, E.G., Unstable plasma oscillations in a magnetic field, *Phys. Rev. Lett.*, 2, 34, 1959.

KARPMAN, V.I., Ju. K. ALEKHIN, N.D. BORISOV, N.A. RJABOVA, Electrostatic electron-cyclotron waves in the multi-temperature plasma, 1974a (preprint).

KARPMAN, V.I., Ju. K. ALEKHIN, N.D. BORISOV, N.A. RJABOVA,
    Electrostatic electron-cyclotron waves in the plasma with a
    loss cone, 1974b (preprint).

KENNEL, C.F. and PETSCHEK, H.E., Limit on stably trapped particle
    fluxes, *J. Geophys. Res.*, 71, 1, 1966.

KENNEL, C.F., F.L. SCARF, R.W. FREDRICKS, J.H. Mc GEHEE and F.V.
    CORONITI, VLF electric field observations in the magnetosphere,
    J. Geophys. Res., 75, 61136, 1970.

LYONS, R.L., Electron diffusion driven by magnetospheric electrosta-
    tic waves, *J. Geophys. Res.*, 79, 575, 1974.

SAGDEEV, R.Z., and V.D. SHAFRANOV, On the instability of a plasma
    with an anisotropic distribution of velocities in a magnetic
    field, *Soviet Physics*, JETP, English Transl., 12 (1), 130-132,
    1961.

SCARF, F.L., R.W. FREDRICKS, C.F. KENNEL and F.V. CORONITI,
    Satellite studies of magnetospheric substorms on August 15,
    1968. OGO-5 plasma wave observations, *J. Geophys. Res.*, 78,
    3119, 1973.

YOUNG, T.S., J.D. CALLEN and J.E. Mc CURE, High frequency electro-
    static waves in the magnetosphere, *J. Geophys. Res.*, 78, 1082,
    1973.

DOUBLE LAYERS

Lars P. Block
Royal Institute of Technology
S-10044 Stockholm, Sweden

An increasing number of observations of charged particle precipitation at auroral latitudes during the last ten or fifteen
years have indicated electrostatic acceleration of these particles
through kilovolt potential drops along the geomagnetic field lines.
Theoretically this seemed difficult to understand at first, although elementary knowledge of plasma experiments prompted Alfvén
(1958) to suggest such acceleration through electrostatic double
layers above or in the upper ionosphere.

Advances in the theory of instabilities in current-carrying
plasmas, together with observations of strong Birkeland (field-
aligned) sheet currents, have now provided several mechanisms by
which field-aligned electrostatic fields can be produced. One such
mechanism, the double layers, will be reviewed here.

1. LABORATORY EXPERIMENTS

Double layers (also called sheaths) have been observed and
studied in laboratory gaseous discharges for at least 50 years
(Langmuir, 1929; Tonks, 1937; Allen and Thoneman, 1954; Torvén,
1968; Babic and Torvén, 1974; and many others).

Visibly, a double layer appears as a sharp boundary across
the positive column. The plasma is differently colored on both
sides of the layer. Many experiments have been run in mercury (for
technological reasons). On the cathode side the plasma has the ordinary bluish mercury color, but on the anode side of the layer it
is more reddish. It should be noted, however, that double layers
can be produced in any gas.

229

A summary of the properties of laboratory double layers is given below. Most of the results are quoted from Andersson et.al. (1969) and Babic and Torvén (1974) but they are representative also of those of a host of other authors.

A double layer appears when the current exceeds a critical current I_c. This is determined by the condition that the average electron drift velocity u_e becomes about equal to the electron thermal velocity:

$$u_e = I_c/(en_eA) \approx (kT_e/m_e)^{\frac{1}{2}} \tag{1}$$

where e, m_e, n_e, and T_e are the electron charge, mass, number density and temperature, respectively, and A is the tube cross section. At sub-critical currents the conductivity is determined by binary particle collisions. Instabilities which may occur at lower drift velocities - e.g. the Buneman instability with threshold at about $u_e = \sqrt{}\ (kT_e/m_i)$ - are observed, but they neither influence the conductivity, nor do they cause double layers.

An increase in discharge current (below I_c) is always associated with an increase in electron density n_e. The drift velocity u_e varies much more slowly with current. This means that over a large range of currents the drift velocity is a substantial fraction of the thermal velocity for electrons (Torvén, 1965).

The slow increase in drift velocity with current cannot go on indefinitely. When a sufficiently large fraction of the initially neutral particles has been ionized the electron density n_e cannot increase very much more. Most of the increase in current must then be accounted for by a corresponding increase in drift velocity u_e. The critical current I_c is then rapidly reached as u_e reaches the electron thermal velocity given by (1).

The exact limit depends on the ionization and other collisional cross sections, and on the typical thickness of double layers (see below). I_c can therefore not be estimated by putting n_e equal to the initial neutral particle number density. In practice I_c is often reached when only a few percent of the particles are ionized. The probability for ionization is then already too low.

Experimental evidence shows that lack of neutral gas is always associated with the formation of double layers (Poletaev, 1951). One cause of this may be absorption of neutral particles by the walls of the discharge tube (Torvén, 1968).

A gaseous discharge acts as a pump in the following way. Charged particles receive momentum from the electric field. This momentum is partly transferred to the neutrals by collisions. The

ions thus give rise to a neutral flux towards the cathode. The
electron induced neutral flux is directed towards the anode. However, the latter is larger, since electrons and ions receive equal
but opposite momenta from the electric field (their lifetimes are
equal in order to preserve charge neutrality) but the electrons
transfer all their momentum to the neutrals at an average collision, compared to half for the ions. Hence, there is a net transport of gas towards the anode. This can be noticed, for example,
in mercury discharges where the pressure is regulated by the temperature of mercury pools at both ends. After some time it can be
seen that the amount of mercury has increased in the anode pool at
the expence of the cathode pool.

This transport can be compensated for by raising the pressure
at the anode. If the tube is long compared to its diameter a measurable neutral density minimum will occur somewhere between anode
and cathode. Neutral particles are streaming both ways towards
this region, where they are ionized and removed as charged particles. At a critical current the double layer always forms at this
minimum (Babic and Torvén, 1974).

A double layer is always associated with an electrostatic potential jump Φ_0 which is larger than kT_e/e. This potential accelerates charged particles both ways, depending on the sign of
charge. At the positive side of the layer an electron beam can be
observed. This is illustrated in Figure 1. The two peaks of the
electron energy distribution correspond, respectively, to electrons accelerated through the layer, and thermal electrons produced on the high potential side that either will be or have been
reflected by the layer. It can be seen from the Figure that the
high energy electrons are gradually dispersed and thermalized,
causing the high energy peak to disappear somewhere beyond 6 cm
from the layer on the anode side. This distance is approximately
equal to a mean free path for the electrons. The thermalization
can hence be explained by simple collision theory. No collective
instabilities need to be invoked.

Another important observation from the same experiment is
that the particles in the undispersed high energy peak obtained
just on the anode side of the layer can account for the entire
discharge current (the tube diameter was 1 cm). This means that
the particle fluxes through the layer can be considered as essentially laminar. This is discussed further in the next section.

The double layer has been described as a boundary between two
distinct regions of plasma. However, this boundary has a finite
thickness which is of the order of several Debye lengths. Since
the potential across the layer changes by more than kT_e/e it means
that quasi-neutrality is not valid in the layer.

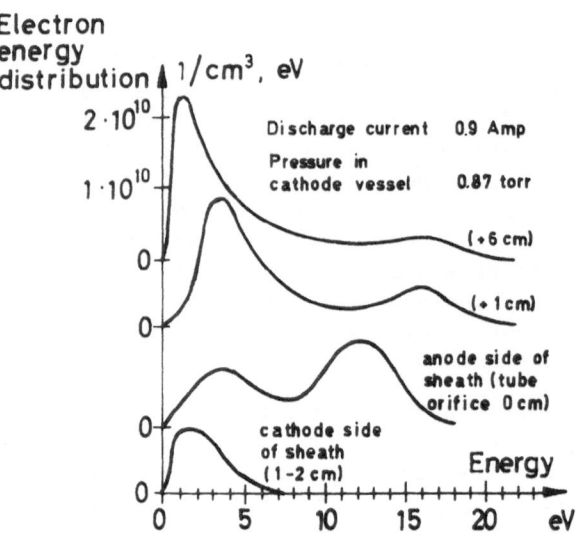

Figure 1. Electron energy distributions observed in a laboratory discharge with a double layer (sheath) according to Andersson et.al. (1969). The big peak around about 12eV in the second lowest curve is produced by the layer, but further down the tube (+ 1 cm and + 6 cm) this peak is gradually dispersed and the beam particles are thermalized.

The potential jump across a double layer may often fluctuate considerably. It may be regarded as having a steady d.c. component with a spectrum of superimposed a.c. components. An example of such fluctuations is shown in Figure 2, obtained by Babic and Torvén (1974). The minimum potential (about 80 volts) is equal to the discharge voltage just below the critical current (about 10 amperes). Apparently double layers are born and die at fairly

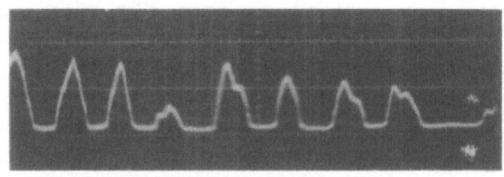

Figure 2. Fluctuating voltage of a laboratory discharge with a double layer (Babic and Torvén, 1974). Horizontal time scale: 5 μs/division. Vertical voltage scale: 100 V/division. Layers exist only during pulses which represent layer potentials. They reach about 100 volts and then fade away.

regular intervals of about 5 μs. This is much longer than the travel time for an electron through the layer but of the same order
as the corresponding time for an ion.

    This behaviour is not necessarily typical of double layers.
For example, a longitudinal magnetic field with maximum strength
somewhere along the discharge may stabilize the layer at the point
of maximum field strength.

    It may be suspected that the occurence of double layers is
always associated with the presence of material walls. However,
the discharge may be detached from the walls by a longitudinal
magnetic field. It then turns out to be easier to create double
layers. The reason is simple. The cross section of the discharge
diminishes with a corresponding increase in current density and
electron drift velocity. In addition the electron temperature may
decrease since the losses of charged particles through diffusion
sideways is inhibited by the field.

                    2. THEORETICAL RESULTS

    The simplest interpretation of the experimental results de-
scribed in the previous section is that a double layer is a potential structure about as shown in Figure 3. It consists of a region
with high electric field, no quasi-neutrality, and two "single"
layers of opposite space charges, one at each side. All this is
sandwiched between two quasi-neutral plasmas where the electric
fields are negligible compared to the field in the layer. The particle flow through the layer is essentially laminar. It is not a
region of anomalous resistivity, caused by a collision frequency
that is effectively increased by collective oscillations or turbulence. Such effects could not possibly produce the observed double
peaked electron energy distribution (Figure 1) from the single
peaked distribution on the low potential side.

    Of course, the laminar flow simplifies the theory consider-
ably. Yet, a fully satisfactory theory still does not exist.

    Next we note that a double layer requires a trapped particle
population, at least on one side of the layer. "Trapped" means
that they are reflected by the layer and by some other reflector
outside the layer. This latter reflection may in practice be obtained through de-ionization, return in the form of neutral particles, and new ionization.

    To demonstrate the need for trapped particles, consider the
potential structure of Figure 3 and assume that there is no trapping on the high potential side, for example. The plasma on this
side then consists of nothing but the electron beam already accel-

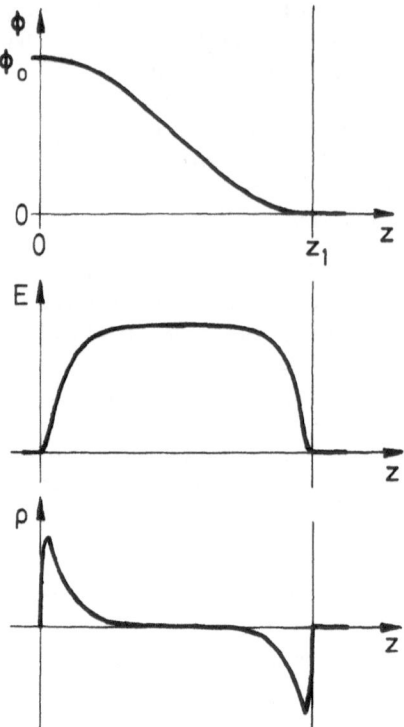

Figure 3. Distribution of potential, electric field, and space
charge in a typical plane double layer.

erated by the layer, and ions to be accelerated by the layer.
These must have equal densities. On the other side of the layer
the electron density must be higher and the ion density lower.
Charge neutrality then requires a trapped ion population on the
low potential side. Conversely, if the free electron and ion densities are equal on the low potential side, trapped electrons are
required on the high potential side.

     In reality there are trapped populations on both sides. This
is evidently unavoidable if non-negligible ionization occurs on
both sides.

2.1. Macroscopic Theory

     With these basic facts it is possible to construct some the-
oretical models. A fully satisfactory model must be based on the
coupled Poisson-Vlasov equations. This is because the thickness

of a double layer is less than a collisional mean free path, so
collisions play no role in the first approximation. However, the
first models were based on macroscopic theory. Langmuir (1929)
presented the first theory based on the assumption that all four
populations (trapped and free electrons and ions, where "free" are
those with sufficient total energy to exist on both sides of the
layer) had zero temperature. It is then easy to construct a self-
consistent model of a double layer with arbitrary potential. The
opposite space charges are held apart by inertia forces. However,
the cold plasmas on both sides are unstable, as is now well known.

The macroscopic theory was extended by Block (1972) to include finite temperatures for all four populations. A number of
self-consistency conditions were derived.

a   The Langmuir condition which is exact in Langmuir's theory
(Langmuir, 1929) and approximate for finite temperatures. This
condition, given by equation (2) below, is due to the fact that
the total charge of the layer is zero since the electric fields
are negligible on both sides. To compensate for the much longer
traversal time of an ion through the layer, compared to that of
an electron, the ratio F_e/F_i of the electron to the ion fluxes
through the layer must be much larger than unity:

$$\frac{F_e}{F_i} = \frac{n_e u_e}{n_i u_i} = (m_i/m_e)^{\frac{1}{2}} \tag{2}$$

If a double layer develops under conditions when (2) cannot be
satisfied in the laboratory frame or in the rest frame of the
plasma, the layer may move with a suitable velocity such that (2)
is satisfied in the rest frame of the layer.

b   The Bohm condition obtained for wall sheaths by Bohm (1949)
for the case when the trapped particles have finite temperature
and the free particles have a delta-function velocity distribution.
This states that the free populations must have some minimum bulk
velocities when they enter the layer before acceleration. The Bohm
conditions were modified by Block (1972) to account for finite
free particle temperatures. They are

$$m_e u_{ef}^2 \geq k(\gamma T_{ef} + T_{it}) \tag{3}$$

for the free electrons when they enter the low potential edge of
the layer and

$$m_i u_{if}^2 \geq k(\gamma T_{if} + T_{et}) \tag{4}$$

for the free ions at the high potential edge. Indices "f" and "t"
refer to free and trapped particles, respectively, and γ is the
ratio of specific heats, which for one-dimensional flow is equal
to three.

It should be noted here that the numerical constants in (3)
and (4) come out differently in Vlasov-theory. The reason for this
is not understood (see below).

The Bohm conditions are approximate also in the respect that
the densities of particles from the other side are neglected. This
is a good approximation for strong layers ($e\Phi_o \gg$ all kT) since
the neglected densities are then very low. Only the most energetic
trapped particles on the other side can reach the edge considered,
and the density of the free particles coming from the other side
is very low due to the large acceleration.

When all particles are included the entering velocities must
be still higher.

The physical reason for these conditions is that the electric
field must grow as one moves into the layer from the edge. Both
reflected (trapped) and free particle densities decay into the
layer, but the Bohm conditions ensure that the reflected particle
densities decay faster than the accelerated. Obviously, this gives
the right sign of the space charge near both edges, since the free
particles must be accelerated away from space charges of their own
sign.

c   Pressure balance. This states that the sums of all pressures
- thermal (nkT) plus kinetic (nmu^2) for the four populations -
must be the same on both sides of the layer. The Langmuir condition immediately follows from this for the case of strong double
layers, since the kinetic pressures of the already accelerated
particles then dominate ($mu^2 = 2e\ \Phi_o \gg$ all kT).

It has been suggested that the formation of double layers
is due to a density instability driven by the current in the
plasma (Alfvén and Carlqvist, 1967; Block, 1972; Carlqvist, 1972).
The threshold as given by macroscopic theory occurs for an electron drift velocity approximately equal to the electron thermal
velocity. It is therefore in agreement with experiments (cf equation 1). A natural consequence is that the layer should appear at
a density minimum.

2.2. Vlasov Theory

Recently Vlasov theory has been employed, but mathematical
difficulties have necessitated either restrictions to simple cases,

approximations, or computer simulation of a great number of individual particles. Nevertheless, important new results have been obtained.

Knorr and Goertz (1974) have studied an inverted problem. Given a potential distribution

$$\Phi(x) = \Phi_o \, \text{tgh}(x/\xi) \tag{5}$$

and the velocity distributions of three particle populations they calculate the fourth distribution. They first obtain the number density of the unknown distribution by inserting equation (5) in the Poisson equation. This density must be everywhere positive, and it can easily be expressed as a function of Φ. This function is put equal to the density integral over the unknown energy distribution, and the resulting integral equation is solved. It is important to note that solutions exist for arbitrary values of the layer potential Φ_0. This agrees both with macroscopic theory and with experiments where Φ_0 can be varied by varying external circuit parameters.

After having constructed some such solutions Knorr and Goertz investigate the stability of the plasmas on both sides of the layer against electrostatic instabilities by applying the Penrose criterion (Penrose, 1960). They find that there exist a large number of physically meaningful energy distributions for which the plasma is Penrose stable at all points in space.

This result is supported by some many-particle computer experiments of Goertz and Joyce (1975). They are able to model the formation of double layers in a one-dimensional particle-in-cell computer simulation. Initially, electrons and ions had Maxwellian velocity distributions with the same temperatures. The electron distribution was then shifted so the electrons were drifting with a certain velocity u_e. At t = 0 the electric field is identically zero. The development in time of the electrostatic potential was then studied for various values of u_e. It was found that double layers spontaneously developed after about an ion plasma period, provided

$$u_e^2 \geq kT_e/m_e \tag{6}$$

Thereafter, the double layer potential Φ_0 could grow to about 100 kT/e. The thickness L of the layer was found to be

$$L \sim (e\Phi_0/kT)^{\frac{1}{2}} \tag{7}$$

The limitation of Φ_0 to about 100 kT/e was probably due to the finite spatial region considered, namely 50 Debye lengths.

It may be concluded that the results of macroscopic theory
have essentially been confirmed by Vlasov-theory. The Langmuir
condition and the pressure balance are identical in both theories.
This is not so for the Bohm conditions, however, unless the free-
particle temperatures are ignored in the macroscopic theory. The
reason for this is certainly not understood.

2.3. Oblique Double Layers

The theories described above have, for reasons of mathematical simplicity, been restricted to one-dimensional, or parallel,
double layers, i.e. the electric field is parallel to the current
and everything is infinitely extended in all directions perpendicular to the current. Another restriction is that only un-magnet-
ized plasmas have been treated.

For geophysical applications these restrictions must, of
course, be removed. If double layers of the kind discussed above
are associated with Birkeland currents the equipotentials must be
perpendicular to the magnetic field. The finite cross section of
Birkeland currents means that the equipotentials must become parallel to the magnetic field outside the current region (Figure 4).
At the transition from perpendicular to parallel they must be
oblique. Hence, a study of a double layer with its normal being
oblique to the magnetic field is imperative. The magnetic field
can, of course, not be ignored then.

Such a two-dimensional study has been made by Swift (1975).
He used Poisson-Vlasov theory. Even here the double layer was an
infinite plane sheet with an arbitrary angle to the magnetic field,

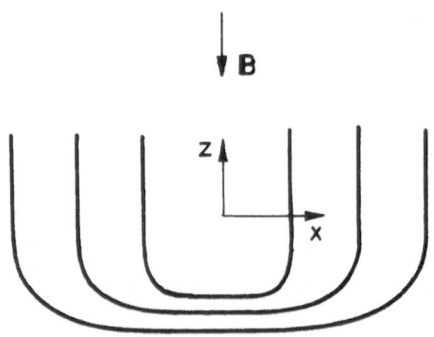

Figure 4. Equipotentials for a double layer turning upwards around
a field-aligned current sheet.

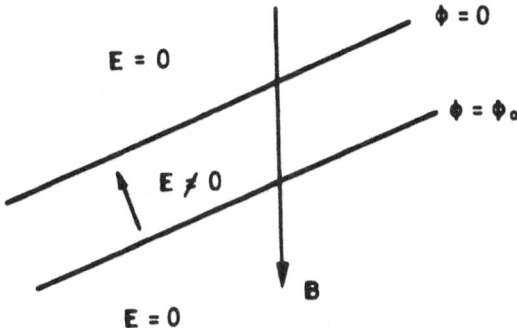

Figure 5. Equipotentials and fields for an oblique double layer.

as shown in Figure 5. The electrons were assumed to move parallel
to the magnetic field, only. The ion number density was calculated
with adiabatic theory, thus accounting for their perpendicular
displacement when entering and leaving the layer. Outside the layer
they have only parallel motion. The plasmas on both sides were assumed homogeneous, and so was the magnetic field. Two plasma models were studied: (1) streaming electrons and a two-temperature
distribution of ions, and (2) streaming and thermal electrons and
ions, just as the four populations of free and trapped particles
in the one-dimensional double layer theories. The first model
seemed less likely to be applicable to geophysical conditions, so
only the second model will be discussed here.

Firstly, Swift demonstrated the existence of self-consistent
solutions with arbitrary potential jumps across the layer. The
thickness was found to be several ion gyro radii, rather than
Debye lengths as in parallel double layers. The ion gyro radius
was considerably larger than the Debye length in the model, in
agreement with magnetospheric and ionospheric conditions. Due to
the obliqueness, the thickness along the magnetic field (the length
of a magnetic field line within the layer) may be extremely large
if the layer normal is nearly perpendicular to the magnetic field.
The parallel component of the electric field may be quite small
even for large layer potentials, although it must of course be
considerably larger than that determined by simple collisional
conductivity if the total potential drop is of the order of kilovolts.

The instability associated with the formation of oblique double layers was identified by Swift to be the electrostatic ion cyclotron instability. The threshold electron drift speed was found to be somewhat less than the threshold for ion acoustic waves in case $T_e \gg T_i$. If $T_e = T_i$ (which is more realistic for ionospheric applications) the threshold speed is comparable to the electron thermal velocity. In fact it is somewhat higher than the threshold for parallel waves.

3. GEOPHYSICAL APPLICATIONS

Double layers have so far been invoked only for explaining the acceleration of auroral precipitation in terms of parallel electric fields. There seem to be no compelling reasons at present to go beyond that, although potential jumps perpendicular to the magnetic field may exist at the magnetopause, for example. Here, we shall confine ourselves to double layers associated with auroral zone Birkeland currents.

3.1. Differences Between Double Layers and Anomalous Resistivity

An important problem is that of the relative importance of anomalous resistivity versus double layers in the auroral acceleration processes. It may be premature to give a definite answer to this question. We shall here only point out the essential differences which must serve as a basis for the judgement.

The double layers are laminar but anomalous resistivity is turbulent. This means that anomalous power dissipation occurs in situ, i.e. at the very place where the anomalous resistivity prevails. This implies in situ heating (mostly of the ions) which can quench the instability producing the resistivity. In double layers the power $I\Phi_0$ certainly produces heat eventually, but in two steps. In situ, the electrical energy is transformed into laminar kinetic energy of the free beam particles. Their energy is then transformed into heat, partly through wave-particle interactions and partly through ordinary collisions, preferably in the much denser ionosphere, provided the beams survive until then. The main point is in any case that the heat is produced, not in the double layers, but somewhere else.

Suppose that along a geomagnetic field line there is a total potential drop of 3kV and a current density of $3 \cdot 10^{-6}$ A/m^2. The total power is 10^{-2} W/m^2. If this is converted in situ into heat in a 10^4 km long column with an average density of $3 \cdot 10^8$ ions/m^3, it implies a heating rate of 30 eV per second per ion. This is excessive, in particular since at times both the potential and current density may be substantially higher, and the particle den-

sity substantially lower (Hagg, 1967). This amount of heat cannot
be carried away by heat transport.

These difficulties obviously do not arise if the parallel potential drop is produced by double layers. The heat is mainly produced in the low ionosphere, where the density is sufficiently
high to limit the heating power per particle to acceptable values.
It may be argued that most of the power in the electron beam should
be dissipated by wave-particle interaction at high altitudes where
the density is low. The counterargument is of course that beams of
nearly monoenergetic, highly field-aligned electron precipitation
are undoubtedly observed at 200 km altitude and higher (see e.g.
Arnoldy et.al., 1974 and many others). Whatever the acceleration
mechanism may be, the beams do survive down to the dense ionosphere
where they produce aurora. They are certainly dispersed somewhat
since they are not as monoenergetic as would perhaps be expected.
This dispersion may be due to wave-particle interactions producing
auroral hiss (Gurnett and Frank, 1972), but the power of this hiss
is not nearly sufficient to account for the above mentioned power.

A way out of this dilemma for those advocating anomalous resistivity may be that the parallel electric field is produced by
a very weak current carried by thermal particles. Other, initially
somewhat more energetic run-away particles acquire the full energy
corresponding to the total potential drop without being impeded
by wave-interaction during the acceleration. However, the control-
ing thermal current must be many orders of magnitude weaker than
the secondary run-away current. This seems very unlikely.

Another process that seems more likely is the following: A
field-aligned energetic electron beam (caused by some unspecified
mechanism) produces waves through interaction with the relatively
cold ambient plasma. Any current carried by the ambient particles
is then subject to anomalous resistivity due to interaction with
the waves produced by the beam.

This process was proposed by Réme and Bosqued (1971) to explain simultaneous bursts of precipitating electrons and ions. The
energetic electron precipitation produces a partly neutralizing
upflow of thermal electrons, associated with a downward parallel
electric field due to anomalous resistivity. This downward electric
field finally accelerates ions.

The drawback of this is, of course, that the cause of the
original electron beam is not specified. But the idea of parallel
electric fields was originally introduced to explain just that.

The velocity distribution of the beam electrons is of course
modified by the beam-plasma interaction. However, field-aligned
keV-electron beams are observed in and above the ionosphere. The

beam properties are therefore essentially retained.

Two explanations for this have been propesed. According to Galeev (1975) the interaction length for dispersion of the beam in the topside ionosphere is very long (about 1000 km). According to Papadopoulos and Coffey (1974) a marginally stable final state is reached, where the beam is somewhat broadened and the tails of the plasma electron velocity distribution have grown many orders of magnitude compared to the original Maxwellian distribution. They claim that this explains observations at 500-800 km altitude by Reasoner and Chappell (1973) of a backscatter ratio of unity for electrons below about 1 keV, simultaneous with a beam of 5-10 keV electrons.

3.2. Power input

A double layer transforms electrical energy into other forms of energy. In this respect it is certainly similar to resistivity, be it anomalous or not. The energy must come from the generator of the current circuit. Suppose there is initially no double layer.

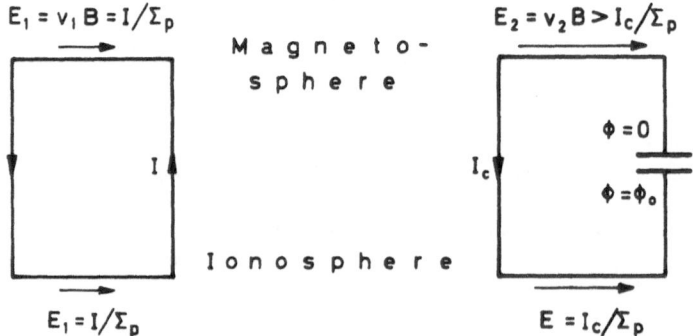

Figure 6. Power input to a double layer. The generator is in the magnetosphere. Circuit to the left: Sub-critical current $(I < I_c)$, no double layer. Right: Critical current, super-critical magnetospheric E-field, double layer. Geometric factors due to magnetic flux tube expansion are ignored.

The potential drop along the magnetic field is negligible. If a
double layer suddenly occurs somewhere along a flux tube carrying
Birkeland current, the generator must immediately increase its
power output, or some other circuit element must reduce its power
consumption.

Figure 6 illustrates these two possibilities. A magnetospheric
generator is connected to the ionosphere via downward and upward
Birkeland currents. The generator could consist of magnetospheric
$\underline{E} \times \underline{B}$-convection. Due to the finite conductivity of the ionosphere
any magnetospheric convection must necessarily be associated with
a similar current system.

Suppose now that the magnetospheric convection velocity increases. The electric field and potential of the generator increases
correspondingly. The current also increases until the threshold for
double layers is reached. No current increase beyond that is possible, but the generator potential can still increase. The power of
the generator increases proportional to the potential if the current
is constant and the power increase goes to the double layer. The
perpendicular electric field is obviously larger above than below
the layer.

The other alternative is that the generator potential is constant but the ionospheric conductivity increases, for some reason.
If the initial current density is just below the double layer
threshold the reduction in potential drop in the ionosphere between
the upward and downward Birkeland currents will appear in the double
layer. The corresponding increase in precipitation energy will further increase the ionospheric conductivity. Thus an explosive growth
of the double layer potential with a sudden formation of an auroral
arc may occur. This process has been proposed by Swift (1975). It
is consistent with observations of weaker perpendicular electric
fields in auroral arcs than outside. In this case, too, $E_\perp$ is larger
above than below the layer.

Everything that has been said about the power input to double
layers is of course equally valid for anomalous resistivity.

3.3. Double Layer Altitudes

Equation (1) gives the necessary condition for the onset of
the instabilities leading to formation of double layers. A consequence of this equation is that double layers occur most easily
where

$$G = A^2 n_e^2 T_e$$

has a minimum. Almost any model of the exospheric plasma exhibits

a minimum somewhere between one half and two or three earth radii.
The table below gives an example of approximate data that may be
valid along an auroral flux tube.

Altitude	$A\ m^2$	$n_e\ m^{-3}$	T_e degrees	G
F-layer	1	10^{11}	10^3	10^{25}
3000 km	3	10^8	10^4	10^{21}
10 R_E	10^3	10^6	10^7	10^{25}

With these values a current density of about $2 \cdot 10^{-5}$ A/m^2 would
induce double layers at 3000 km altitude.

It was noted in Section 1 that a low neutral gas density fa-
vours formation of double layers. The normal ionization processes
maintaining the plasma cannot work. Certainly, there is very little
neutral gas at the high altitudes where double layers are expected
in space.

It is conceivable that a large number of double layers are
induced in the same magnetic flux tube over a range of altitudes
if the G-minimum is very flat (Block, 1972). If these layers are
fluctuating as shown in Figure 2 their combined effects may look
similar to anomalous resistivity. However, this is not so since
they are still laminar. No particles falling through the layers
are impeded by waves. The current is limited to the random current
determined by the thermal speed, and not by "collisions" with particles or waves.

3.4. Equipotentials

We conclude by showing possible electric field or equipotential distributions for parallel and oblique double layers. Figures
7 and 8 illustrate two possibilities for parallel layers. They of
course always imply oblique parts as pointed out before. In figure
7 there is an $E_\perp$ reversal above the layer and no electric field
below. The maximum precipitation energy occurs at the field reversal. This agrees with the observations of inverted V-events (Gurnett
and Frank 1973) if there are many double layers, both above and below the satellite orbit. The maximum precipitation energy observe-
red on the satellite is equal to the maximum potential drop above
the satellite, and the integrated $E_\perp$ along the orbit up to the reversal is equal to the potential drop below the satellite. This

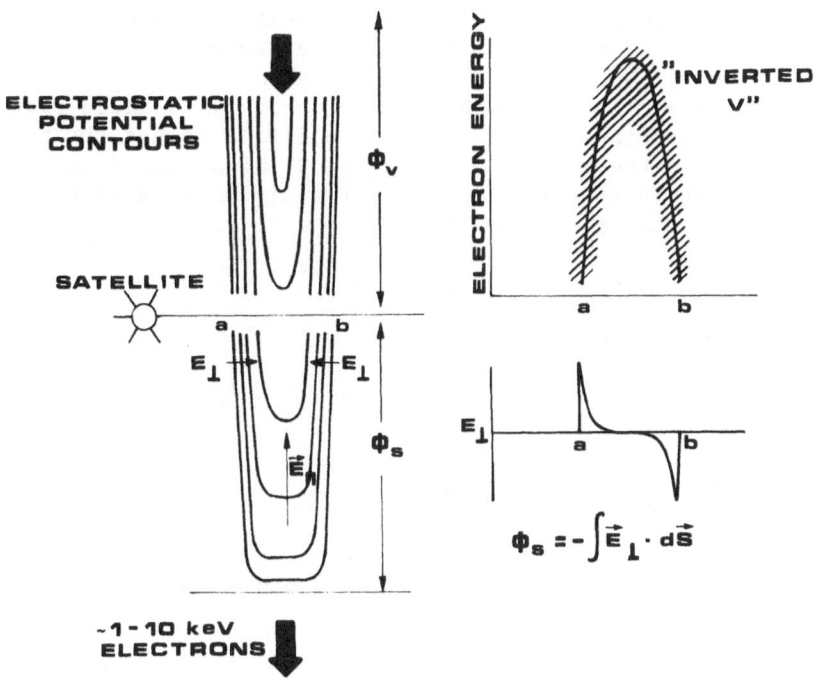

Figure 7. Suggested electrostatic potential distribution with
parallel electric fields (left) at inverted V-events (top right).
The total parallel potential drop below the satellite orbit is the
integral of $E_\perp$ as shown at bottom right (Gurnett, 1972).

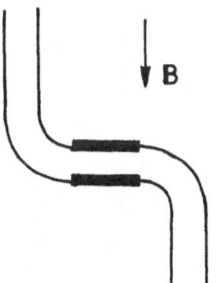

Figure 8. Equipotentials of a double layer with its normal parallel
to $\vec{B}$ and with no reversal of $E_\perp$.

type of electric field distribution was originally proposed by
Block (1969) and has been invoked by Carlqvist and Boström (1970)
to explain observed shear motion of auroral forms.

     The configuration in Figure 8 does not imply a field-reversal.
If this configuration is stretched out a little we get the truly
oblique layer in the top panel of Figure 9. If a mirror image of
this field distribution is added we get again a field reversal as
shown in the bottom panel of Figure 9.

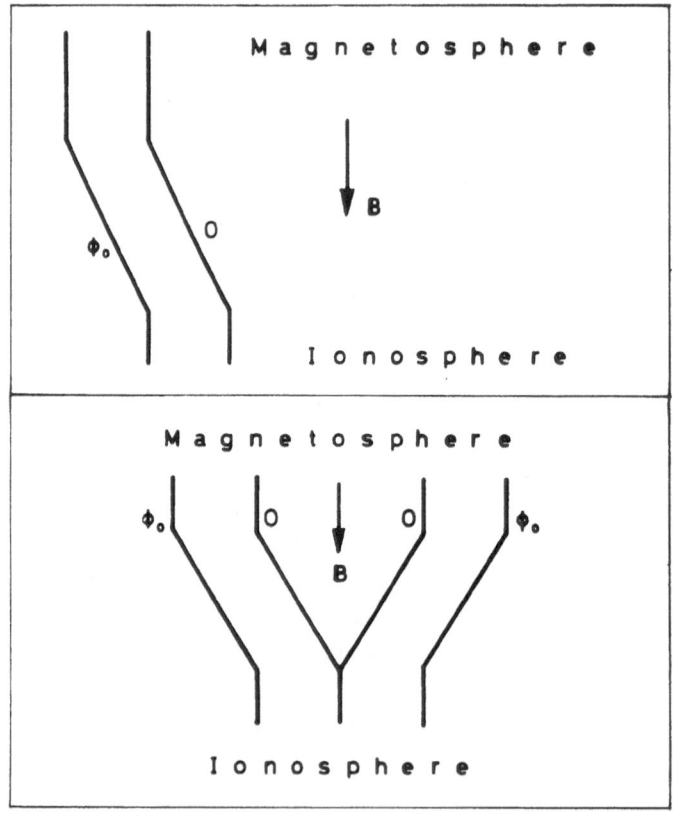

Figure 9. Equipotentials for oblique double layers without (top
panel) and with (bottom panel) electric field reversal.

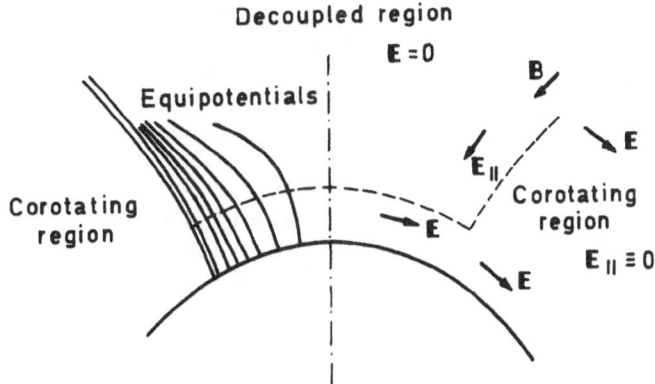

Figure 10. Decoupling the plasma on polar magnetic field lines
requires parallel electric fields. These may be produced by
oblique double layers.

Figure 10 shows a configuration with oblique layers in the polar regions as proposed by Fälthammar and Block (1971). This configuration should result because the earth's rotation must be de-
coupled from the plasma in the remote parts of the open polar geomagnetic flux tubes.

It should be noted that a simple transformation between two
frames of reference moving with a constant velocity relative to
each other in a direction perpendicular to the plane of the Figures
can transform Figure 7 to something qualitatively similar to Figure
9, i.e. an entirely oblique layer. Quantitatively the difference is
great though, since the electric field in the thin layer differs so
much with that outside. It is therefore important to use the right
reference frame, namely the one in which the average perpendicular
plasma velocity is minimized.

SUMMARY

The properties of double layers as observed in laboratory gaseous discharges is reviewed. Theories based on macroscopic equations are compared with those based on the coupled Maxwell-Vlasov
equations. Only minor differences in their results are found. The
necessary existence criteria, namely the Langmuir and Bohm conditions are essentially the same in the two theories. Important recent results are the proof by Knorr and Goertz (1974) that the
plasmas adjacent to double layers is Penrose stable, and the demonstration by Swift (1975) that double layers with the electric field
oblique to the magnetic field can exist for current densities comparable to those required for parallel double layers.

Since double layers are laminar, in contrast to anomalously resistive plasmas, they do not imply the excessive heating at high altitudes that seems to be necessarily associated with anomalous resistivity. Instead, the energy is converted into kinetic energy of electron precipitation, which then produces aurora and heat in the dense E-layer.

Two mechanisms for the power supply to double layers are described, and possible electrostatic field configurations in the auroral ionosphere-magnetosphere system are demonstrated.

REFERENCES

Alfvén, H.: 1958, Tellus, 10, 104
Alfvén, H. and Carlqvist, P.: 1967, Solar Phys., 1, 220
Allen, J.E. and Thonemann, P.C.: 1954, Proc. Phys. Soc. B, 67, 768
Andersson, D., Babic, M., Sandahl, S. and Torvén, S.: 1969, Ninth
    Int. Conf. on Ionized Gases, Bucharest, Rumania, p. 158
Arnoldy, R.L., Lewis, P.B. and Isaacson, P.O.: 1974, J. Geophys.
    Res. 79, 4208
Babic, M. and Torvén, S.: 1974, Tech. Rep. TRITA-EPP-74-02, Dept.
    of Electron Phys., Royal Inst. of Tech., Stockholm, Sweden
Block, L.P.: 1969, Ninth Int. Conf. on Ionized Gases, Bucharest,
    Rumania, Available as Tech. Rep. 69-30, Dept. of Plasma Phys.,
    Royal Inst. of Tech., Stockholm, Sweden
Block, L.P.: 1972, Cosmic Electrodyn., 3, 349
Bohm, D.: 1949, The Characteristics of Electrical Discharges in
    Magnetic Fields, Mc Graw Hill (Editors: A. Guthrie and R.K.
    Wakerling), p. 77
Carlqvist, P.: 1972, Cosmic Electrodyn. 3, 377
Carlqvist, P. and Boström, R.: 1970, J. Geophys. Res., 75, 7140
Fälthammar, C.G. and Block, L.P.: 1971, ESRO Colloqvium on Wave-
    Particle Interactions in the Magnetosphere, Orléans, France,
    ESRO SP-72, p. 147
Galeev, A.A.: 1975, This book
Goertz, C.K. and Joyce, G.: 1975, Astrophys. and Space Sci., 32, 165
Gurnett, D.A.: 1972, COSPAR-IAGA-URSI Symp. on Critical Problems of
    Magnetospheric Phys., Madrid, Spain, p. 123
Gurnett, D.A. and Frank, L.A.: 1972, J. Geophys. Res., 77, 172
Gurnett, D.A. and Frank, L.A.: 1973, J. Geophys. Res., 78, 145
Hagg, E.L.: 1967, Can. J. Phys., 45, 27
Knorr, G. and Goertz, C.K.: 1974, Astrophys. and Space Sci., 31, 209
Langmuir, I.: 1929, Phys. Rev., 33, 954
Papadopoulos, K. and Coffey, T.P.: 1974, J. Geophys. Res., 79, 1558
Penrose, O.: 1960, Phys. Fluids, 3, 258
Poletaev, I.A.: 1951, Zh. Tekh. Fiz., 21, 1021
Reasoner, D.L. and Chappell, C.R.: 1973, J. Geophys. Res., 78, 2176
Reme, H. and Bosqued, J.M.: 1971, J. Geophys. Res., 76, 7683

Swift, D.W.: 1975, On the formation of auroral arcs and acceleration
     of auroral electrons, (accepted for publ.) J. Geophys. Res., 80
Tonks, L.: 1937, Trans. Electrochem. Soc., 72, 167
Torvén, S.: 1965, Ark. Fys., 29, 533
Torvén, S.: 1968, Ark. Fys., 35, 513

PLASMA TURBULENCE IN THE MAGNETOSPHERE
WITH SPECIAL REGARD TO PLASMA HEATING

A.A. GALEEV

Institute of Space Research

Academy of Sciences, Moscow, USSR

1. INTRODUCTION

In the absence of collective effects plasma heating occurs due to Joule dissipation of the electric currents and due to adiabatic acceleration of particles in
slowly varying magnetic and electric fields. Adiabatic
heating due to plasma convection from weak magnetic
field regions to regions with stronger magnetic fields
provides the right order of magnitude for the energy of
the ring current and plasma sheet particles. Anisotropic plasma instabilities such as the loss-cone instability (Rosenbluth, 1965) and subsequent quasilinear particle diffusion in velocity space towards the loss-cone
results in a temperature increase of the plasma which
is left outside the loss-cone and therefore still trapped in the magnetosphere (Galeev, 1967). Although this
effect could change the quantitative estimates for adiabatic plasma heating it would not change the situation
principally. That is why we restrict our attention to

the problem of cold magnetospheric plasma heating.
Bearing in mind that the problems of anomalous plasma
resistivity due to longitudinal currents and magnetic
field line reconnection in the neutral sheets are the
subjects of special reports at this symposium we consider two other mechanisms:

1. Cold plasma heating due to linear and non-
linear absorption of electromagnetic waves, excited due
to instabilities in the hot magnetospheric plasma.

2. Heating of the ionosphere by electron plasma
waves, excited due to beam instability of the electrons
precipitating into the ionosphere.

2. TURBULENT DISSIPATION OF RING CURRENT
   ENERGY AT THE PLASMAPAUSE

Experiments on board the S^3 satellite (Fritz et
al., 1974; Williams and Lyons, 1974) showed that the
interaction of the cold plasmasphere and hot ring current predicted by Cornwall et al. (1970) really takes
place at the plasmapause. The physics of the interaction is that the phase velocity of the ion cyclotron
waves decreases in a dense plasmasphere and becomes comparable to the velocity of the ring current protons penetrating into the plasmasphere, and this results in
strong resonant wave-particle interaction. Analyses
show (Cornwall et al., 1970; Kennel and Petchek, 1966)
that such an interaction in a plasma with an anisotropic velocity distribution leads to intense growth of
oscillations with frequencies

$$\omega < \omega_{max} = A_+ \Omega_c / (1 + A_+) \qquad (1)$$

due to resonance with ions moving along the magnetic
field lines with a velocity

$$ v_{\|R}^2 > V^2 = \frac{B^2}{4\pi(n_{RC} + N_{PS})M} \; A_+^{-2}(1+A_+)^{-1} \quad (2) $$

where B is the magnetic field, N_{PS} and n_{RC} are the densities of the plasmasphere and ring current respectively, A_+ is the plasma anisotropy defined by Kennel and
Petchek (1966), ω and $\underline{k}$ are the frequency and wave vector of the disturbance, Ω_c is the ion cyclotron fre-
quency, $v_{\|R} = (\omega - \Omega_c)/k_\|$ is the longitudinal velocity of
the resonant ring current ions.

Under the influence of growing ion-cyclotron waves
the particle distribution relaxes towards the stable
one. Wave energy resulting from such a quasilinear relaxation can be found as the difference of the particle
energy in the initial and final states. In Figure 1, taken from Sagdeev and Galeev (1969), the particle distribution is shown for both states. With the help of this
drawing the wave energy in the final state can be estimated as

$$ \frac{1}{4\pi}\int\frac{d^3k}{(2\pi)^3}\,|\underline{B}_{\underline{k}}|^2 \; \simeq \; MN\overline{\eta(v_{\|R})\,v_{\|R}^2}\,A_+^2 \quad (3) $$

where $\eta(v_{\|R}) = 2\pi\,v_{\|R}\,N^{-1}\int_o^\infty v_\perp dv_\perp\,f(v_\perp, v_\| = v_{\|R})$ is the
number density of resonant particles, $N = N_{PS} + n_{RC}$ is the
total density and $f(v_\perp, v_\|)$ is the distribution func-
tion for the ring current protons.

The described relaxation leads to quenching of the
instability only if we neglect anisotropy recovery due
to new portions of plasma entering the interaction re-

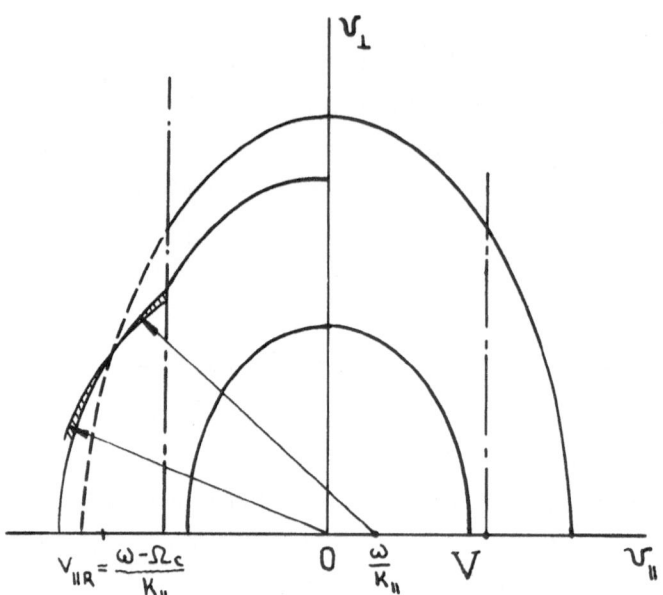

Fig.1. Ring current proton distribution contours.
In the left part of the resonant region ($v_{\|R} < -V$, V is
defined by Eq. (2)) the solid line represents the level
contour for the relaxed distribution and the dashed
line represents the same for the initial anisotropic
distribution. The level contour for the case of only
pitch-angle diffusion in the vicinity of the resonant
velocity for a given wave mode is shown to estimate
the portion of particle energy transferred to waves.

gion and particle precipitation through the loss-cone.
Both these effects keep the plasma in a weakly unstable
state very close to the stable one, resulting from quasilinear relaxation. Small deviations from the stable
state can be described by the following quasilinear
equation:

$$c\frac{\underline{E} \times \underline{B}_0}{B^2} \cdot \nabla f_{QL} + V_{\parallel} \frac{\partial}{\partial S} f_{QL} = \Omega_c^2 \int \frac{d^3 \underline{K}}{(2\pi)^3} \left[\left(\frac{\partial}{U_{\parallel} \partial U_{\parallel}} - \frac{\partial}{V_{\perp} \partial V_{\perp}} \right) + \frac{\omega}{\omega - \Omega_c} \frac{\partial}{V_{\perp} \partial V_{\perp}} \right] \cdot$$

$$\cdot \left(\frac{\omega - \Omega_c}{K_{\parallel}} \right)^2 V_{\perp}^2 \pi \delta \left(\omega - \Omega_c - K_{\parallel} V_{\parallel} \right) \left| \underline{B}_{\underline{K}} \right|^2 B_0^{-2} \left[\left(\frac{\partial}{V_{\parallel} \partial V_{\parallel}} - \frac{\partial}{V_{\perp} \partial V_{\perp}} \right) + \frac{\omega}{\omega - \Omega_c} \frac{\partial}{V_{\perp} \partial V_{\perp}} \right] f^{(1)}$$

(4)

where $f^{(1)} = f - f_{QL}$ is the deviation from the stable
distribution for the ring current protons, f_{QL}, reached
at the end of the quasilinear relaxation, $\underline{E}$ is the qua-
sistationary electric field in the magnetosphere, S is
the coordinate along the magnetic field lines and
$\left| \underline{B}_{\underline{K}} \right|^2 / 8\pi$ is the spectral energy density of the
wave magnetic field. Equation (4) is one of two equa-
tions we need to find the growth rate of the strongly
suppressed instability and the turbulence level. In a
number of papers (Cornwall et al., 1970; Kennel and
Petchek, 1969; Cornwall et al., 1971) the wave energy
density is estimated by equating the time of quasilinear diffusion into the empty loss-cone and the time of
particle flight between the magnetic mirrors. But we
see that the reduction of particle diffusion into the
loss-cone can be achieved either by the suppression of
excited electromagnetic noise or by the relaxation of
the particle distribution towards the stable one. As a
result the requirement of balance between the rate of
quasilinear diffusion into the loss-cone and the rate of
particle loss from it gives only one equation for two

the spectral wave energy density varies slowly over a
frequency range of the order of the ion sound $\left(\omega_s = |\underline{K} - \underline{K}'| c_s \right)$ or Doppler $\left(\omega_D \sim |\underline{K} - \underline{K}'| v_{TH} \right)$ frequencies:

$$\left(\frac{\partial}{\partial t} + \frac{\partial \omega}{\partial K_{\parallel}} \frac{\partial}{\partial S} - \frac{\partial \omega}{\partial r} \frac{\partial}{\partial K_r} \right) |\underline{B}_{\underline{K}}|^2_+ =$$

$$= 2 \left[\gamma_{QL} - \frac{m}{M \sqrt{\pi}} \frac{\omega^3}{\Omega_c^2} \frac{K_{\perp}^2}{K_{\parallel}^2} \left(\frac{\omega}{K_{\parallel} v_{THe}} \right)^3 \exp \left(- \frac{\omega^2}{K_{\parallel}^2 v_{THe}^2} \right) \right] |\underline{B}_{\underline{K}}|^2_+$$

$$+ 2\omega K_{\parallel} |\underline{B}_{\underline{K}}|^2_+ \frac{\partial}{\partial K_{\parallel}} K_{\parallel} \int |\underline{B}_{\underline{K}'}|^2_- B_o^{-2} \frac{d^2 K'_{\perp}}{(2\pi)^2} \Big|_{K'_{\parallel} = K_{\parallel}}$$

$$\tag{6}$$

The left hand side of this equation describes ion-
cyclotron wave propagation in a geometrical optics approximation. The wave refraction represented by the
last term on the left hand side of (6) leads to a deviation of the wave path from the magnetic field line and,
therefore, results in electron Landau damping (Kennel
and Wong, 1967). We will use the Landau damping decre-
ment in a form found by Cornwall et al. (1971). The instability growth rate can be estimated from Eq. (5).
The nonlinear term is given for the low frequency limit
$(\omega < \Omega_c)$ when it can be reduced to a differential form
appropriate for the low β-plasma. It should be noted
that the interaction takes place only between waves propagating in opposite directions, and the intensity of
the corresponding waves is labelled by $\pm$ signs.
        In a uniform plasma approximation, therefore ne-
glecting refraction, this equation has a solution in
the form of a "stream line" in phase space:

parameters: the noise intensity and the growth rate of
the instability which characterizes the small deviation
from the stable state. Taking into account both effects
supporting the instability (convection and loss-cone
erosion) we write this equation in the approximate form

$$\frac{1}{4\pi} \int \frac{d^3 K}{(2\pi)^3} |\underset{\sim}{B}_{\underset{\sim}{K}}|^2 \underset{\underset{\sim}{K}}{\gamma} \simeq MN\eta \overline{(V_{\parallel R})} V_{\parallel R}^2 \left(A_+^2 \Big/ T_E + L^{-3} T_{min}^{-1} \right) \quad (5)$$

where $T_E^{-1} \sim cE_\perp / B_0 R_E$ is the rate of convection and
$T_{min}^{-1} \sim v_{\parallel R} / L^3 R_E$ is the rate of particle loss
through the loss-cone, R_E is the Earth radius and the
factor L^{-3} characterizes the amount of particles within the loss-cone.

It should be noted that protons from the plasma
sheet can reach the region of stable confinement of the
ring current without essential loss only if $T_E < T_{min}$
(Kennel, 1969; Thorne and Kennel, 1971). This means
that in the outer part of the ring current the instability is supported mainly by anisotropic plasma convection. But at the plasmapause we still consider for the
sake of simplicity the particle removal from the loss-
cone as the dominant mechanism for anisotropy recovery.
The results can easily be changed for the opposite case.

The spectral wave energy density can be found by
considering the nonlinear effects which are responsible
for wave growth saturation. It is known that in a low
β plasma with hot electrons the low frequency electromagnetic waves (such as Alfvén and ion-cyclotron waves)
are strongly coupled to the ion sound waves. If the ion
temperature is comparable with that of the electrons,
this coupling takes the form of induced wave scattering
on ions. The wave kinetic equation is simplified when

$$\left| \underset{\sim}{B}_{\kappa} \right|_{\pm}^{2} = (2\pi)^{2} \delta(\underset{\sim}{\kappa}_{\perp}) B_{o}^{2} \kappa_{\parallel}^{-1} \int_{\kappa_{\parallel}}^{\omega_{max}/V_A} \frac{\gamma_{QL}}{\omega \kappa_{\parallel}} d\kappa_{\parallel} \qquad (7)$$

The quasilinear estimate (5) of the integral energy input into the waves and the nonlinear estimate of the
saturated wave intensity as a function of the growth
rate now permit us to find the turbulence level as a
function of plasma parameters

$$\frac{1}{8\pi} \int \frac{d^3\kappa}{(2\pi)^3} \left| \underset{\sim}{B}_{\kappa} \right|^{2} \simeq \left(\frac{7}{\omega^7 L^3 T_{min}} \right)^{1/2} \frac{B_o^2}{8\pi} \qquad (8)$$

Let us now turn to the question of what fraction
of the ion-cyclotron wave energy is dissipated in the
plasmasphere and what part of it enters the ionosphere.
To answer this question we should compare the rate of
three processes: induced wave scattering leading to
ion heating, electron Landau damping of oblique waves
appearing as the result of wave refraction at the plasmapause, and wave energy flux along the field lines to
the ionosphere. For existing parameters of the plasma
and the magnetic field the induced wave scattering is
the dominant process in Eq. (6). Therefore, most of the
wave energy is dissipated into the ion component. The
ion temperature growth is saturated when the divergence
of the ion heat flux to the ionosphere becomes comparable with the energy input due to the instability. For
typical parameters of the ring current (n_{RC}=10 cm^{-3},
$n_{RC}Mv^2$=10^{-7}-10^{-6}erg cm^{-3}, N=300 cm^{-3}, B=5·10^{-3} g) the
plasmaspheric ion temperature at the plasmapause is so
high that the ion mean free path exceeds the magnetic
field line length from the equatorial plane to the ionosphere. In this case the balance between the energy

input due to the instability and the energy transfer to the ionosphere by freely moving ions can be written as

$$\frac{1}{4\pi}\int \gamma_{QL}\,|\underset{\sim}{B}_{\underset{\sim}{K}}|^2\frac{d^3\underset{\sim}{K}}{(2\pi)^3} \simeq \frac{N_{PS}\,T_i^{3/2}}{R_E\,\sqrt{M}} \tag{9}$$

Using Eq.(5) to estimate the left hand side of this equation we find the ion temperature

$$T_i \sim M\,\overline{V_{\shortparallel R}^2}\;\eta^{2/3}\,L^{-4} \tag{10}$$

For the above-mentioned plasma parameters this gives an ion temperature rise towards the plasmapause, which has been observed experimentally (Frank, 1970; Serbu and Maier, 1970) and which can give rise to additional heating and evaporation of the heavy ionospheric ions below the plasmapause. (see Fig.2.) Because of their shorter mean free path, the plasmaspheric ions heat the rare upper ionosphere more efficiently than the precipitating ring current protons.

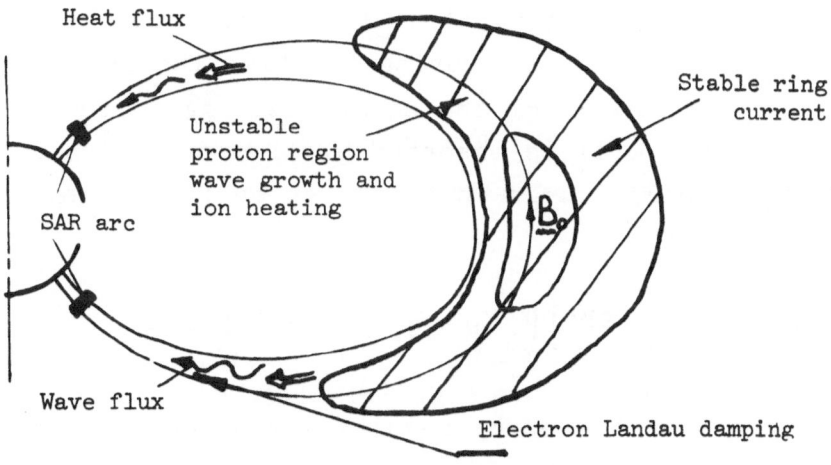

Fig.2. The energy transfer processes at the plasmapause.

Wave refraction due to the plasma density gradient
at the plasmapause leads to the deviation of the wave
path from the field line and the Landau damping of these
oblique waves results in heating of the electrons. If
the plasma density gradient at the plasmapause is not
very sharp, strong Landau damping suppresses the amplitude of the oblique waves and the solution of Eq. (6)
still has the form of a "stream line" in phase space.
The balance between the growth of oblique waves due to
refraction and their suppression due to strong electron
Landau damping defines the finite width of the stream
line $(K_r \leq 0)$

$$|\underset{\sim}{B}_K|^2 = 12\pi^2 \delta(K_\varphi) B_0^2 \frac{\exp\left(K_r^3/|K_{\parallel}|^3 tg^3\Theta\right)}{\Gamma(1/3)\, K_{\parallel}^2\, tg\,\Theta} \int_{K_{\parallel}}^{\omega_{max}/V_A} \frac{\gamma_{QL}\, dK_{\parallel}}{\omega K_{\parallel}} \quad (11)$$

where $\Theta = arctg \left| \frac{3\sqrt{\pi}M}{m} \frac{\partial \ln\omega}{K_{\parallel}\partial r} \frac{\Omega_c^2}{\omega^2} \left(\frac{K_{\parallel} U_{THe}}{\omega}\right)^3 \right|^{1/3}$ is the
angular width of the stream line. For the above-mentioned
plasma parameters this width is very large. But if
$tg\,\Theta < L$,   then during the time the oblique waves propagate from the equatorial region to the
ionosphere they will be absorbed by electrons. The
electron temperature can in this case be found from the
balance equation

$$\frac{1}{4\pi} \int \frac{d^2K}{(2\pi)^2} \frac{\partial\omega}{\partial r} |\underset{\sim}{B}_K(K_r=0)|^2 \simeq \frac{V_{THe}\, N_{PS}\, T_e}{R_E} \quad (12)$$

Here, the electron heating resulting from wave absorption is balanced by the electron heat flux along the
field lines to the ionosphere. Using Eq. (10) for the

ion-cyclotron wave intensity we obtain

$$T_e \simeq (tg\theta)^{-2/3} \left(\frac{m}{M\omega} \frac{\eta}{T_{min} L^3} \right)^{1/3} M \overline{U_{\parallel R}^2} \tag{13}$$

As has been shown by Cornwall et al.(1971) the heat flux of such electrons is sufficient to cause SAR arcs (see Fig.2).

3. HEATING OF THE IONOSPHERE DURING ELECTRON PRECIPITATION

Precipitation of hot magnetospheric electrons with a field-aligned pitch-angle distribution into the iono-sphere can be represented as the entrance of an elec-tron beam with thermal velocity spread into a cold plas-ma. The interaction of such a beam with the plasma leads to the excitation of a wide spectrum of Langmuir waves with phase velocities close to the beam velocity

$$\omega/\kappa \simeq V_b \tag{14}$$

In the case where the beam velocity does not very much exceed the thermal spread of the electron distribution, the main effect leading to wave growth saturation is the quasilinear relaxation of the electron distribu-tion resulting in plateau formation in velocity space (Vedenov et al., 1962). In the opposite case, the wave-lengths of the excited oscillations are so small accord-ing to Eq. (14) that the stability criterion for self-modulation of the spatially uniform intensity of Lang-muir waves is violated (Vedenov and Rudakov, 1964)

$$\int \frac{d^3\kappa}{(2\pi)^3} \left| \underline{E}_{\underline{\kappa}} \right|^2 / 4\pi n_0 T_e < \kappa^2 \lambda_D^2 \qquad (15)$$

where $\left| \underline{E}_{\underline{\kappa}} \right|^2 / 4\pi$ is the spectral energy density
of the Langmuir waves, $\lambda_D = \sqrt{T_e / m \omega_p^2}$ is the De-
bye length, T_e is the temperature of the bulk electrons.
Since quasilinear diffusion in velocity space, spreading
the beam velocities over the velocity interval Δv_b
results in wave energy growth up to the value

$$\frac{1}{4\pi} \int \frac{d^3\kappa}{(2\pi)^3} \left| \underline{E}_{\underline{\kappa}} \right|^2 \simeq n_b m v_b \Delta v_b \qquad (16)$$

we find that, for typical parameter values of the precipitating electrons and the ionospheric plasma, the stability criterion (15) is strongly violated.

Thode and Sudan were the first who considered Langmuir wave-growth saturation due to a parametric instability of the excited wave packet. Application of this
idea to the case of electron burst precipitation into
the ionosphere (Papadopolous and Coffey, 1974) gave
some rough estimates of the ionospheric electron heating. However, it is difficult to see how to improve such
a theory since the mode coupling approach fails in the
case of strong parametric interaction. On the other
hand, in the case of weak parametric interaction, induced wave scattering is the dominant nonlinear process
and it gives rise to wave energy flow towards the long
wavelength region, where there is no energy sink. This
creates the problem of the plasmon condensate (see discussion in the survey by Galeev and Sagdeev (1973)).

As was proposed by Galeev et al. (1972) the self-
modulational instability of the plasmon condensate can

be an effective mechanism to produce short wavelength
oscillations and, therefore, to remove all the problems
with the plasmon condensate. The rigorous theory providing the wave energy flux towards the short scales
and subsequent wave absorption by electrons (Galeev et
al., 1972) is based on Zakharov's idea about the Langmuir wave collapse (Zakharov, 1972).

   The mechanism of the wave collapse is the following. In the case of a large plasmon energy density the
electric field pressure forces plasma outwards. This
results in the creation of a plasma cavity. The plasma
cavity in its turn serves as a potential well for the
plasmons and gives an additional concentration of wave
energy. A very simple argument of Sagdeev gives a hint
whether a collapse is possible or not. The argument is
based on the use of the following expression for the
wavelength of plasmons trapped in the density cavity:

$$K^2 \lambda_D^2 \sim \delta n / n_o \qquad (17)$$

and of the conservation of plasmon number in the cavern:

$$\int d^s r \, |E|^2 / 4\pi\omega_p = \text{const} \qquad (18)$$

According to Eq. (18) the electric field pressure in-
creases with a decrease of the cavern volume as $|E|^2 \sim$
$\sim \ell^{-s}$ where S is the dimensionality of the cavern.
At the same time, to force plasma out of the cavern we
need to create a pressure which is larger than the plasma pressure $\delta n \, T_e \sim \ell^{-2}$ Therefore,
in the one-dimensional case with a given initial plasmon energy concentration, the narrowing of the cavern
stops when the pressure balance is reached (see Fig.3.)

In the two-dimensional case the collapse can not be
stopped if the initial wave concentration is sufficiently strong. And, finally, in the three-dimensional case
the plasmon collapse seems unavoidable.

Energy transfer towards the short wavelength region can be represented as a wavelength reduction of

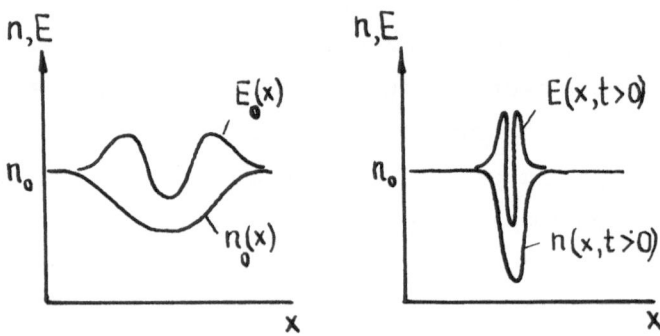

Fig.3. Evolution of the plasma cavern and the trapped
       waves.

plasmons trapped in collapsing caverns. For a given production rate of the large scale caverns it is possible
to find the number of caverns with different scales and
thus to solve the problem of the wave energy distribution in k-space.

If we want to apply this theory to ionospheric phenomena we should generalize it for the case of a strong
influence of the Earth's magnetic field on the plasma
wave dispersion. In the limit of weak magnetic field
$(\omega_c < \omega_p)$ the generalization of the theory

is simple, since plasmons still can be trapped efficiently within the plasma cavern. This can be seen from
the wave dispersion relation:

$$\omega = \omega_p \left(1 + \tfrac{3}{2} \, \kappa_{\parallel}^2 \lambda_D^2 \right) + \left(\omega_c^2 / 2\omega_p \right) \kappa_{\perp}^2 / \kappa_{\parallel}^2 \tag{19}$$

Using this equation we can write the nonlinear wave
equations for the Langmuir waves and for the density
disturbance under the influence of the wave electric
field pressure:

$$\left(i\frac{\partial}{\partial t} + \frac{3}{2}\,\omega_p \lambda_D^2 \frac{\partial^2}{\partial z^2} \right) \frac{\partial^2}{\partial z^2} E + \frac{\omega_c^2}{2\omega_p}\Delta_\perp E = \frac{\omega_p}{2n_0}\frac{\partial^2}{\partial z^2}\delta n E \tag{20}$$

$$\left(\frac{\partial^2}{\partial t^2} - \frac{T_e}{M}\frac{\partial^2}{\partial z^2} \right)\delta n = \frac{1}{16\pi M}\frac{\partial^2}{\partial z^2}|E|^2 \tag{21}$$

where z is the coordinate along the magnetic field line,
and the plane (x,y) is perpendicular to it; $\underline{E}(z,\underline{r}_\perp,t)$
is the z-component of the wave electric field. From Eq.
(21) we see that in a supersonic collapse regime the
wave electric field behaves as $E \sim (t_0 - t)^{-1}$
since $n_1 \sim \partial^2/\partial z^2$ in accordance with Eq. (17).
Therefore, we can try to find a selfsimilar solution of
Eq. (20) in the form

$$E \sim (\tau_0 - \tau)^{-1}\,\psi\left(Z\sqrt{M(\tau)}\,,\, R\,M(\tau) \right) \tag{22}$$

where $\tau = \omega_p t \dfrac{\langle E^2\rangle}{8\pi n_0 T_e}$, $Z = \dfrac{z}{\lambda_D}\sqrt{\dfrac{\langle E^2\rangle}{12\pi n_0 T_e}}$, $R = \dfrac{r}{\lambda_D}\dfrac{\sqrt{2}\,\omega_p}{\sqrt{3}\,\omega_c}\dfrac{\langle E^2\rangle}{8\pi n_0 T_e}$

With the help of the conservation law for the plasmon
number in the cavern we obtain the dependence of μ
on τ

$$\mu(\tau) = (\tau_0 - \tau)^{-4/5} \tag{23}$$

The rate of short wave production is defined by the
rate of the collapse. Assuming the wave energy flux towards the short scales to be constant (as in hydrodyna-
mic turbulence) we write (Galeev et al., 1975)

$$\frac{W_k \, dk_z}{d\,t\,(k)} \simeq \mathrm{cons}\,t \tag{24}$$

where $d t(k) = \frac{15}{4} \omega_P^{-1} \left(\frac{\langle E^2 \rangle}{12\pi n_0 T} \right)^{-7/2} k_z^{-7/2} \lambda_D^{-5/2} dk_z$
is the time which the cavern spends within the wave
number interval $(k_z, k_z + dk_z)$. This equation gives the
following turbulence spectrum:

$$W_k \sim k_z^{-7/2} \tag{25}$$

Equation (24) is not valid in the short wavelength region where electron Landau damping becomes important.
Resonant interaction of the short wavelength oscillations with the ionospheric electrons results in the
creation of a non-Maxwellian electron tail. Selfcon-
sistent solutions for the electron distribution within
the non-Maxwellian tail and the turbulence spectrum in
the short wavelength region can be found from the quasilinear equation for the electrons

$$\frac{\partial f_e}{\partial t} = \frac{\partial}{\partial V_z} \, 4\pi^2 \int \frac{dk_z}{2\pi} \, W_k \, \delta(\omega_P - k_z V_z) \frac{\partial f_e}{\partial V_z} \tag{26}$$

and the wave kinetic equation including electron Landau damping and wave energy flux in k-space:

$$\frac{\partial}{\partial t} W_k = -\frac{\partial}{\partial k_z}\left(\frac{dk_z}{dt} W_k\right) + \frac{\pi \omega_p^2}{k_z^2} \frac{\partial f_e}{\partial v_z}\Big|_{v_z=\omega_p/k_z} \cdot W_k \qquad (27)$$

A selfsimilar solution of these equations based on the assumption of constant particle flux into the non-Maxwellian tail is:

$$W_k \sim k_z^{-11/2}, \qquad f_e \sim v_z^{-7/2} \qquad (28)$$

Finally we can find the rate of electron-beam energy-absorption in the ionospheric plasma. This rate is defined by the linear growth rate of the beam instability and by the wave intensity in the long wavelength region:

$$\frac{\partial}{\partial t} \tfrac{1}{2} n_b m v_b^2 \simeq -\gamma_b |E_0|^2/4\pi \qquad (29)$$

where $\gamma_b = \pi \dfrac{\omega_p}{k_z^2} \dfrac{\partial f_b}{\partial v_z}\Big|_{v_z=\omega_p/k_z} \sim \omega_p \dfrac{n_b}{n_0}\left(\dfrac{v_b}{\Delta v_b}\right)^2$

The level of excited plasma turbulence, $\langle E^2 \rangle /4\pi$, can be found from the balance of the energy input to the long wave oscillations with the dissipation of this energy due to wave scattering on caverns:

$$\gamma_b \frac{E_0^2}{4\pi} \simeq \nu_{eff} \frac{E_0^2}{4\pi} \qquad (30)$$

where $\nu_{eff} = \omega_p \langle E^2 \rangle / 8\pi n_0 T_e$ is the effective collision frequency of the long wave oscillations on caverns with plasmon energy density equal to $\langle E^2 \rangle /4\pi$

(Galeev et al., 1975). The wave intensity at large wave-
lengths can be found from the balance of the wave energy
trapped in the caverns (the loss is due to modulational
instability and the source is the long wavelength oscillations excited by the beam)

$$\nu_{eff} \frac{E_o^2}{4\pi} \simeq \gamma_{md} \frac{\langle E^2 \rangle}{4\pi} \tag{31}$$

where $\gamma_{md} = \sqrt{\frac{m}{M} \frac{\langle E^2 \rangle}{8\pi n_o T_e}}$ is the growth rate of
the modulational instability.

With the help of the derived Equations (29)-(31)
we can estimate the rate of absorption of precipitating
electron energy by the ionospheric plasma:

$$\frac{\partial}{\partial t} \frac{1}{2} n_b m V_b^2 \simeq \omega_p n_b m V_b^2 \frac{T_e}{m \Delta V_b^2} \left(\frac{m}{M}\right)^{3/4} \left(\frac{n_b}{n_o} \frac{V_b^2}{\Delta V_b^2}\right)^{1/2} \tag{32}$$

For typical ionospheric parameters and flux of precipitating electrons this rate is insufficient to result in
complete absorption of the beam energy.

REFERENCES

Bezrukich V.V., Breus T.K., and Gringauz K.I., Kosmicheskie Issledovania. 5, 798, 1967.

Cohen R.H., in "Solar Wind Three" ed. by C.T. Russell, Univ. of Calif., L.-A., 1974, p. 281.

Cornwall J.M., Coroniti F.V., and Thorne R.M., J.Geophys.Res. 75, 4699, 1970.

Cornwall J.M., Coroniti F.V., and Thorne R.M., J.Geophys.Res. 76, 4428, 1971.

Frank L.A., J.Geophys.Res. 75, 1263, 1970.

Fritz T.A., Smith P.H., Williams D.J., Hoffman R.A., and Cahill L.J., in "Correlated Interplanetary and Magnetospheric Observations", 1974, D. Reidel Publishing Co., Dordrecht-Holland pp. 485-506.

Galeev A.A., J.Plasma Phys. 1, 125, 1967.

Galeev A.A., Oraevskii V.N., and Sagdeev R.Z., Zh.Eksper. Teor.Fiz.Pis'ma Red. 16, 194, 1972.

Galeev A.A., and Sagdeev R.Z., Nuclear Fusion 13, 603, 1973.

Galeev A.A., Sagdeev R.Z., Shapiro V.D., and Shevchenko V.I., Fiz.Plasmi, 1, 11, 1975.

Kennel C.F., Rev. of Geophys. 7, 379, 1969.

Kennel C.F., and Petchek H.J., Geophys.Res. 71, 1, 1966.

Kennel C.F., and Wong H.V., J.Plasma Phys. 1, 81, 1967.

Papadopolous K., and Coffey T., J.Geophys.Res. 79, 674, 1974.

Rosenbluth M.N., and Post R., Phys.Fluids 8, 547, 1965.

Sagdeev R.Z., and Galeev A.A., Nonlinear Plasma Theory. Benjamin, Amsterdam, N.-Y., 1969.

Serbu G.P., and Maier E.J.R., J.Geophys.Res. 75, 6102, 1970.

Thode L.E., and Sudan R.N., Phys.Rev.Lett. 30, 732, 1973.

Thorne R.M., and Kennel C.F., Comments on Astrophysics and Space Physics 3, 115, 1971.

Vedenov A.A., Velichov E.P., and Sagdeev R.Z., Nuclear Fusion. Suppl., Part 2, p. 465, 1962.

Vedenov A.A., and Rudakov L.I., Doklady AN SSSR. 159, 767, 1964.

Williams D.J., and Lyons L.R., J.Geophys.Res. 79, 4195, 1974; 79, 4791, 1974.

Zakharov V.E., Zh.Eksper.Teor.Fiz. 62, 1745, 1972.

Zakharov V.E. et al., Zh.Eksper.Teor.Fiz.Pis'ma Red. N° 1, 1975.

CHARACTERISTICS OF INSTABILITIES IN THE MAGNETOSPHERE DEDUCED FROM

WAVE OBSERVATIONS

Frederick L. Scarf

Space Sciences Department, TRW Systems Group, One Space
Park, Redondo Beach, California 90278 U.S.A.

1.   INTRODUCTION

In almost all regions of the magnetosphere, the collisionless
plasma distribution functions are significantly non-Maxwellian and
they are either continuously unstable with respect to generation of
plasma waves, or readily perturbed into unstable states so that enhanced wave levels develop during geomagnetic storms and substorms.
The plasma waves act back on the particle distributions to produce
pitch angle scattering, spatial diffusion, acceleration, or heating.
The plasma turbulence also provides an important mechanism for energy dissipation in collisionless shocks and field annihilation regions, and it can lead to anomalous resistivity effects that allow
macroscopic parallel DC electric fields to be maintained. It appears that plasma waves also play a very important role in providing
the coherence that leads to intense electromagnetic radiation from
the magnetospheric plasma.

The characteristics of a magnetospheric instability are as-
sessed by examining both the local plasma distribution function and
the local plasma turbulence spectrum. This separation is never a
clear one since, by definition, the waves associated with any significant plasma instability will already have acted to perturb the
distribution function. Thus, for instance, the observer of a
strongly interacting system must try to work backwards to identify
the initial source of the observed wave turbulence in terms of conventional instability dispersion relations and initial growth rates.
In the time-varying large scale plasma laboratory of the magnetosphere, this task is rarely a simple one. Nevertheless, there has

been much progress in this area during the last decade; this report
contains a very general summary of the types of unstable plasma distributions encountered in the magnetosphere, along with an assess-
ment of the present state of knowledge concerning instability characteristics based on wave observations.

2. VARIETIES OF UNSTABLE PLASMA DISTRIBUTION FUNCTIONS

Even during the initial phase of exploration, it was recognized that some plasma distribution functions with "free energy"
sources capable of yielding growing plasma waves had to be present
in the magnetosphere. It was anticipated long ago that some kind
of two-stream instability would play an important role at the solar
wind-magnetosphere interface. Moreover, it was noted at an early
stage that, as in any magnetic mirror device, the existence of a
loss-cone already implies a thermal anisotropy that can yield wave
amplification.

In the past decade an enormous amount of detailed information
on local particle characteristics has been gathered, and it is now
very evident that many mechanisms operate to produce a complex variety of non-Maxwellian and unstable plasma distributions in the
earth's ionized environment. Within the magnetosphere thermal
anisotropies with $T_\perp > T_\parallel$ develop because of: a) inward plasma diffusion and convection with conservation of μ; b) cyclotron and betatron acceleration effects; and, c) selective pitch angle scattering
into the loss cone. The observed magnetospheric distributions with
$T_\parallel > T_\perp$ are probably associated with inward motion and conservation
of the longitudinal action invariant, while similar distributions
in the solar wind are thought to be connected with conservation of
μ, as the interplanetary field strength in the expanding plasma
continuously declines. Other thermal anisotropies that are significant with respect to plasma instabilities involve heat flux and
higher order moments, and these perturbations can be associated
with parallel electric fields, currents, particle beams, and heat
conduction.

It has also become clear in recent years that the natural
plasmas are locally non-Maxwellian in the sense that there are significant peaks and dips in the energy distributions. These fluctuations arise because: a) the solar wind and the ionosphere are
more or less independent sources of warm and cool magnetospheric
plasma; b) varying gradient drifts selectively remove trapped particles in restricted energy ranges from the magnetosphere and they
leave residual quasi-trapped distributions with deep flux ripples;
c) collisionless acceleration processes apparently provide enhanced
fluxes at certain energies, and; d) resistive dissipation and heat

conduction (and perhaps runaway) contribute to the observed non-thermal tail populations.

The original fairly narrow application of the two-stream insta-bility has also been greatly generalized in recent years. Supra-thermal flows are commonly detected within the magnetosphere (i.e., in the polar cusp and in the tail during large substorm events), and it is known that the equivalent electron-proton drift speed in the solar wind and magnetosheath has a significant contribution asso-ciated with electron heat conduction. Current-driven instabilities are found to be important at field-merging regions in the magneto-sheath, as well as at the bow shock, and the field-aligned currents associated with pressure gradients in the magnetosphere, the cusp boundary and the high latitude ionosphere can involve streaming in-stabilities.

Some remarkable examples of extreme non-Maxwellian plasma dis-tributions at synchronous orbit are described in McIlwain's[1] paper on the ATS-6 measurements. Although ATS-6 does not have an on-board plasma wave investigation to provide correlative data on plasma in-stabilities, it now seems clear that natural magnetospheric process-es provide enough variety in distribution functions to allow growth of almost any conceivable plasma wave mode.

3. GYRORESONANT INTERACTIONS

The initial suggestions by Dungey[2] and Cornwall[3] that electro-magnetic whistler mode turbulence can lead to precipitation of ra-diation belt particles were soon followed by the classic self-con-sistent theory of Kennel and Petschek[4], which is based on use of a natural gyroresonance instability mechanism for the wave growth (amplification associated with trapped particle $T_\perp > T_\parallel$ pitch angle distributions) and the concept of turbulent pitch angle diffusion arising from the local wave-particle interactions. The Kennel and Petschek theory predicted overall stable trapping limits and preci-pitation patterns that agreed fairly well with available observa-tions, and this pioneering effort has since stimulated an enormous amount of more detailed analytical activity; as described in a recent review by Fredricks[5], improved self-consistent theories for electron whistler and ion cyclotron mode turbulence were developed, these ideas were applied to analyze lightning whistler amplification processes, numerical simulation studies were performed, and the basic concepts were used in attempts to explain localized plasma-pause phenomena as varied as ring current decay, SAR arc formation and the development of an energetic electron "slot."

It is difficult to verify details of the generalized Kennel-Petschek theory for a number of reasons: a) whistler mode waves propagate readily in the dispersive magnetospheric plasma, and they are reflected and refracted so that the identification of the source region and the initial wave spectrum is a complex and frequently uncertain project; b) the particles that interact resonantly with these waves have relatively high energies (more than tens of kilovolts) and the ions, in particular, drift in and out of the interaction regions; c) there are generally significant local sources of resonant particles, and these sources mask the effects of the loss mechanisms.

In fact, only a few aspects of the whistler mode instability theory have been directly verified in some isolated studies. Rosenberg et al.[6] analyzed the arrival characteristics of bursts of precipitating $E \simeq 30-100$ keV electrons and propagating chorus emissions at Siple Station, Antarctica, and they provided convincing evidence that the chorus interaction with trapped particles near the geomagnetic equator gives rise to pitch-angle scattering that leads to the precipitation. In another area, Kivelson and Russell[7] showed directly that enhanced whistler mode turbulence develops in a region with a cold density enhancement, where the theoretical resonant electron energy (proportional to $B^2/8\pi N$) decreases so that more electrons are available to feed the wave growth. However, Kivelson and Russell could not find evidence of enhanced pitch-angle scattering during these events.

During a large substorm event when betatron acceleration produced a sudden local increase in $T_\perp/T_\parallel$, Kivelson et al.[8] showed conclusively that the anisotropy enhancement was associated with chorus growth. The measurements are displayed in Figure 1, adapted, in part, from an OGO-5 report by Scarf et al.[9]. The shaded region in the bottom panel marks the onset of a $T_\perp > T_\parallel$ distribution for 79 keV electrons, and it can be seen that this change is associated with the change in field orientation (second panel) and with an increase in 467 Hz search coil output (third panel). The top two panels in Figure 1 contain simultaneous broadband dynamic spectra from the short electric antenna and the magnetic search coil. It can be seen that the B-field chorus onset at 0732:15 to 0732:40 UT is indeed very well correlated with the anisotropy enhancement. Since the measured plasma parameters give a resonant electron energy ($B^2/8\pi N$) range near 30-90 keV at this time, it may be concluded that the chorus onset is directly associated with development of a suitable anisotropy in the resonant electron population. It should be noted, however, that this gyroresonant interaction does not lead to anything approaching strong diffusion. During the preceeding period with near-isotropy the chorus was absent, but large amplitude electrostatic waves with $f \simeq 3\, f_c^e/2$ were detected; we shall discuss this point in more detail below.

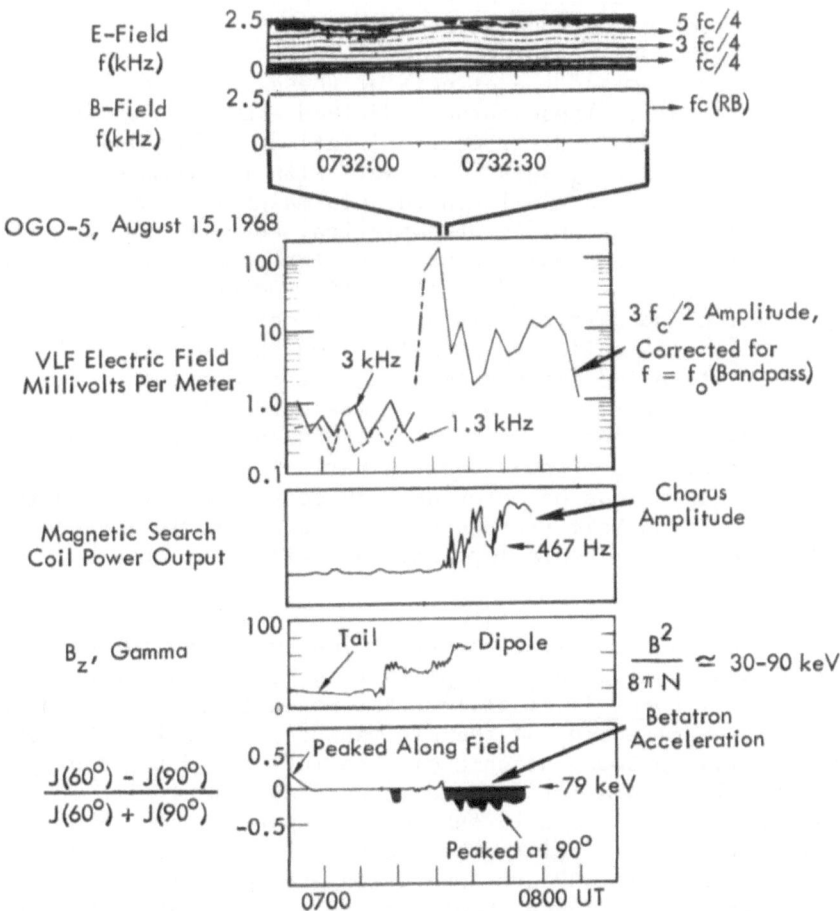

Figure 1. Wave-particle interaction phenomena detected near local midnight during a substorm expansion event. The top panels show the characteristic dynamic spectra for 3 $f_c^e/2$ noise (E-field) and chorus (B-field). The next panel down gives the corrected $f \simeq 3$ $f_c^e/2$ amplitude for t > 0727 UT, but, for the earlier interval, uncorrected center frequency amplitudes in the 1.3- and 3-kHz channels are plotted. The lowest two panels show that chorus emissions and loss cone (shaded) particle distributions were detected after 0733 UT, when the $f \simeq 3$ $f_c^e/2$ wave intensity decreased markedly.

The most convincing verifications of the Kennel-Petschek ideas
on radiation belt structure involve analytical treatments based on
use of generally measured wave turbulence spectral characteristics.
In a series of papers, Lyons et al.[10] and Lyons and Thorne[11] assumed a distribution of oblique waves with amplitude and spectral
shape that match many measurements of ELF hiss within the plasmapause, and they examined the effects of these waves on the trapped
particles in detail. These authors derived energy-dependent and
L-dependent pitch-angle distribution lifetimes and equilibrium distribution functions that agree very well with the complex profiles
measured on OGO-5 and S^3-A within the plasmasphere. Recently,
there have also been significant analytical advances in treating
the other half of the problem concerning the origin of the observed
gyroresonant emissions. Etcheto et al.[12] studied the problem of
wave-generation and particle diffusion self-consistently, although
they did confine attention to parallel propagating modes. These
authors used iterative techniques to derive a wave turbulence spectrum that agrees fairly well with the measured ELF hiss spectrum,
and they also clarified concepts of the general stable trapping
limit and the effects of cold density variations. These semi-analy-
tical papers taken together provide very convincing evidence that
the inner radiation belt trapped electron characteristics are controlled by an internal whistler mode instability associated with
the thermal anisotropy.

A very elegant parallel theory of ion cyclotron turbulence has
been developed and applied to explain ring current decay and proton
precipitation. Cornwall et al.[13] first noted that the sudden decrease in resonant energy at the plasmapause should lead to wave
growth, greatly enhanced pitch-angle scattering and rapid ring current decay. These predictions are in good agreement with energetic
particle observations, and in subsequent reports[5] associated theories were developed to relate the presumed ion cyclotron turbulence
at the plasmapause to SAR arc generation and formation of an energetic electron slot. These theoretical ideas are very appealing
and, as summarized by Williams[14], many aspects of the equatorial
and low altitude proton measurements are in agreement with the theoretical predictions. However, there appears to be a persistent
theoretical or experimental problem in connection with the model
because there is little evidence that significant ion cyclotron turbulence develops at the plasmapause. Taylor et al.[15] have detected
only a few low frequency electromagnetic events on S^3-A, and
Gurnett has reported (private communication) that only once has any
1.7 Hz signal been detected near the plasmapause with the high sensitivity Hawkeye-1 search coil.

We conclude that it is now known with certainty that gyroresonant interactions play an important role in magnetospheric dynamics,
but only for trapped electrons in the plasmasphere has it been
demonstrated that these interactions actually control the dynamics.

There is still no complete theory to account for whistler mode emissions observed at the magnetopause, or the surprisingly strong (up to 200 milligamma amplitude) whistler mode turbulence detected on IMP 7 in the distant plasma sheet at frequencies above the lower hybrid resonance. Comprehensive explanations of chorus fine structure and whistler mode triggered emissions are also not yet at hand.

4. ELECTROSTATIC INSTABILITIES NOT DRIVEN BY CURRENTS

The most intense waves detected in the magnetosphere with frequencies related to the local electron gyrofrequency, f_c^e, are the $(n + 1/2) f_c^e$ emissions first discussed by Kennel et al.[16]. Subsequent analyses by Fredricks and Scarf[17] and Scarf et al.[9] suggested that these strong emissions (which very frequently have amplitudes as high as 10 millivolts/meter) are substorm related, and Gurnett and Shaw[18] verified that the modes are electrostatic. The most common observation of this mode is at $f \simeq 3 f_c^e/2$, and the broadband E-field dynamic spectrum at the top of Figure 1 shows the typical diffuse and irregular character of the emission (the narrow tones in this f-t diagram represent telemetry output from the OGO-5 Rubidium magnetometer at harmonics of $f_c^e/4$; these narrow lines appear only because the two OGO-5 experiments share the broadband telemetry, and they provide a convenient high resolution indicator of the local electron gyrofrequency).

The VLF electric field panel in Figure 1 shows that the $3 f_c^e/2$ amplitude has a very large peak (approximately 100 mV/meter during this large substorm) in the region where the energetic electron flux is essentially isotropic; Scarf et al.[9] crudely estimated the associated diffusion coefficient for this event and they suggested that on August 15, 1968 the unusually strong wave-particle interactions might yield strong diffusion and near-isotropy for the 80 keV electrons. A much more definitive analysis of the effects of $(n + 1/2) f_c^e$ emissions on magnetospheric particle distributions was carried out later by Lyons[19], who demonstrated that the commonly observed turbulence amplitude of 1-10 mV/m is sufficient to put electrons with energies of several kilovolts in strong diffusion. Lyons' calculations also verified that 100 mV/m waves would indeed lead to strong diffusion for 100 keV electrons, as indicated by the data of Figure 1. In addition, Lyons showed that some detailed low altitude (rocket) measurements of 1-20 keV electron energy spectra in the loss cone can be very well explained on the assumption that moderate amplitude (a few mV/m) $3 f_c^e/2$ waves produce strong diffusion near the equator. It now seems certain that the $(n + 1/2) f_c^e$ modes play a very important role in providing the source of high-latitude electron precipitation fluxes, especially during geomagnetically active periods.

The precise origin of the $(n + 1/2)$ f_c^e mode is presently somewhat obscure. The recent review by Fredricks[5] contains fairly detailed discussions of linear stability analyses carried out by
Fredricks, Young, Karpman and others, and comparison with a wave-
wave interaction theory proposed by Oya. There is little direct
information on the specific local plasma distribution characteristics that can be associated with this mode. On OGO-5 no plasma information at all was available (the probe failed soon after launch)
and complete IMP-6 results are not yet available. However, Gurnett
and Frank[20] have reported that the $(n + 1/2)$ f_c^e electrostatic emissions appear to be quenched whenever the density of very low energy
thermal plasma exceeds a small fraction (on the order of 5 percent)
of the suprathermal plasma density. This suggests that the generation mechanism for these modes (which are also occasionally detected near the magnetopause) involves a hump in the energy distribution as well as a thermal anisotropy.

A different kind of electrostatic instability was recently discussed in connection with the problem of ring current decay and proton precipitation. Coroniti et al.[21] showed that for an initial
ring current proton distribution peaked near 90° with an almost
empty loss cone, there will be sufficiently rapid growth of electrostatic waves with $\omega \approx \omega_p$ to provide significant pitch-angle
scattering and strong precipitation. Coroniti et al. noted that
the marginal stability state could have enough residual anisotropy
to provide some electromagnetic ion cyclotron turbulence, but the
model calculations indicated that the strongest waves would be electrostatic, with $\omega \approx (0.1-0.5)\ \omega_p$. The calculations also indicated
that strong proton precipitation would readily develop beyond the
plasmapause. As summarized by Williams[14], strong high latitude
precipitation events are observed, and it may well be that more than
one mechanism accounts for proton scattering into the loss cone.

The OGO-5 electric field experiment had no wideband measurements below 1 kHz, and poor bandpass channel coverage below 560 Hz,
so that essentially no information about this mode (nominal frequency range 20-300 Hz) could be obtained. On S^3-A, Anderson and
Gurnett[22] reported on the infrequent detection of 20-500 Hz electrostatic wave turbulence just beyond the plasmapause. However, it appears that on Hawkeye-1 moderate amplitude ($\approx$ 1 mV/M) electrostatic
waves with $f \approx$ 50 Hz are commonly detected beyond the plasmapause
(D. Gurnett, private communication). Thus, it now seems that some
proton precipitation events may be associated with an important new
low frequency wave-particle interaction effect that requires much
more study in the future.

5.   CURRENT-DRIVEN INSTABILITIES

Currents strong enough to drive plasma instabilities are clearly detected at the bow shock and along the polar cusp and plasma sheet boundaries. During storms and substorms the high-latitude, field-aligned magnetospheric current systems appear to drive especially strong plasma instabilities in the region extending from the topside ionosphere (say 1000 to 2000 km altitude) to a geocentric distance of (2-4) R_e. From a plasma physics viewpoint, space measurements of strong wave-particle interactions associated with current-driven instabilities are particularly valuable because the interactions are highly localized and the scale lengths, although small in terms of magnetospheric scales, are large in comparison with the dimensions of the spacecraft measuring platform; the measurements are also of major importance for magnetospheric physics because current-driven instabilities produce very important physical effects such as dissipation at the bow shock, anomalous resistivity, and a possible microscopic mechanism for field-line merging.

The most extensive measurements in space are related to studies of the collisionless bow shock upstream from the magnetopause, and Figure 2 shows some typical features of the wave-particle-field interaction phenomena detected during early OGO-5 crossings of the bow shock. We first direct attention to the left side of Figure 2, and discuss the various panels in order, starting from the top.

The magnetic field profile at the shock frequently exhibits ULF magnetic turbulence that can, depending on the interplanetary field orientation, spread into the entire upstream and downstream region. The data displayed here are relatively clean, however, and a clear signature of a strong fairly well-defined current layer parallel to the shock surface is shown in the region centered around 2248:45 UT. It is apparent that the enhanced VLF electric field turbulence shown in the next panel is associated with this current. At this time the upstream proton density was 15.1 cm^{-3} (see Fredricks et al.)[23] and the proton plasma frequency was 0.82 kHz, so that the waves detected in the 1.3 kHz and 3.0 kHz channels had frequencies somewhat higher than the upstream f_p^+-value in the spacecraft frame of reference.

The search coil panel shows essentially no activity in the 1000 Hz channel, indicating that the kiloHertz electric fields involve detection of electrostatic waves; this OGO-5 conclusion of Fredricks et al.[23] was recently verified in great detail by Rodriguez and Gurnett[24] on the basis of IMP-6 measurements. However, it can be seen that electromagnetic waves were detected at lower frequencies; the comprehensive IMP-6 capabilities have been

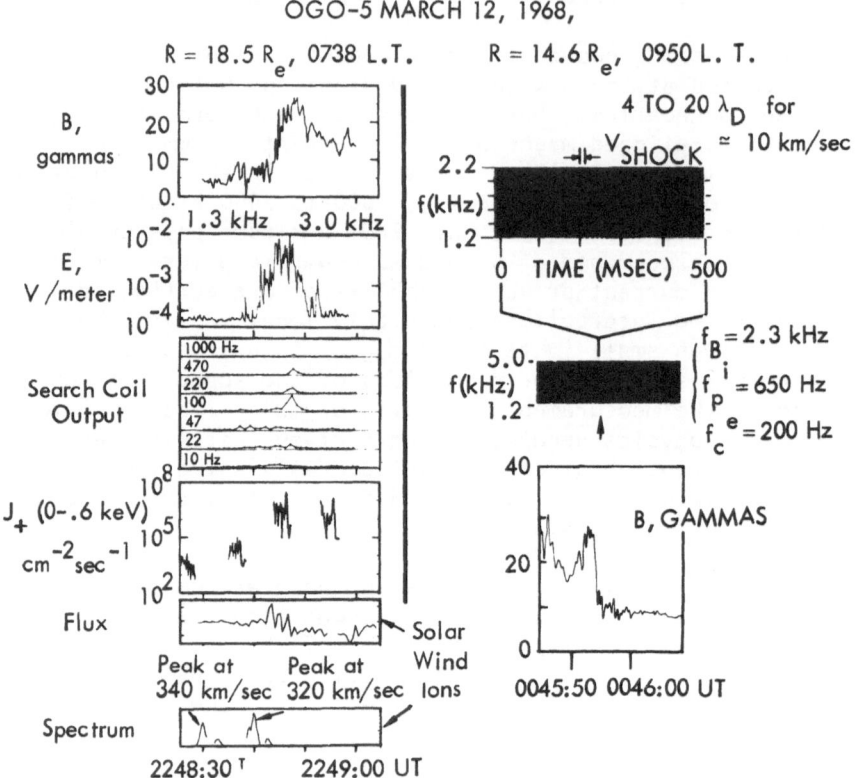

Figure 2. Typical wave-particle interaction phenomena detected at quasi-laminar shocks. These early OGO-5 observations indicate that electrostatic instabilities provide shock dissipation, although whistler mode turbulence also develops (left-hand side). The characteristic features of the E-field signals at the shock are indicated in the dynamic spectra on the right-hand side.

used by Rodriguez and Gurnett to demonstrate that the E to B ratio for the waves with $f < f_c^e$ is consistent with expectations for whistler mode turbulence.

The 0-600 eV ion measurements shown in Figure 2 were derived from the OGO-5 light ion spectrometer pointing 61° away from the sun, and the data thus provide relative information on thermalization and deflection of solar wind flow. The lowest panels contain

measurements from the OGO-5 ion plasma probe pointing directly toward the sun. Frequently, as discussed by Neugebauer,[25] it is possible to detect a shift toward lower streaming speed just before
the shock indicating that charge separation electric fields start
to decelerate the upstream ions; typically, as shown here, there
are also flux oscillations at the shock, and the local flux spikes
may be (2-4) times the upstream flux value.

The left-hand side of Figure 2 clearly shows that the solar
wind thermalization, deflection and flux compression started well
before the whistler mode turbulence was detected, and it is commonly agreed that electrostatic wave interactions provide the dominant dissipation mechanism for this type of laminar shock.

The right-hand side of Figure 2 shows some other aspects of
the typical bow shock electric field turbulence. The data from
the wideband telemetry link provided the E-field dynamic spectrum
in the central panel, and this f-t diagram illustrates the basic
spectral characteristic of the E-field turbulence associated with
the current-driven instabilities measured in space; in every case
the observed electrostatic waves consist of a number of impulsive
or rapidly rising noise bursts covering a relatively broad range
in frequency. The rest frame electron cyclotron, ion plasma and
Buneman mode frequencies associated with the upstream parameters
are shown next to the f-t diagram, and it can be seen that the measured bow shock turbulence covers a range extending down to $f_p{}^i$,
or below, and up to about f_B. In terms of $f_c{}^e$, this noise would
be related to relatively high harmonics (i.e., $nf_c{}^e$, with n up to
10 or more).

In the initial discussions of these bow shock measurements,
Fredricks et al.[23] interpreted the VLF electric field turbulence
in terms of ion waves associated with a two-stream instability.
The disparity between the measured wave frequencies and the anticipated range of 0 to $f_p{}^i$ (based on the upstream density value)
was noted and it was speculated that Doppler shifts play important
roles for ion sound waves. In fact, the flux spikes observed
within shock structures represent density enhancements, and the
local values of $f_p{}^i$ may exceed the upstream values by factors of
2 or more. However, there does appear to be some problem in explaining the observed E-field frequency range in terms of simple
linear ion sound waves, since Rodriguez and Gurnett[24] show that
the noise spectral density customarily remains quite high up to
about 2 kHz, while $f_p{}^i$(NOMINAL) $\simeq$ 500-700 Hz in the plasma rest
frame.

Another apparent difficulty with the ion acoustic two-stream
instability is related to observation of limited ranges for $\theta =$
T_e/T_i within the shock structure. Although Montgomery et al.[26]
did report a local θ-value as high as 9.3 for one shock crossing,

the ions generally appear to be heated much more readily than anticipated in the earlier papers, so that (on the basis of calculations by Fried and Gould[27]) electron-proton drift speeds many times
higher than the ion thermal speed are needed to provide an ion
sound instability.

For these reasons, other bow shock instability theories have
recently been discussed. As summarized by Greenstadt and
Fredricks,[28] various authors have invoked ion-ion streaming and
cyclotron drift instabilities in order to explain the E-field shock
turbulence, but there is little direct evidence in favor of these
interpretations. For instance, the very high resolution dynamic
spectrum at the top right-hand side of Figure 2 shows that the turbulence has a complex structure consisting of many single modes
with short duration, but there is no evidence indicating that these
tones are related to harmonics of the electron cyclotron frequency.

Mode identification can be achieved if precise polarization
measurements are made but, to date, no really conclusive measurements within the shock are available. OGO-5 was stabilized and
instantaneous polarization was not measured, while IMP-6,8 both
spin slowly in comparison with the time needed to transit through
a laminar shock. However, incomplete polarization studies reported
by Rodriguez and Gurnett[24] are consistent with detection of longitudinal electrostatic oscillations aligned parallel to the local
B-field.

One very important problem connected with analysis of the bow
shock plasma instability concerns the motion of the shock. Since
it is necessary to deduce the actual current profile from the field
variation measurements, it is also necessary to determine the true
shock thickness in the shock rest frame, and this requires use of
more than one spacecraft. In a few cases coordinated multispace-
craft bow shock observations have been made (see Greenstadt
et al.)[29] and Figure 3 contains the OGO-5 data, along with an outline of some preliminary analysis, for one bow shock crossing on
February 12, 1969. The lower left panel shows a very sharp magnetic gradient and a wideband (1 to 22 kHz) electric field profile that has a well-confined maximum centered within the B-field
shock ramp. At this time the upstream density was 2 cm^{-3}, the
wind speed was 418 km/sec, T_+ was $5.8 \times 10^{4}°K$, the Alfvén Mach number was near 2, and the interplanetary field angle with respect to
the shock normal was near 80°.

The upper left panel in Figure 3 shows the relative locations
of HEOS-1 and OGO-5 on February 12, 1969. Greenstadt et al.[29]
were able to determine the bow shock motions for a series of crossings by comparing data from the two spacecraft and for Case 3
(shown here) they found V = 11 km/sec, so that the shock rest frame
shock thickness and current magnitude could be evaluated, as indicated in the upper right side of Figure 3.

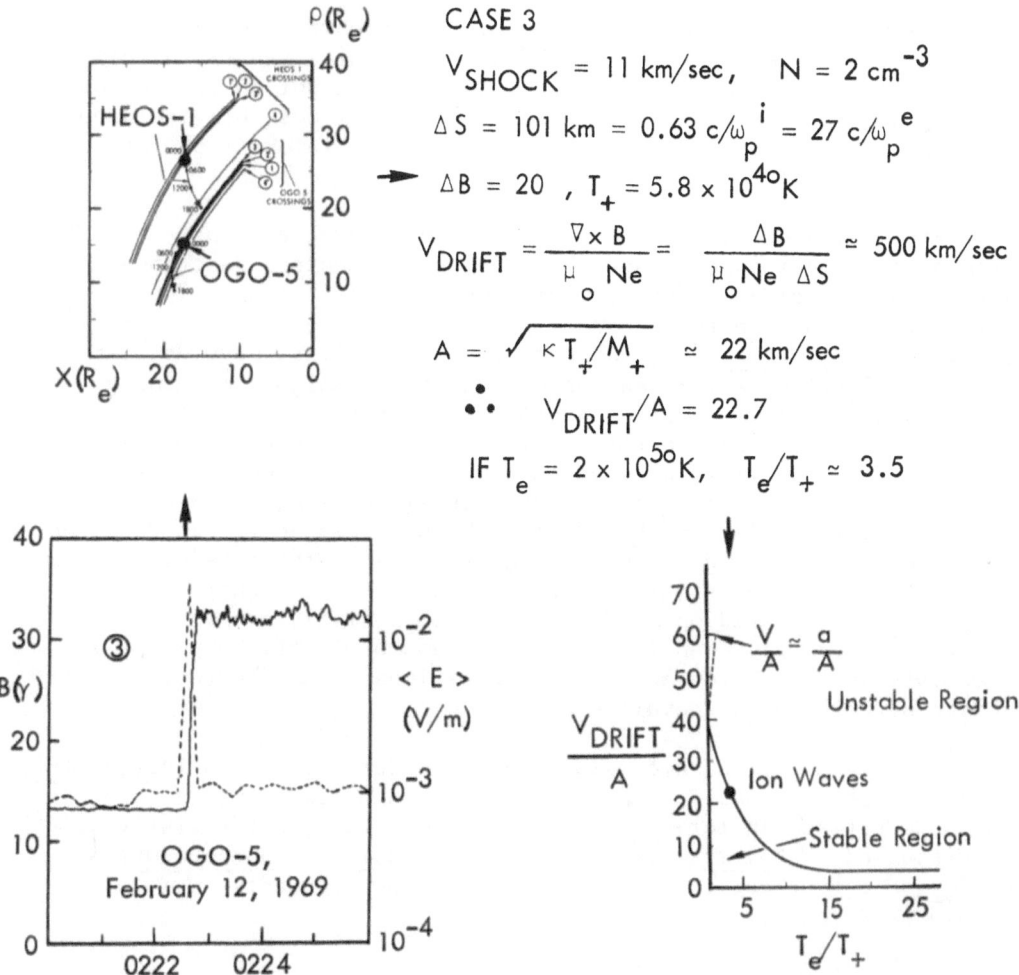

CASE 3

$$V_{SHOCK} = 11 \text{ km/sec}, \quad N = 2 \text{ cm}^{-3}$$

$$\Delta S = 101 \text{ km} = 0.63 \, c/\omega_p^{\,i} = 27 \, c/\omega_p^{\,e}$$

$$\Delta B = 20 \, , \, T_+ = 5.8 \times 10^{4 \circ}K$$

$$V_{DRIFT} = \frac{\nabla \times B}{\mu_o \, Ne} = \frac{\Delta B}{\mu_o \, Ne \, \Delta S} \approx 500 \text{ km/sec}$$

$$A = \sqrt{\kappa \, T_+/M_+} \approx 22 \text{ km/sec}$$

$$\therefore \quad V_{DRIFT}/A = 22.7$$

$$\text{IF } T_e = 2 \times 10^{5 \circ}K, \quad T_e/T_+ \approx 3.5$$

Figure 3. Instability analysis for a laminar shock crossing for which multi-spacecraft data are available to give information on the rest frame shock thickness.

        This analysis leads to a surprisingly high value for V(DRIFT)/A within the shock. Precise local measurements of T_e were not available on OGO-5, but even if a nominal value of $2 \times 10^{5}°K$ is used, the lower left panel shows that the ion wave possibility, based on the Fried and Gould[27] calculations, cannot be ruled out. If more realistic non-Maxwellian shock electron distribution functions having $T_e \approx 6\text{-}7 \times 10^{5}°K$ (see Montgomery et al.)[26] are considered, along with the effect of electron heat conduction (see Forslund)[30],

it is even easier to understand how ion sound waves can be driven
by currents in the shock layer. Of course, real advances in this
area of bow shock plasma instability analysis cannot be carried out
on the basis of chance multi-spacecraft observations, but the pend-
ing ISEE mission is designed to provide precise information for
this type of study.

Field-aligned currents are also of considerable importance
within the magnetosphere, and the OGO-5 experimenters described the
detection of field-aligned currents together with strong electrostatic turbulence in the polar cusp during a large storm (Fredricks
et al.[31], Scarf et al.[32]), and near the low altitude dayside cusp
boundary (Scarf et al.[33]) and inner edge of the nightside plasma
sheet (Scarf et al.[34]) during substorms. Figure 4 shows some of
the characteristic wave-particle-field relations observed within
the high latitude field-aligned current regions of the magnetosphere. It can be seen that strong impulsive rising E-field noise
bursts, quite similar to the spectra shown in Figure 2, are detected
near the field-aligned currents. However, as in the case of the bow
shock, there can be no unambiguous estimate of the magnetospheric
current density based on measurements from a single moving spacecraft, and a complete analysis of this type of important plasma instability must be deferred until observations from a suitable multi-
spacecraft mission are available.

The plasma turbulence associated with current-driven instabilities produces changes in the distribution functions, and the process
can be described in terms of an anomalous or turbulent resistivity.
As discussed by Kadomtsev,[35] Sagdeev,[36] Biskamp,[37] Papadopolous and
Coffey,[38] and others, the effective collision frequency, υ_{eff},
should be given by

$$\upsilon_{eff} \simeq \sqrt{\frac{\pi}{2}} \, \omega_p{}^e \, (\varepsilon_o E^2/2N\kappa T_e) \qquad (1)$$

and, in principle, the anomalous resistivity can be evaluated directly by measuring the local plasma wave turbulence amplitude and
the electron moments. For instance, with the shock of Figure 3,
υ_{eff} turns out to be approximately 30-60 collisions per second
using Equation (1) with $T_e = 2 \times 10^5 °K$, and the anomalous conductivity, $\sigma = Ne^2/m \, \upsilon_{eff}$, can account for the shock dissipation.

Recently Fredricks et al.[31] used wave and particle measurements from OGO-5 to argue that during storms anomalous resistivity
near the boundaries of the earth's polar cusp allows large scale
parallel electric fields to develop. Other aspects of the local
measurements are consistent with an anomalous resistivity interpretation. For instance, the electron spectra plots in the bottom
panel of Figure 4 show a depletion in low energy population and
appearance of suprathermal particles as the spacecraft encounters

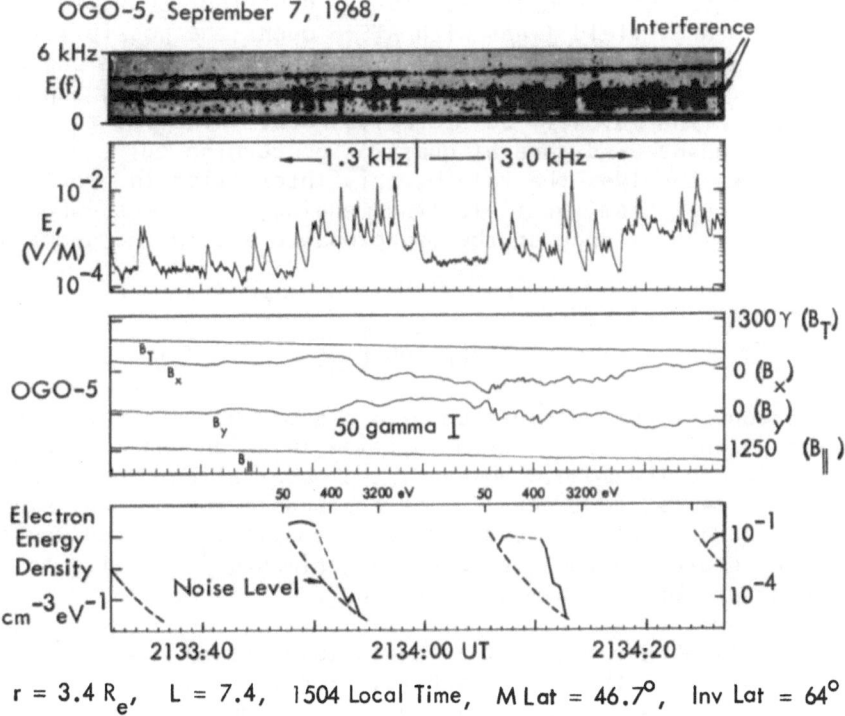

Figure 4. Observations of strong, low-altitude field-aligned currents, magnetospheric noise bursts that are typically associated with current-driven plasma instabilities, and simultaneous changes in the low energy electron distribution functions. These data, taken during a substorm period, can be interpreted in terms of local particle energization associated with the development of anomalous resistivity just beyond the trapping boundary.

the current system and its associated turbulence; these changes could be explained by resistive heating associated with the wave-particle interactions. However, there are still a fair number of important open questions concerning this microscopic process, particularly as it relates to the substorm observations of field-aligned current phenomena and the role of the plasma instability in regulating line-tying and magnetosphere-ionosphere coupling. One of the biggest gaps in this area involves the measurement of the complete particle distribution function in the appropriate

unstable current-carrying high latitude region; McIlwain's[1] description of the very complex equatorial distributions detected
on closed auroral field-lines with ATS-6 opens the possibility that
extremely non-Maxwellian distributions are present all along auroral
field-lines, and local acceleration, streaming, runaway, and similar
complex phenomena may well be operative here. Finally, the most
significant unanswered general question concerning current-driven
instabilities involves the role of this interaction in providing a
dissipation mechanism for field-line merging, a process expected to
be of great significance at the magnetopause and in the geomagnetic
tail.

6. MODE COUPLING

The conventional discussion of a plasma instability in space
involves essentially a linear dispersion relation and the concept
of a single growing mode. However, the observations show that such
simple treatments are generally not applicable for natural signals,
and the greatest value of space science to plasma physics may turn
out to be related to the spacecraft experimenter's ability to study
non-linear processes and mode coupling.

There is abundant direct evidence that strong wave-wave interactions develop in the magnetosphere. For instance, Coroniti and
Kennel[39] noted that chorus amplitudes are frequently modulated during micropulsation events, and they attempted to explain this in
terms of a micropulsation modulation of the particle distribution
that affects the instantaneous chorus growth rate. Helliwell[40]
recently summarized many of the other complex wave-wave interactions
that involve a variety of magnetospheric signals propagating in the
whistler mode, and he presented some of the current non-linear ideas
concerning feedback effects. There are a number of other prominent
cases where mode coupling effects are probably quite important; for
instance, the current-driven instabilities discussed in the previous
section clearly involve coupling of ULF electromagnetic pulses and
higher frequency electrostatic waves. This coupling persists even
when large amplitude MHD waves are found in the upstream solar wind
(Scarf et al.[41], Fredricks et al.[42]), and the non-linear wave-wave
interactions can lead to dissipation so that the magnetic pulses
steepen into collisionless shocks.

Other important wave-wave interactions are assumed to be operative because the two modes are generally observed together. The
whistler mode turbulence at the shock ramp and at magnetosheath
field gradients ("lion roars") appears to involve current-driven
modifications of electromagnetic growth mechanisms, and a similar
mechanism may be associated with the generation of the strong
whistler mode noise detected with IMP-7 near the neutral sheet in
the distant tail (Scarf et al.[43]).

The most striking example of a non-linear interaction in the earth's plasma environment is the wave-wave coupling that apparently leads to the remarkably intense kilometer-wave radiation from the regions above auroral arcs. Gurnett[44] showed that at peak intensity the total power radiated is about 10^9 watts. This is approximately one percent of the maximum energy dissipated by the precipitating auroral particles. It appears highly unlikely that any incoherent radiation mechanism can account for the tremendous power radiated above the auroral arcs, and it seems to be necessary to invoke mode coupling in which local plasma waves organize the radiating particles to provide a high degree of coherence; similar wave-wave interactions may well be operative at Jupiter.

7. OUTLOOK

During the past decade it has become clear that plasma instabilities and wave-particle interactions play a central role in the magnetosphere. Much progress has been made in identifying specific instabilities and in verifying that magnetospheric particles interact strongly with local plasma turbulence. However, certain very important problems (such as the merging dissipation problem and the question of the acceleration of O_2^+ ions to kilovolt energies) are still essentially completely open, with respect to local observations. In dealing with phenomena involving runaway and the interrelation of field-aligned currents, parallel electric fields and anomalous conductivity, the observational status is still highly preliminary. It seems clear that much better measurements of magnetospheric plasma distribution functions are needed, and it is also clear that multi-satellite missions are required to solve many of the outstanding problems.

ACKNOWLEDGMENTS

I wish to thank Drs. F. V. Coroniti, R. W. Fredricks, D. A. Gurnett, C. T. Russell, and D. J. Williams for helpful discussions. This work was supported by the National Aeronautics and Space Administration under Contract NASW-2659.

REFERENCES

1. McIlwain, C. E., This volume.
2. Dungey, J. W., Plan. Space Sci., 11, 591, 1963.
3. Cornwall, J. M., J. Geophys. Res., 69, 1251, 1964.
4. Kennel, C. F., and H. Petschek, J. Geophys. Res., 71, 1, 1966.
5. Fredricks, R. W., Wave-Particle Interactions in the Outer Magnetosphere, Magnetospheres of Earth and Jupiter (Edited by V. Formisano), D. Reidel, Dordrecht-Holland, in press.
6. Rosenberg, T. J., R. A. Helliwell and J. P. Katsufrakis, J. Geophys. Res., 76, 8445, 1971.

7. Kivelson, M. G., and C. T. Russell, Radio Sci., **8**, 1035, 1973.
8. Kivelson, M. G., T. A. Farley and M. P. Aubry, J. Geophys. Res., **78**, 3079, 1973.
9. Scarf, F. L., R. W. Fredricks, C. F. Kennel, and F. V. Coroniti, J. Geophys. Res., **78**, 3119, 1973.
10. Lyons, L. R., R. M. Thorne and C. F. Kennel, J. Geophys. Res., **77**, 3455, 1972.
11. Lyons, L. R., and R. M. Thorne, J. Geophys. Res., **78**, 2142, 1973.
12. Etcheto, J. R., Gendrin, J. Solomon and A. Roux, J. Geophys. Res., **78**, 8150, 1973.
13. Cornwall, J. M., F. V. Coroniti and R. M. Thorne, J. Geophys. Res., **75**, 4699, 1970.
14. Williams, D. J., This volume.
15. Taylor, W. L., B. Parady and L. J. Cahill, Jr., Univ. of Minnesota, Tech. Report, June 1974.
16. Kennel, C. F., F. L. Scarf, R. W. Fredricks, J. H. McGehee, and F. V. Coroniti, J. Geophys. Res., **75**, 6136, 1970.
17. Fredricks, R. W., and F. L. Scarf, J. Geophys. Res., **78**, 310, 1973.
18. Gurnett, D. A., and R. R. Shaw, J. Geophys. Res., **78**, 8136, 1973.
19. Lyons, L. R., J. Geophys. Res., **79**, 575, 1974.
20. Gurnett, D. A., and L. A. Frank, J. Geophys. Res., **79**, 2355, 1974.
21. Coroniti, F. V., R. W. Fredricks and R. White, J. Geophys. Res., **77**, 6243, 1972.
22. Anderson, R. R., and D. A. Gurnett, J. Geophys. Res., **78**, 4786, 1973.
23. Fredricks, R. W., G. M. Crook, C. F. Kennel, I. M. Green, F. L. Scarf, P. J. Coleman, and C. T. Russell, J. Geophys. Res., **75**, 3751, 1970.
24. Rodriguez, P., and D. A. Gurnett, J. Geophys. Res., **80**, 19, 1975.
25. Neugebauer, M., J. Geophys. Res., **75**, 717, 1970.
26. Montgomery, M. D., J. R. Asbridge and S. J. Bame, J. Geophys. Res., **75**, 1217, 1970.
27. Fried, B. D., and R. W. Gould, Phys. Fluids, **4**, 139, 1961.
28. Greenstadt, E. W., and R. W. Fredricks, Magnetospheric Physics (Edited by B. M. McCormac), D. Reidel Publishing Company, Dordrecht, Holland, 281, 1974.
29. Greenstadt, E. W., C. T. Russell, F. L. Scarf, V. Formisano, and M. Neugebauer, J. Geophys. Res., **80**, 502, 1975.
30. Forslund, D. W., J. Geophys. Res., **75**, 17, 1970.
31. Fredricks, R. W., F. L. Scarf and C. T. Russell, J. Geophys. Res., **78**, 2133, 1973.
32. Scarf, F. L., R. W. Fredricks, I. M. Green, and C. T. Russell, J. Geophys. Res., **77**, 2274, 1972.

33. Scarf, F. L., R. W. Fredricks, C. T. Russell, M. Neugebauer, M. Kivelson, and C. R. Chappell, J. Geophys. Res., 80, in press, 1975.

34. Scarf, F. L., R. W. Fredricks, C. T. Russell, M. Kivelson, M. Neugebauer, and C. R. Chappell, J. Geophys. Res., 78, 2150, 19 1973.

35. Kadomtsev, B. B., Plasma Turbulence, Acad. Press, London, N. Y., 1965.

36. Sagdeev, R. Z., Proc. Symp. Applied Math., 18, 281, 1967.

37. Biskamp, D., Nucl. Fusion, 13, 719, 1973.

38. Papadopoulos, K., and T. Coffey, J. Geophys. Res., 79, 1558, 1974.

39. Coroniti, F. V., and C. F. Kennel, J. Geophys. Res., 75, 1279, 1970.

40. Helliwell, R. A., Space Sci. Rev., 15, 781, 1974.

41. Scarf, F. L., R. W. Fredricks, L. A. Frank, C. T. Russell, P. J. Coleman, Jr., and M. Neugebauer, J. Geophys. Res., 75, 7316, 1970.

42. Fredricks, R. W., F. L. Scarf, C. T. Russell, and M. Neugebauer, J. Geophys. Res., 77, 3598, 1972.

43. Scarf, F. L., L. A. Frank, K. Ackerson and R. Lepping, Geophys. Res. Lett's., 1, 189, 1974.

44. Gurnett, D. A., J. Geophys. Res., 79, 4227, 1974.

SOME EXPERIMENTALLY DETERMINED CHARACTERISTICS

OF THE TURBULENCE IN THE MAGNETOSPHERE

Bengt Hultqvist

Kiruna Geophysical Institute

S-981 01 Kiruna 1, Sweden

INTRODUCTION

The magnetospheric turbulence - by which we here mean all processes which give rise to scattering in configuration or velocity
space of particles and waves - affects the hot plasma
strongly in several ways: it redistributes energy between the
particles themselves and between particles and waves, it redistributes the hot plasma spatially by giving rise to diffusion and it
redistributes the particles in pitch angle. The last mentioned
effect is in many ways the most important one. This is because
the pitch angle diffusion probably is the main loss mechanism for
most of the hot plasma in the magnetosphere. The particles which
are scattered into the loss cone disappear into the atmosphere
and produce excitations, ionizations and heat. Although acceleration processes near the atmosphere certainly play a major role for
the production of particle precipitation in discrete auroral forms
(see e g Evans in this volume), most of the worldwide loss of hot
plasma to the atmosphere is probably caused by pitch angle diffusion.

The redistribution of energy within the hot plasma has the
interesting characteristic that it creates a high-energy tail in
the distribution function. It seems likely that the particles
with energies of one hundred to several hundred keV, which appear
so quickly in the inner magnetosphere during substorms and which
are so difficult to understand in terms of adiabatic acceleration
processes, constitute the high-energy tail produced by energy
diffusion in turbulent regions of the magnetosphere.

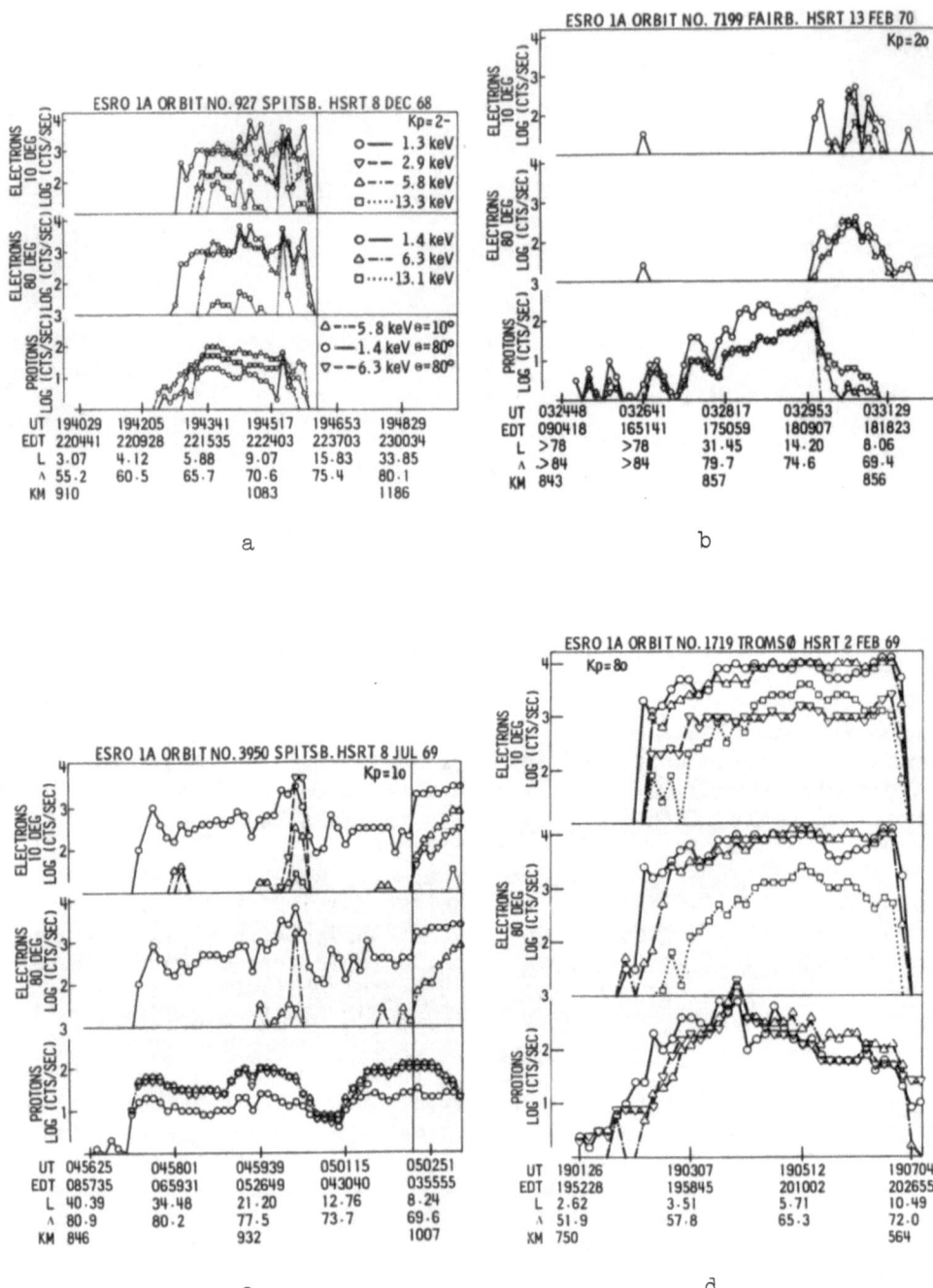

Figure 1 (caption on p. 293)

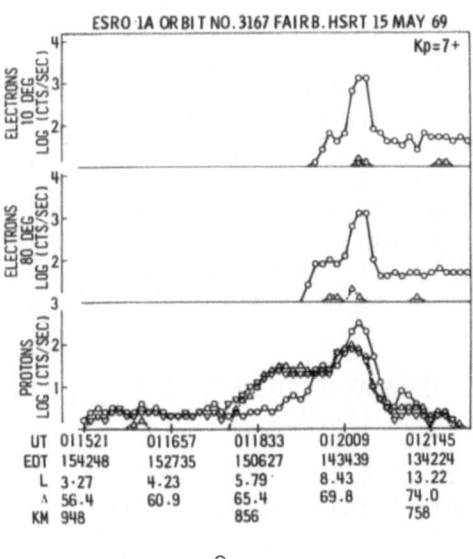

Figure 1. KeV electron and ion data from ESRO 1A passes in different local time sectors and at different disturbance levels. Each data point represents measurements during 8 sec. The center energy and approximate pitch angle of the various detectors are shown in a). EDT means excentric dipole time, Λ is the invariant latitude and the satellite altitude is given in km. The straight vertical lines over some of the diagrams indicate the location of a trapping boundary for >40 keV electrons as determined on ESRO 1A by Page and Shaw (1972). The factors for conversion from count rates to differential fluxes (dimension: part/cm^2 s sr keV per counts/s) have the following values for the various detectors:

10° pitch angle		80° pitch angle	
ions	5.8 keV: $0.37 \cdot 10^4$	ions	1.4 keV: $13.5 \cdot 10^4$
electrons	1.3 keV: $12.7 \cdot 10^4$	ions	6.3 keV: $0.74 \cdot 10^4$
electrons	2.9 keV: $50 \cdot 10^4$	electrons	1.4 keV: $10.6 \cdot 10^4$
electrons	5.8 keV: $0.48 \cdot 10^4$	electrons	6.3 keV: $0.94 \cdot 10^4$
electrons	13.3 keV: $0.50 \cdot 10^4$	electrons	13.1 keV: $4.5 \cdot 10^4$

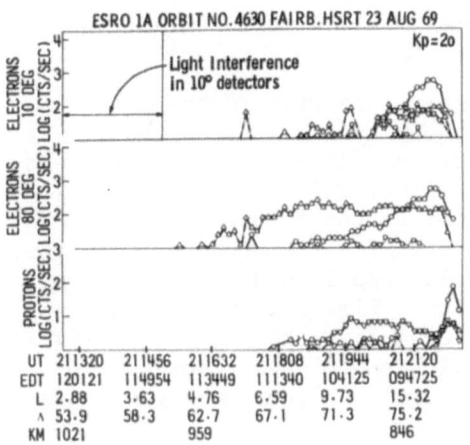

Figure 2. KeV electron and ion data from an ESRO 1A pass through the forenoon sector showing anisotropic electron and ion pitch angle distributions. For the meaning of the various symbols and the values of the conversion factors see Figure 1.

In the present report we will concentrate on what observations of the hot plasma in and near the loss cone may tell us about the turbulence. It is thus mainly the turbulence effects on the pitch angle distribution that will be discussed and the reason for this choice is simply that it is the easiest part of the problem from an experimental point of view. Most of the observations dealt with below were made by means of the ESRO 1A and B satellites. The reader is referred to Riedler et al. (1971) for a description of the satellites and their hot plasma experiments and to Bernstein et al. (1974) and Hultqvist et al. (1974) for more complete data presentations and discussions of several of the subjects treated here.

ON THE OCCURRENCE OF STRONG PITCH ANGLE DIFFUSION IN THE HOT MAGNETOSPHERIC PLASMA

The dominating characteristic of the keV ion pitch angle distribution, as indicated by the near 0° and near 90° pitch angle fluxes measured by ESRO 1A and B, may be expressed as follows: the ion pitch angle distribution is roughly isotropic (within a factor of two or three) in most of the particle precipitation zone. This is generally true for all local times and at all disturbance levels - from completely quiet to strongly disturbed conditions.

At the equatorward edge of the precipitation zone the loss cone is strongly depleted. At the poleward edge, the distribution is more variable. Among other things, the field-aligned distributions discussed earlier by Hultqvist et al. (1971) and Hultqvist (1971) are most commonly seen there on the night side of the earth. Figure 1 shows a number of examples of latitude profiles at approximately 10° and 80° pitch angle for keV electrons and ions as measured in the upper ionosphere by ESRO 1A at various local times and different disturbance levels (after Hultqvist et al., 1974). Only in the central parts of the dayside have exceptions from roughly isotropic ion pitch angle distributions been seen. An example is shown in Figure 2.

The keV electrons are also roughly isotropic in the main part of the precipitation zone, as shown by e g ESRO 1A (Hultqvist et al., 1974). There is, however, more variability in this respect in the keV electrons than in the keV ions. This is illustrated in Figures 1b and 2 and also in Figure 3. As can be seen there the structure in the 10° electron fluxes is most pronounced at the highest energy (13 keV). On the dayside the loss cone is found to be depleted of keV electrons most of the time at low and medium K_p values in the entire region equatorward of the cusp. The roughly isotropic regions are mostly those with the highest fluxes. The flux dependence is much more pronounced for higher energy electrons, as shown already by O'Brien (1964).

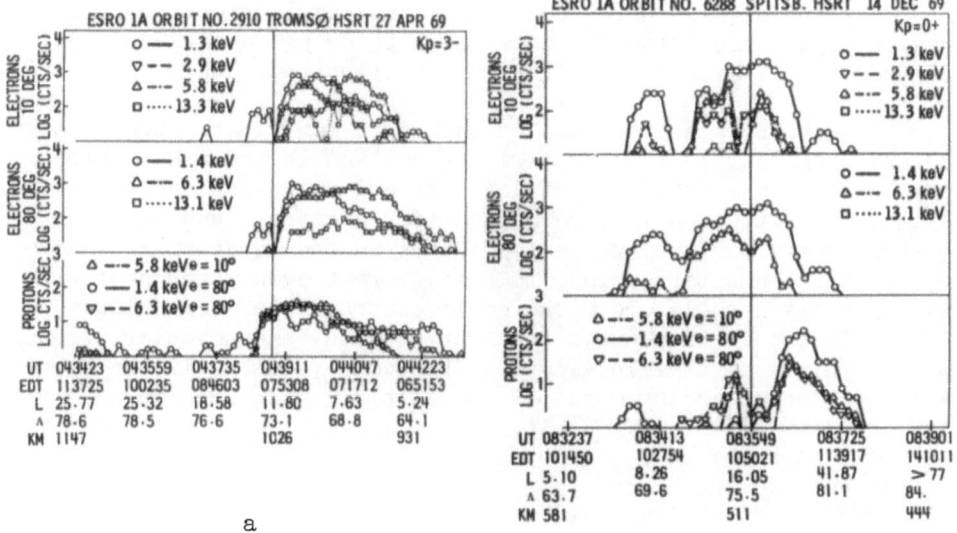

a

b

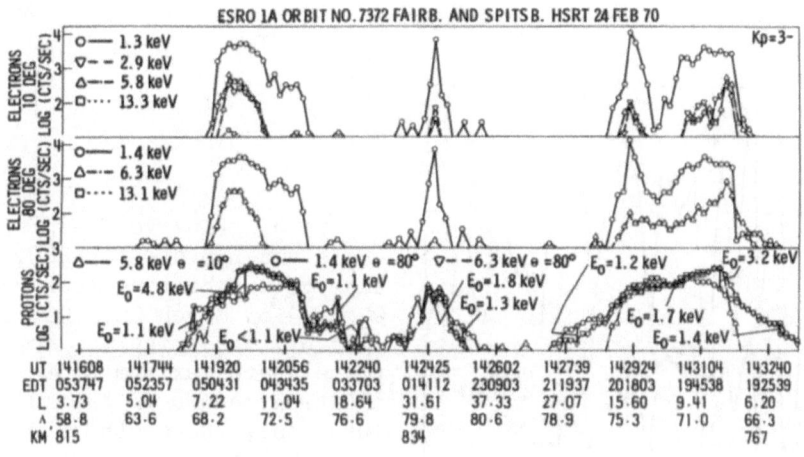

c

Figure 3. KeV electron and ion data from a number of ESRO 1A passes showing regions with anisotropic electron pitch angle distributions even at the lowest energies. About the vertical straight lines in a) and b) see caption of Figure 1. Factors for converting from count rates to differential fluxes can also be found in the caption of Figure 1.

For the higher energy ions (>100 keV) an isotropic central region in the precipitation zone seems to be permanently present (Søraas, 1972, Mizera, 1974). Figure 4 illustrates this.

At keV energies the roughly isotropic ion fluxes appear to be quite similar in the equatorial plane and in the upper ionosphere along a magnetic field line (cf e g Hultqvist et al., 1974). The situation is more unclear for electrons and also for higher energy protons, where the fluxes in the equatorial plane and in the ionosphere appear to be different even when the low-altitude pitch angle distribution is isotropic (cf e g Hultqvist, 1975). It should be pointed out that no simultaneous observations at two points of a field line have been made, but still the above statement about the keV protons seems fairly well founded.

The isotropic keV ion and electron precipitation takes place mainly outside the plasmasphere in a wide region. As illustrated by e g Figures 1a, b and 3a, c the roughly isotropic proton region generally does not reach lower L values than 5 and mostly extends upward to above L = 10 during undisturbed and moderately disturbed conditions. Figure 5 shows the dependence of the latitudinal location (L) of the point where the 6 keV precipitated (10° pitch angle) proton flux, as observed by ESRO 1A, falls below the detection threshold on geomagnetic activity (K_p). Also shown are plasmapause locations determined by Chappell et al. (1970) with OGO 5.

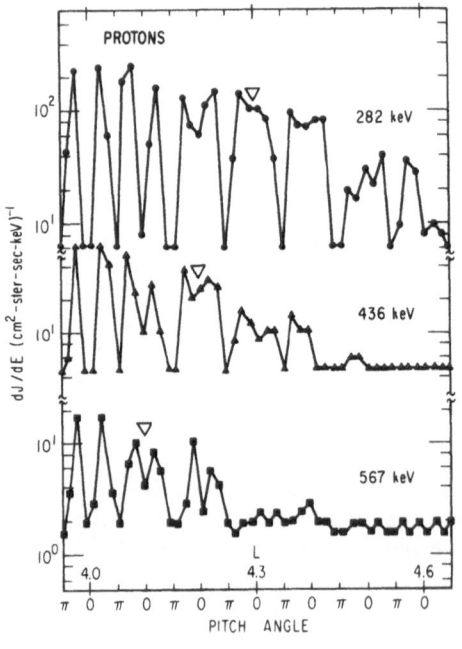

Figure 4. Flux versus pitch angle and L plots for 282, 436 and 567 keV protons measured by OV 1-19 on March 20, 1969 at 0625 UT. The triangles mark the place where maximum proton precipitation occurs in the 100 km loss cone, the width of which is illustrated by the width of the symbols (after Mizera, 1974).

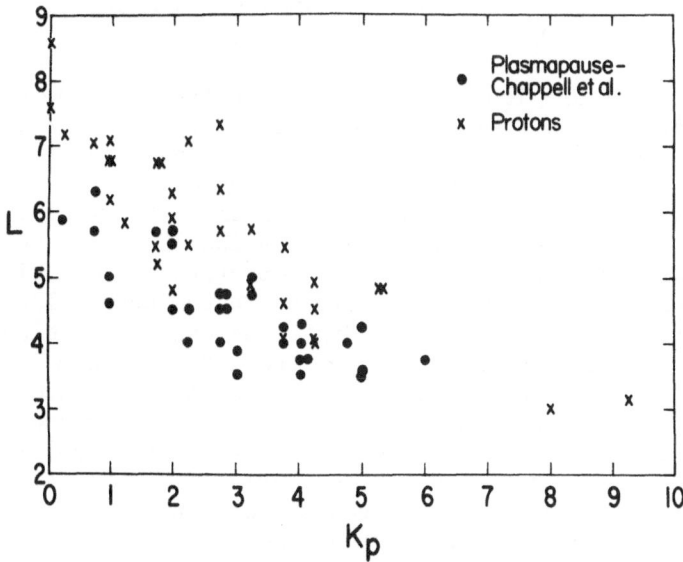

Figure 5. Dependence of the latitudinal location (L) of the
point where the 6 keV precipitated (10° pitch angle) proton flux
falls below the detection threshold on geomagnetic activity (K_p).
Also shown are plasmapause locations determined by Chappell et al.
(1970) for equivalent geomagnetic conditions (after Hultqvist et
al.,1974).

The proton observations were made in the evening sector whereas
the plasmapause determinations were made in the early morning
sector. If these differences in local time are taken into account,
the two sets of points will coincide still better. In spite of the
fact that the two kinds of data originate in different spacecraft,
they clearly indicate that the keV proton precipitation takes
place outside the plasmasphere. There are no systematic indications
of an intensification of the precipitation just inside the plasmapause as suggested by e g Cornwall et al. (1971) on theoretical
grounds. Satellite observations of >100 keV protons by Søraas
(1972), Mizera (1974) and Søraas and Berg (1974) show that the
latitude of maximum precipitated flux and of the low latitude
limit of the isotropic zone in general are lower in L the higher
the energy.

COINCIDENCE OF REGIONS OF STRONG PITCH ANGLE
DIFFUSION FOR KEV ELECTRONS AND IONS

KeV electrons and ions precipitate into the atmosphere in
closely coinciding zones except sometimes in the noon and afternoon sectors, where keV ions but no keV electrons often are found.
At the edges of the precipitation zone one species or the other
frequently is found to extend farther than the other (Hultqvist
et al., 1974). In the polar cusp (cleft) regions ESRO 1A and B
mostly saw only protons as they had no detectors for energies
below 1 keV where the electrons are found in the cusps. The
characteristics of the latitudinal distributions of keV electrons
and ions in the noon sector - i e with the keV ions reaching
several latitude degrees farther towards the pole than the keV
electrons - is found also at the flanks of the magnetosphere with
diminishing difference between ion and electron latitude profiles
the farther from noon one observes (Hultqvist et al., 1974;
Hultqvist, 1975). An extreme case can be seen in Figure 1b. But
even when such differences are present the conclusion that keV
electrons and ions are precipitated in roughly coinciding regions
is generally valid. Exceptions from this are found, apart from in the
cusps, over the polar caps, where soft electron precipitation, "polar
rain", seems to occur mostly, without any measurable fluxes of precipitated keV ions (Hultqvist et al., 1974), and in the afternoon,
where "pure" ion precipitation without any measurable keV electron
fluxes is sometimes observed at auroral latitudes (Hultqvist et al.,
1974). The polar cap precipitation will not be dealt with in this
review, but the "pure" proton pitch angle diffusion will be discussed
somewhat later in this section as well as in the Discussion section.
The coincidence of keV electron and ion zones is illustrated by the
examples of ESRO 1A passes shown in the previous section (Figures 1-
3). Therefore, only two additional ESRO 1A passes showing more or
less identical regions with measurable fluxes of keV electrons and
ions in the upper ionosphere are reproduced here (Figure 6).
Figure 6a has been included because of the very good coincidence of
the boundaries of the main electron and ion regions. Poleward of
the >40 keV trapping boundary there are no measurable fluxes of
keV particles. In Figure 6b, on the contrary, there are higher keV
electron fluxes poleward of the trapping boundary than equatorward
of it, while the keV ions show the opposite relation. But still the
keV ions and electrons are found in closely coinciding regions.

Thus, all examples of ESRO 1A passes shown hitherto give support
to the hypothesis that keV electrons and ions are affected by
pitch angle diffusion in roughly identical regions of the magnetosphere. However, in the afternoon sector there are frequently no
measurable fluxes of keV electrons in the upper ionosphere but there
are always keV ions (Hultqvist et al., 1974). Examples of observations

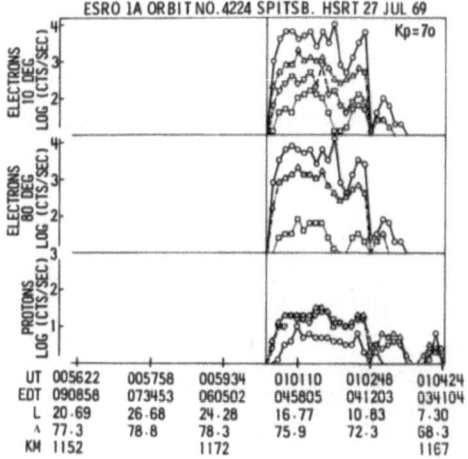

Figure 6. KeV electron and ion data from ESRO 1A passes illustrating the coincidence of the regions in which keV electrons and ions are found in the upper ionosphere. For the meaning of the symbols as well as about the vertical straight lines and the values of the factors for conversion from count rate to differential flux the reader is referred to Figure 1.

a

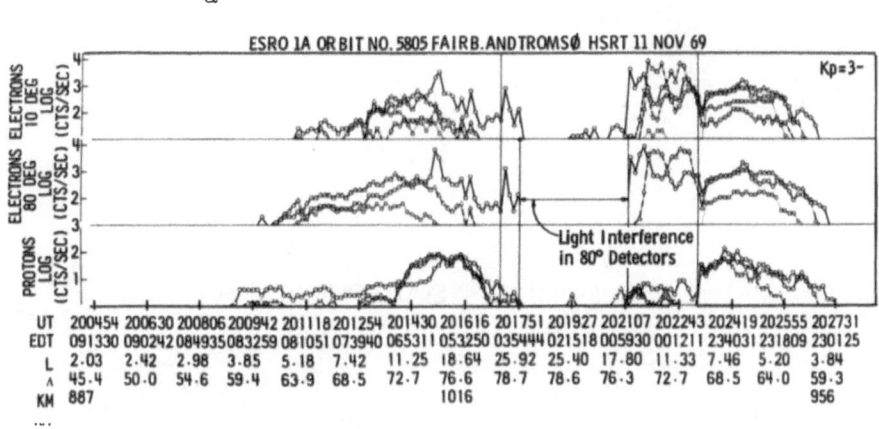

b

of such "pure" ion precipitation are shown in Figure 7. We will discuss their interpretation later.

RELATIONS BETWEEN SPATIAL/TEMPORAL VARIATIONS IN THE KEV ION AND ELECTRON PRECIPITATION RATES

The relations between spatial/temporal variations in the precipitation rates of keV electrons and ions are extremely variable. This is illustrated by several passes in Figures 1, 3 and 6 and by those in Figure 8. Figure 1a shows at UT ~194530 a flux peak for both the electrons and the ions, but whereas the peak is most pronounced for the highest electron energies the lowest ion channel

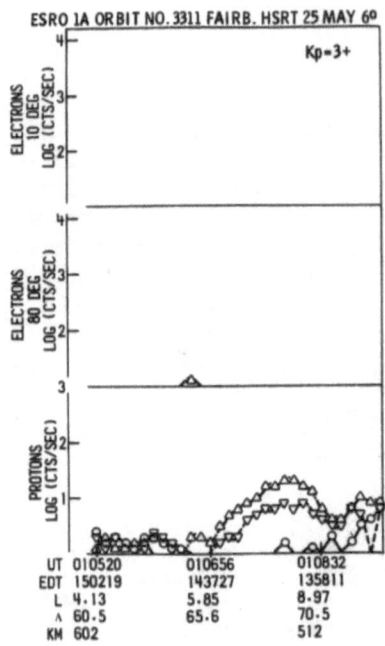

a

Figure 7.   ESRO 1A passes through
the afternoon sector illustrating
"pure" ion precipitation. About
the vertical straight lines and
values of conversion factors, see
caption of Figure 1.

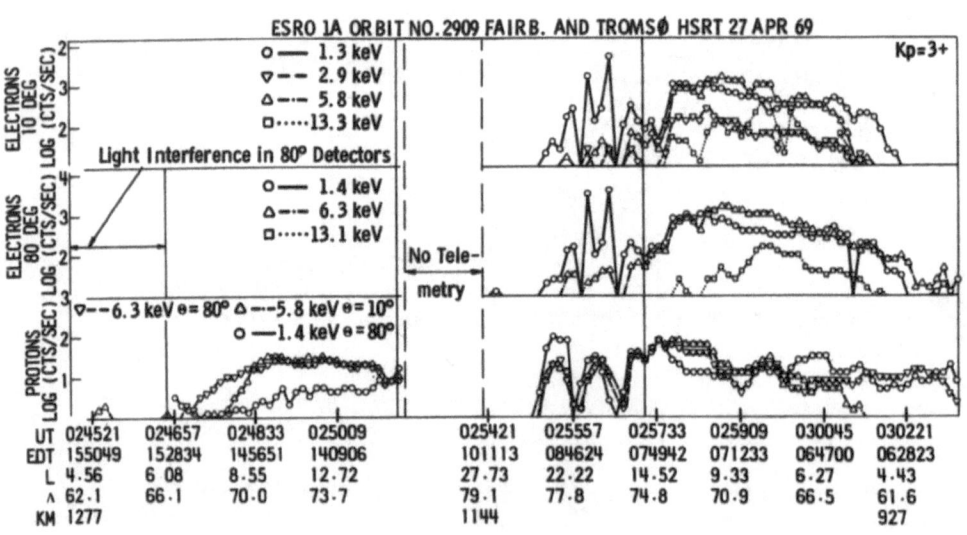

b

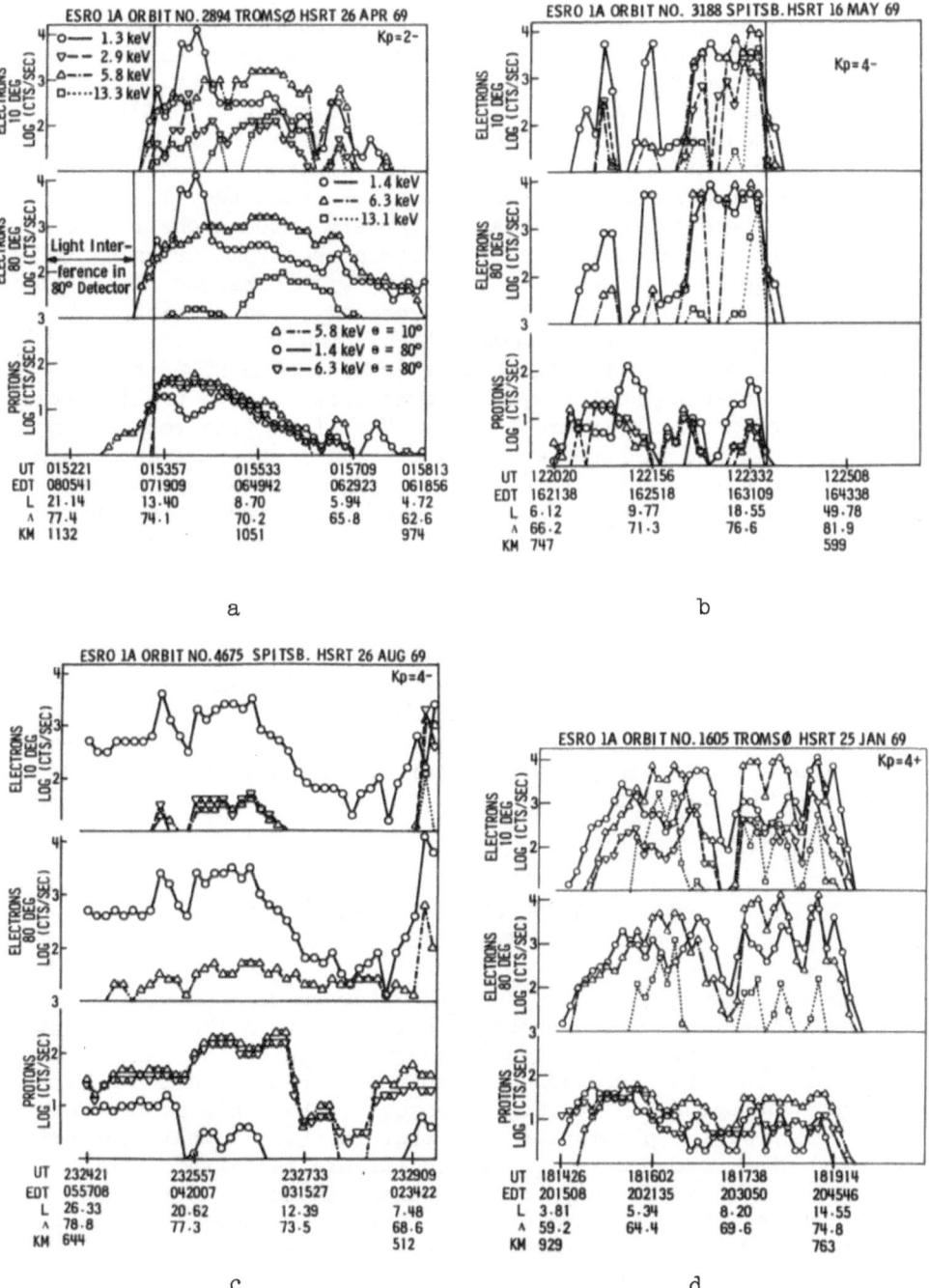

Figure 8.   ESRO 1A passes illustrating various relations between flux variations in keV electrons and ions. About the vertical straight lines and the values of the conversion factors see caption of Figure 1.

shows the strongest enhancement. Figure 8a contains a region of
enhanced low energy electron flux coincident with diminished ion
flux at the lowest energy. Figure 8b shows exactly the opposite
relation between 1 keV electron and ion fluxes at UT ~122130.
Figure 8c shows at about UT 2327 an enhancement of the keV ion
fluxes coincident with a decrese of the electron fluxes. Whereas
the ion spectrum hardens in the 1 to 6 keV range the electron
spectrum softens there. This case is a very unusual one in the
ESRO 1A data. The association of electron spectrum hardening in
the keV range and ion spectrum softening with enhanced fluxes of
both ions and electrons is the most commonly seen combination,
(although any combination is possible to find in the ESRO 1 data).
This is also observed in Figure 8d which shows an ESRO 1A pass
through the evening sector with several peaks in the precipitated
fluxes which most likely are inverted V's. (The time resolution
of the ESRO 1 plasma experiment was generally not fast enough to
resolve the narrower inverted V's). In Figure 8d an exact correspondence of enhancements of 13 keV electrons, indicating the
center of the inverted V's, and of 1 keV ions can be seen. In the
most equatorward broad region with an inverted-V-like structure
there is also a small peak in the 1 keV ion flux (at UT ~181620).

A further conclusion that the ESRO 1A data invite you to draw
is that the rate of flux variation in a pass is generally greater
in the keV electrons than in the keV ions. This is illustrated in
most of the passes shown above. Figure 8c (at UT ~2327) is a rare
exception to that rule. We will return to the interpretation of
these observations later.

PRECIPITATION OF RING CURRENT IONS
WITHIN THE PLASMASPHERE

We have concluded above that the ring current particles
penetrate into the plasmasphere to some extent in the equatorial
plane, but precipitation takes place only outside the plasmapause.
This means that the loss cone is empty within the plasmapause.
Williams and Lyons (1974a) found, however, that this is true only
in a zone near the plasmapause and that farther in weak scattering
into the loss cone occurs. Their results are illustrated in
Figure 6 of Williams (this volume), which demonstrates the L
dependence of the pitch angle distribution for energies between 1
and 390 keV in the early recovery phase of a magnetic storm. The
matrix shows a flat pitch angle distribution for the higher L
values. Explorer 45 had fields of view of the particle detectors
which were wider than the loss cone. At the lowest pitch angle the
measured flux was generally so much lower in the region with flat
pitch angle distribution that it was consistent with an empty loss

cone. At a certain L value there is a rapid transition from a flat to a rounded pitch angle distribution. The latter is formed by scattering particles near the loss cone into it. The transition from flat to rounded distribution takes place at higher L value the higher the energy.

In the region with rounded pitch angle distribution deep in the plasmasphere we expect to see ring current protons in the loss cone and, in fact, we do (Hultqvist et al., 1974b). Figure 9 shows the latitudinal distribution of keV protons of 10° and 80° pitch angles measured by ESRO 1A at an altitude of about 1400 km in the late evening. The magnetosphere was almost completely quiet (K_p = 1$_+$, Dst = -3γ) when the measurements were made. Figure 9 shows the isotropic precipitation zone outside the plasmasphere, which on this occasion was unusually narrow, between L = 5.9 and 8.7. Equatorward of it there is a region with highly anisotropic proton pitch angle distribution (no protons seen in the loss cone). At L = 4.2 loss cone protons appear again and they can be seen to below L = 3. In the same region the 80° fluxes are enhanced. It may be pertinent to state here that the count rates in this region in our view cannot be due to highly penetrating MeV electrons.

The observations in the upper ionosphere illustrated in Figure 9 seem to correspond very well to the observations in the equatorial plane in the previous figure, although the two sets of data refer to quite different levels of disturbance: next to the main precipitation zone there is a region with empty loss cone and farther in a region with weak pitch angle diffusion is found.

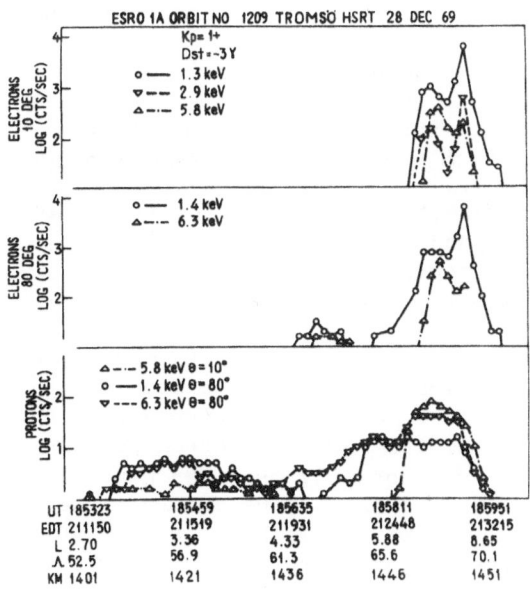

Figure 9. An example of latitudinal profiles for keV proton and electron fluxes observed in the upper ionosphere with a highly anisotropic zone equatorward of the main precipitation zone (after Hultqvist et al., 1974b).

    ESRO 1A, B data from sufficiently low subauroral latitudes
were obtained only in a limited number of passes. The available
information is, however, sufficient to conclude that this kind
of subauroral precipitation occurs most of the time. The precipitation below L = 4 has been seen in all·local time sectors. The
locations of the boundaries between the isotropic zone and the
highly anisotropic region equatorward of it, between the highly
anisotropic and the moderately anisotropic zones and the equatorward boundary of the region with weak pitch angle diffusion have
been found to depend on the magnetic disturbance level in different
ways. Whereas the low L boundary of the isotropic zone follows the
plasmapause inward with increasing K_p (Bernstein et al.,1974;
Hultqvist et al., 1974), the boundaries of the plasmaspheric region
with weak pitch angle diffusion are not influenced by moderate
magnetic activity. Magnetic storms do, however, move also these
boundaries, but ESRO 1A, B have not provided any data of this kind
from the nightside for magnetic storm periods. The magnetic storm
data from the dayside is also very limited. It indicates, however,
that during storms the fluxes increase very much both in the intermediate zone with empty loss cone and in the most equatorward
region with weak diffusion. The interpretation of the weak pitch
angle diffusion will be discussed later.

                        DISCUSSION OF THE TURBULENCE

    A number of conclusions about the characteristics of the
magnetospheric turbulence outside and inside of the plasmapause
may be drawn from the observational results described above.
Some conclusions are first listed below and are then discussed
one by one. The turbulence characteristics we want to emphasize here
are the following:
  - The hot plasma is permanently in a turbulent state in a wide
    region outside the plasmapause.
  - The turbulence outside the plasmapause seems to be due to a
    complex set of instabilities which are coupled in complex and
    variable ways.
  - Cold plasma appears to inhibit the turbulence outside the plasma-
    pause rather than.to enhance it and to determine thereby the
    inner boundary of the turbulent region.
  - The turbulence which isotropizes the keV ions appears to be
    located near the equatorial plane.
  - The ion cyclotron instability is not likely to be of importance
    for the strong pitch angle diffusion of keV ions outside the
    plasmapause. Loss cone instabilities may be of importance.
  - The turbulence may be the cause of the harder electron spectrum
    and softer ion spectrum which are frequently observed near the
    atmosphere in regions of flux enhancement in the auroral zone.
  - The observed "pure" keV ion precipitation (without measurable
    fluxes of keV electrons) observed equatorward of the polar cusps

in the afternoon sector and sometimes in the noon sector appears
to be due to the absence of electrons of keV energies in this
part of the magnetosphere.
The observations of keV ion pitch angle diffusion within the
plasmapause are consistent with the hypothesis that the ion
cyclotron instability causes the diffusion in the plasmasphere.

Permanent Turbulent State:

A fairly well filled loss cone is found in a wide region outside the plasmapause at all local times irrespective of the disturbance level of the magnetosphere. It is characteristic of the
central part of the precipitation zone in completely quiet as well
as in highly disturbed situations. This is true for ions of energies
up to at least several hundred keV but for electrons mostly only
below 10 keV of energy, with the loss cone more completely filled
with electrons the lower the energy and the higher the fluxes. The
diffusion process thus seems to be velocity dependent rather than
energy dependent.

The persistence of the strong pitch angle diffusion indicates
that the turbulence is due to some basic, always present characteristics of the magnetosphere in a region outside the plasmasphere.
As soon as plasma is brought there by the influence of the
magnetic and electric fields instabilities are excited. A process
which has such a permanence is the increase of the anisotropy of
the hot plasma when it moves towards the earth and there may be
others too.

Before leaving this subject it should be emphasized that on the
dayside the pitch angle diffusion of keV ions and electrons quite
frequently is weak instead of strong equatorward of the cusp and
that the weak diffusion on the dayside is more frequently seen in
the keV electrons than in the ions. It should also be remembered
that the roughly isotropic pitch angle distribution is characteristic of the major part of the zone outside the plasmasphere in
which keV particles reach close to the atmosphere but never for
the equatorward edge of this zone, i e closest to the plasmapause,
where the loss cone is strongly depleted. In addition, field aligned
pitch angle distributions of both electrons and ions are seen
fairly frequently in limited regions, indicating the existence of
acceleration processes near the atmosphere in addition to the
equatorial plane processes (see e g Evans in this volume concerning
field aligned electrons and Hultqvist et al., 1971, and Hultqvist,
1971, about field aligned ions). In this report only processes which
scatter particles into the loss cone are discussed, however.

A Complex Set of Scattering Processes:

The observational result that the strong pitch angle diffusion
regions largely coincide for ions and electrons outside the plasmapause, in combination with the observed wide variety of relations
between spatial/temporal variations (one spacecraft cannot separate
spatial and temporal variations) of keV ion and electron characteristics indicate that no single physical process that affects only a
small number of variables of the magnetospheric plasma is the
cause of the pitch angle diffusion. Instead, a complex set of
physical processes seems to affect simultaneously both ions of
energies up to hundreds of keV and keV electrons. The processes
affecting ions and electrons are strongly coupled but appear to
be so in complex and variable ways. In general, however, the 6 keV
ion flux variations seem to be more closely related to the structure
in the 1 keV electron flux than to the 6 keV electron variations.

Swift (1968, 1970) has suggested that one single process, an
electrostatic ion loss cone instability associated with ring
current protons, may precipitate both protons and electrons, the
former by pitch angle scattering in the interaction between the
protons and the low frequency waves and the latter through acceleration of the electrons along the magnetic field lines by the waves,
which have a phase normal almost perpendicular to the magnetic
field lines and therefore a high phase velocity along the field
lines (assuming quasiplane waves). In Swift's model the instability
is located in the upper ionosphere. That this is not the case can
be concluded from the observation that the height variation of
proton fluxes between 1450 and 250 km altitude is consistent with
the hypothesis that the only significant modification in the
characteristics of isotropic keV proton precipitation introduced
at low altitudes is attributable to charge exchange and charge
stripping reactions with the atmosphere (Hultqvist et al.,1974).
However, this is not an argument against Swift's mechanism
affecting the hot plasma near the equatorial plane. A major
difficulty with such a mechanism located near the equator is that
the main effect on the electron population from an energy point of
view is one of deceleration along the field lines (a high energy
tail of accelerated electrons is also produced). This would mean
that the process would contribute to increasing the anisotropy at
lower energies and decreasing it at higher, which is contrary to
what has been observed. In addition, the observations show a very
high degree of complexity and variability in the relations between
changes of electron and ion parameters which seems very difficult
to interprete in terms of one single process. We, therefore,
conclude that presently available observational results appear to
be difficult to interpret in terms of any known single instability
affecting both ions and electrons in wide energy ranges. They
rather indicate that a well developed turbulence with strong

coupling between several kinds of waves is permanently present
outside the plasmapause. Many of these waves have certainly not
been identified experimentally yet. Some of them may be electrostatic with frequencies in the kHz range (Lyons, 1974) as well as
of the order of magnitude unity (Hultqvist et al., 1974).

Effects of Cold Plasma on the Turbulence
Outside the Plasmasphere:

The observation that the strong pitch angle diffusion of ions
and keV electrons occurs outside the plasmapause may be due either
to the plasmapause formation and the hot plasma distribution being
determined by the same process, or to the cold plasma in the
plasmapause region determining the distribution of the hot plasma.
The common cause case may e g be one of the convection bringing
the hot plasma in to the forbidden region of each energy and in
so doing also removing the cold plasma (of ionospheric origin)
from outside its forbidden region. The cold plasma may produce the
inner boundary of the pitch angle diffusion region either by precipitating all particles into the atmosphere at the edge of the
plasmasphere and thereby keeping the region within this boundary
free of hot plasma, although the convection electric field reaches
farther in, or by quenching the instabilities so that no pitch
angle diffusion occurs within the plasmapause even when the hot
plasma reaches well within it. The mechanism of emptying of the
field tubes by precipitation just in the plasmapause region has
been proposed by Kennel (1969). It is not consistent with the
observations that the energy spectrum softens in the poleward
direction for both electrons and ions and that keV electrons and
ions are precipitated largely together. If the hot plasma reaches
well within the plasmapause but pitch angle diffusion occurs only

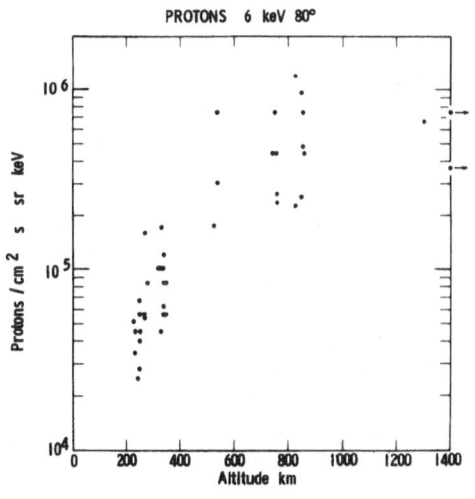

Figure 10. Maximum observed flux
values of 6 keV, 80° protons
observed in a number of ESRO 1A
and 1B passes through the particle
precipitation zone at different
altitudes (after Hultqvist et al.,
1974).

outside it, the convection alone cannot be the determining
process. Some additional mechanism is required and the quenching
of the instabilities by the cold plasma is a likely candidate.

The measurements in fact show that the ring current reaches
well within the plasmapause in most cases (Frank, 1971; Williams
and Lyons, 1974a; see also Hultqvist, 1975 for a discussion). So the
observations are consistent with the cold plasma having a strong
quenching effect on those instabilities which give rise to the
pitch angle diffusion of ions and electrons in a wide region outside the plasmapause.

The Location of the Turbulence
which Affects the keV Ions:

The peak differential flux of 6 keV ions observed in a pass
is shown for a number of ESRO 1A and 1B passes in Figure 10. Above
600 km altitude, say, the peak flux varies by about an order of
magnitude from orbit to orbit. The high altitude values in Figure 10
are found in the range from $(2-3) \cdot 10^5$ to slightly above 10^6 ions/cm^2
s sr keV. The highest 6 keV ion flux observed by the ESRO 1
satellites was only slightly higher than the highest values shown
in Figure 10. The fluxes of 6 keV ions near the atmosphere appear
to be quite close in magnitude to those observed near the equatorial plane. Figure 11a contains an ion energy spectrum from ATS 5
for which the differential number flux at 6 keV amounts to $8 \cdot 10^5$
ions/cm^2 s sr keV. Figure 11b, c show some data taken at high
pitch angles in the equatorial plane by means of the S^3 satellite
during various levels of magnetic disturbance (Smith and Hoffman,
1974; Williams et al., 1974). They show 6 keV fluxes in the range
of Figure 10. Considering the limited accuracy of the absolute
flux values, it seems reasonable to conclude that the keV ion fluxes
observed at high pitch angles near the equatorial plane differ
from those in the magnetically conjugate region of the upper
ionosphere, i e the fluxes in and near the loss cone, by only a
fraction of an order of magnitude. This may not be true at higher
ion energies in spite of the isotropy in and near the loss cone
(see e g Hultqvist, 1975).

The similarity of the differential fluxes of keV ions in the
loss cone and at large pitch angles near the equatorial plane shows
that the isotropization of the pitch angle distribution by pitch
angle diffusion takes place fairly close to the equator. For if it
did not, the fluxes which were affected by the turbulence at some
distance along the field lines would be lower than the flux at
large pitch angles near the equator (disregarding here the field
aligned pitch angle distributions due to non-stochastic effects
which are also seen near the equatorial plane by ATS 6 according to
McIlwain in this volume).

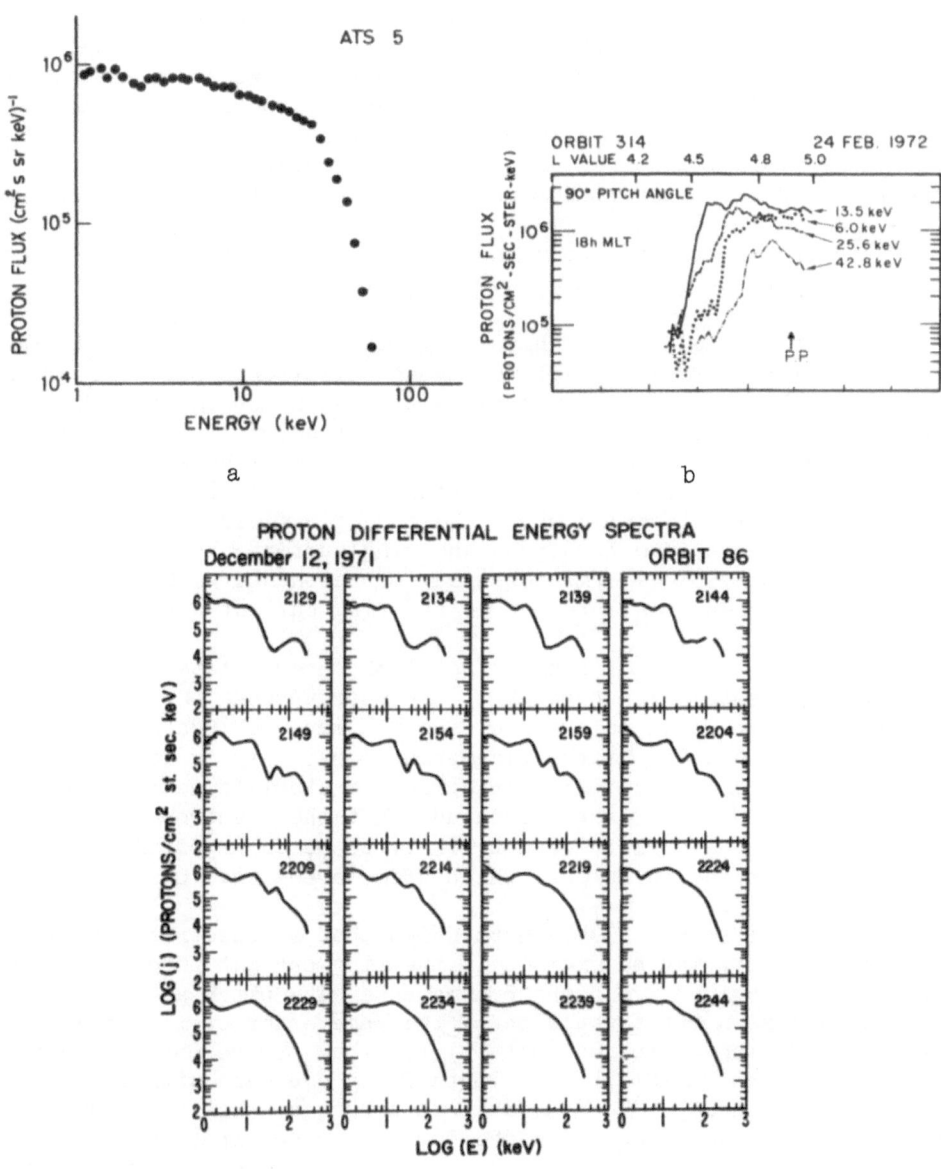

Figure 11. Ion flux data obtained near the equatorial plane (a) Ion energy spectrum measured by DeForest and McIlwain with ATS 5. (b) Explorer 45 data from the storm of Feb. 24, 1972 (after Smith and Hoffman, 1974). (c) Five-minutes averages of ion spectra taken by Explorer 45 before, during, and after the substorm of 2200 ±10 UT on Dec. 12, 1971. Times given are in UT (after Williams et al., 1974).

With the presently available observational data, it seems
difficult to draw any corresponding conclusions about the location
of the unstable regions where the ions and electrons of higher
energies are scattered. The strong coupling between the processes
affecting various species over various energy ranges,which has been
emphasized earlier in this report,makes it natural to assume that
all the instabilities are located near the equatorial plane. The
observations do, however, not provide any proof for this assumption.

On the Roles of Some Individual Instabilities in
the Turbulent Region Outside the Plasmapause:

Cornwall et al. (1971) have suggested that the electromagnetic
ion cyclotron instability, driven by an anisotropic particle distribution in the equatorial plane plays a role in determining the
spatial distribution of the pitch angle diffusion. Their theoretical
model suggests that the ring current protons will be subject to
enhanced pitch angle and energy diffusion just within the plasmapause and at high L shells where the magnetic field intensity is
small. At intermediate locations the ring current should be stable
because the particle energy required for resonance with the cyclotron waves exceeds that present in the ring current distribution.

The observations of roughly isotropic precipitation observed
by low orbit satellites throughout the energy range one to several
hundred keV extending poleward of the plasmapause independently of
the specific location of the plasmapause obviously do not sub-
stantiate the existance of an intermediate stable zone for low
ion energies. We, therefore, conclude that the electromagnetic
ion cyclotron instability is not likely to be an important
constituent in the turbulence which affects the keV ions.

Coroniti et al. (1972) have shown that an electrostatic loss
cone instability may be excited in the ring current throughout the
low density region beyond the plasmapause. The pitch angle and
energy diffusion rates would be of the same order of magnitude as
those of the ion cyclotron instability. They furthermore suggest
that the growth rates for this instability decrease with increasing
cold plasma density. The available data suggest that a process with
such characteristics provides the dominant precipitation loss
process for the ring current ions. That the loss cone permanently
is roughly filled up to within a factor of two, say, of the 90° flux
may not mean that there is not sufficient free energy available for
the instability. Even quite small deviations from isotropy may
suffice in this respect (Galeev, personal communication). Electrostatic loss cone instabilities are, therefore, strong candidates
to be those constituents in the complex turbulence which cause
diffusion of the ions and electrons in the keV energy range. But
even if these loss cone instabilities are important for pitch

angle diffusion of the low energy (keV) particles, it still remains
to be understood why the instabilities affecting electrons and
ions are so strongly coupled as observations show. A single
microinstability which can affect both ions and electrons would be
a good explanation but that instability remains to be identified
if it exists, as pointed out earlier.

Before leaving the matter of the identities of important
microinstabilities in the magnetosphere it may be noted that drift
wave instabilities are not likely to play a significant role for
the large scale fairly diffuse precipitation of hot plasma but
may be the more important in discrete auroral forms (Haerendel,
personal communication).

Possible Effect of the Turbulence on the
Energy Distribution of the Hot Plasma:

As shown above, the ESRO 1 observations demonstrate that when
there are enhancements in the precipitated keV electron fluxes
there is usually also an enhancement in the keV ion flux in and
near the loss cone, but a smaller enhancement than in the electrons.
The energy spectrum in the 1-10 keV range most frequently is harder
for the electrons and softer for the ions in these regions with
enhanced fluxes than in the surroundings. This is the most commonly
observed relation between flux variations and spectral variations
in the keV electrons and ions in and near the loss cone, but it
should be emphasized that all sorts of combinations of variations
can be found in the data. The combination of correlated flux
increases with correlated spectral changes in the electrons and
ions appears to be the most difficult one to find in the ESRO 1
data (see Hultqvist et al.,1974).

Because the pitch angle diffusion is quite strong already when
the precipitated flux is at a fairly low level, the flux increases
in and near the loss cone cannot in general be due only to increased diffusion rate out of a constant population near the
equatorial plane, but most of the enhancements that are seen in
the upper ionosphere have to originate in regions of space and/or
time with increased plasma density. Even so it seems likely that
the mentioned relations between variations of ion and electron
populations in and near the loss cone may be produced by the
turbulence. This is mainly because it appears to be difficult to
find any alternative process which can give rise to the large
variety of relations observed and, in particular, can produce
opposite spectral changes in keV ions and electrons in regions of
enhanced fluxes. A well developed turbulence offers both the
variability and the possibility that energy is transferred from
ions to electrons in some yet unknown way. We do not claim, however, that the available data provide a proof for this suggestion.

"Pure" Ion Precipitation:

The almost "pure" keV ion precipitation in the afternoon
hours, which is demonstrated in Figure 7, has been seen by ESRO 1A
at as low K_p values as 1o (Hultqvist et al.,1974), but the
phenomenon becomes more common with increasing K_p and seems to be
rule rather than exception at high disturbance level. Exceptions
with keV electrons present are, however, seen also at very high
K_p values.

The observations of keV ions without any associated keV
electrons at ionospheric heights in the afternoon may indicate
that precipitation mechanisms that affect keV ions but not keV
electrons are frequently operative in the noon and afternoon
sectors, particularly during strongly disturbed conditions. The
"pure" ion precipitation may, however, also be interpreted as
caused by the absence of keV electrons in the magnetospheric
particle populations from which the keV particles are diffused
into the upper ionosphere. The latter alternative is identical to
the assumption that the keV ions are precipitated from a ring
current where the electrical neutrality is brought about by
electrons of energies below the keV range. If this is the case,
basically the same mechanisms as in the other local time sectors,
where both ions and electrons are precipitated, may be responsible
for the ion precipitation (and not species specific processes
like those proposed by Kennel and Petschek, 1966; Eather and
Carovillano, 1971; Cornwall, 1972 and others). This seems the
more likely since the electron-free precipitation is the one
extreme of a whole spectrum of ratios between ion and electron
fluxes observed.

It also appears that the size of the local time sector where
only keV ions but no keV electrons are found in the upper ionosphere
may sometimes extend into the noon sector and even to before noon.
The growth of the electron-free sector into the forenoon hours
seems to be associated with the expansive phase of substorms.
Unfortunately we have ESRO 1 data from the noon sector for only
one substorm (or small magnetic storm). Figure 12a shows the AE
index for 26-27 July 1969 with three ESRO 1A passes marked on it.
Figure 12b demonstrates that in the very early phase there were
both keV ions and electrons present at 1000 km altitude south of
the cusp as well as north of it. During the expansive phase
(Figure 12c) the keV electrons disappeared completely and the keV
ion fluxes disappeared north of the cusp, and they decreased and
their spectrum softened in the subcusp region. At the center of
the disturbance 100 min later (Figure 12d) no single keV electron
count was recorded by ESRO 1A in passing through the invariant
latitude range 53° to 79° in the noon sector, whereas the ion

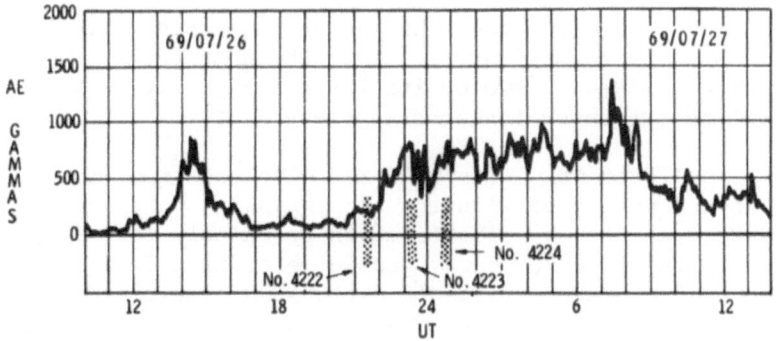

a

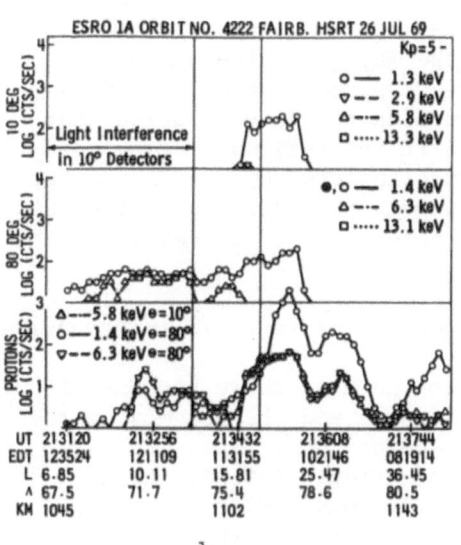

b

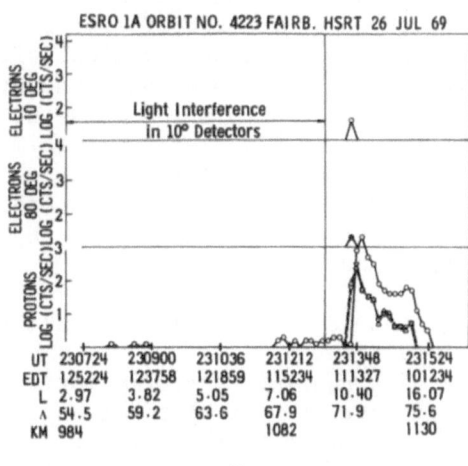

c

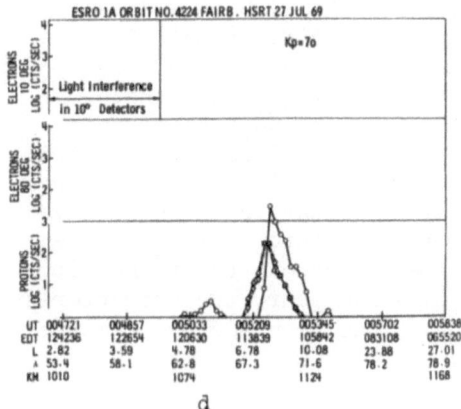

d

Figure 12. Variations of keV electron and ion fluxes in the noon sector during a strong substorm (or small magnetic storm). (a) shows the AE index with the ESRO 1A passes marked (after Hultqvist et al., 1974).

fluxes in the cusp were unusually high. At this stage electrons
of energies above 14 keV had arrived into the noon sector
(Hultqvist et al., 1974). Two orbits later keV particles were
again present in the upper ionosphere poleward of the cusp.

The substorm evidently wiped out keV electrons from the upper
atmosphere in the noon sector during its expansive phase and
electrons were fed into this sector again presumably from the
nightside, with fairly high-energy particles of high drift speeds
arriving first. The disappearance from the 1000 km altitude of keV
electrons and the decrease of keV ion fluxes equatorward of the
cusp indicate that not only the loss cone but also the neigh-
bouring regions of the angular distribution were emptied by the
disturbance and that no measurable diffusion into this angular
region occurred thereafter for several hours. It seems difficult
to combine these observations with the fact that the magnetosphere
was heavily disturbed and, therefore, supposedly highly loaded
with waves and turbulence, without coming to the conclusion that
there were no keV electrons available in the relevant regions of
the magnetosphere for scattering into the loss cone in this later
phase of the disturbance. In other words, the substorm emptied at
least part of the dayside of the magnetosphere of keV electrons
in the early part of it.

At least the following mechanisms may be involved in producing
these effects:
- disappearing by precipitation into the atmosphere because of
  increased rate of pitch angle diffusion and thereby moving the
  boundary of the "empty" sector to before noon;
- modification of drift orbits to the extent that particles
  temporarily do not penetrate into the noon sector any more;
- the cusp moving equatorward through the region where keV
  particles were found before the storm.
The last of these effects is unable to explain the difference
between the situations in Figures 12b and 12c with regard to keV
electrons at the lowest latitudes. As the modification of the
convection pattern has to be a severe one and as it does not
eliminate the keV ions from the noon sector it appears that
increased rate of precipitation into the atmosphere at present
seems most likely to be the process that determines the distribution of the keV electron population in the dayside magnetosphere
together with the drift into the dayside of particles which have
been injected on the nightside during the substorm. Whereas the
time scale for the disappearance of keV electrons by strong pitch
angle diffusion appears to be less than one hour, the characteristic
time for feeding new keV particles into the noon sector from the
nightside is several hours.

Weak Pitch Angle Diffusion of keV
Ions within the Plasmasphere:

Weak pitch angle diffusion of keV ions has been found in a
region between L ~2.7 and ~4 during quiet and moderately disturbed
conditions as described earlier. The stable location of this
region may be interpreted as being a consequence of the wave particle
interaction that causes the pitch angle diffusion requiring at
least several hundred cold ions per cubic centimeter to operate.
As has been demonstrated by Chappell et al. (1970), the density
distribution in the plasmasphere at levels above say 500 ions/cm^3
is hardly affected at all by variations in K_p up to about four,
whereas the outermost edge with lower density is very much affected.

The large width of this weak diffusion region appears to be
interpretable in terms of the relative variation with L of the
growth rate for the ion cyclotron instability (Hultqvist et al.,
1974b). A minimum cold plasma density of about 500 ions/cm^3 for
6 keV ions to participate in this instability is in agreement with
the Explorer 45 (S^3) results from near the equatorial plane
(Williams, personal communication).

Most of the ESRO 1 observations of plasmaspheric pitch angle
diffusion of keV ions were obtained during fairly quiet conditions.
They appear to be consistent with the S^3 results from the equator
(Williams and Lyons, 1974a, b; Williams, this volume) but supplement
these by showing that the intermediate stable region with empty
loss cone is present between the outer precipitation region with
strong pitch angle diffusion and the inner region with weak diffusion not only in the recovery phase of magnetic storms but always
(at least on the nightside). We thus end up with a model where the
dominating precipitation of the hot magnetospheric plasma, occurring
outside the plasmapause due to yet unidentified instabilities, is
inhibited by cold plasma already at low densities, but cold plasma
of fairly high densities drive the instability causing weak
diffusion of keV ions within the plasmasphere (probably the ion
cyclotron instability). At higher ion energies lower cold plasma
densities are needed for the plasmasphere instability to run, as
shown by Williams and Lyons (1974a, b) and Williams (this volume).

SUMMARY

    The magnetospheric turbulence is discussed on the basis of
hot plasma measurements in and near the loss cone, mainly by means
of the ESRO 1A and 1B satellites. The observational results may be
summarized as follows:
  - The hot plasma is permanently in a turbulent state in a wide
    region outside the plasmapause.
  - The turbulence outside the plasmapause seems to contain a
    complex set of instabilities which are coupled in complex and
    variable ways.
  - Cold plasma appears to inhibit the turbulence outside the plasma-
    pause rather than to enhance it and to determine thereby the
    inner boundary of the turbulent region.
  - The turbulence which isotropizes the keV ions appears to be
    located near the equatorial plane.
  - The ion cyclotron instability is not likely to be of importance
    for the strong pitch angle diffusion of keV ions outside the
    plasmapause. Loss cone instabilities may be of importance.
  - The turbulence may be the cause of harder electron spectra and
    softer ion spectra which are frequently observed near the
    atmosphere in regions of flux enhancement in the auroral zone.
  - The "pure" keV ion precipitation (without measurable fluxes of
    keV electrons), observed equatorward of the polar cusps in the
    afternoon and sometimes in the noon sector, appears to be due
    to the absence of electrons of keV energies in this part of
    the magnetosphere.
  - The observations of keV ion pitch angle diffusion within the
    plasmapause are consistent with the hypothesis that the ion
    cyclotron instability causes the diffusion in the plasmasphere.

ACKNOWLEDGEMENT

    The research on which this report is based has been supported
economically by the Swedish Natural Science Research Council, the
Swedish Board for Space Activities and the European Space Research
Organization. Part of the investigations was done when the author
was on leave at NOAA, ERL Space Environment Laboratory, Boulder,
Colorado, USA as a National Research Council - National Academy of
Sciences senior research associate.

REFERENCES

Bernstein, W., B. Hultqvist and H. Borg, Some implications of low altitude observations of isotropic precipitation of ring current protons beyond the plasmapause, Planet. Space Sci., 22, 767, 1974.

Chappell, C.R., K.K. Harris and G.W. Sharp, A study of the influence of magnetic activity on the location of the plasmapause as measured by OGO-5, J. Geophys. Res., 75, 50, 1970.

Cornwall, J.M., F.V. Coroniti and R.M. Thorne, A unified theory of SAR arc formation at the plasmapause, J. Geophys. Res., 76, 4428, 1971.

Cornwall, J.M., Precipitation of auroral and ring-current particles by artificial plasma injection, Rev. Geophys. and Space Phys., 10, 993, 1972.

Coroniti, F.V., R.W. Fredericks and R. White, Instability of ring current protons beyond the plasmapause during injection events, J. Geophys. Res., 77, 6243, 1972.

Eather, R.H. and R.L. Carovillano, The ring current as the source region for proton auroras, Cosmic Electrodynamics, 2, 105, 1971.

Frank, L.A., Relationship of the plasma sheet, ring current, trapping boundary, and plasmapause near the magnetic equator and local midnight, J. Geophys. Res., 76, 2265, 1971.

Hultqvist, B., On the production of magnetic field-aligned electric field by the interaction between the hot magnetospheric plasma and the cold ionosphere, Planet. Space Sci., 19, 749, 1971.

Hultqvist, B., H. Borg, P. Christophersen and W. Riedler, Observations of magnetic field-aligned anisotropy for 1 and 6 keV positive ions in the upper ionosphere, Planet. Space Sci., 19, 279, 1971.

Hultqvist, B., H. Borg, P. Christophersen, W. Riedler and W. Bernstein, Energetic protons in the keV energy range and associated keV electrons observed at various local times and disturbance levels in the upper ionosphere, NOAA, Technical Report ERL 305 - SEL 29, Boulder, Colorado, 1974.

Hultqvist, B., W. Riedler and H. Borg, Ring current protons in the upper ionosphere within the plasmasphere, Kiruna Geophysical Institute, Preprint 74:305, 1974b.

Hultqvist, B., The ring current and particle precipitation near the plasmapause. Paper presented at the 2nd Annual Meeting of the European Geophys. Soc., Trieste, 1975; Ann. Geophys. (in print).

Kennel, C.F. and J.E. Petschek, Limit of stably trapped particle fluxes, J. Geophys. Res., 71, 1, 1966.

Kennel, C.F., Consequences of a magnetospheric plasma, Rev. Geophys., 7, 379, 1969.

Lyons, L.R., Electron diffusion driven by magnetospheric electro-static waves, J. Geophys. Res., 79, 575, 1974.

Mizera, P.F., Observations of precipitating protons with ring
    current energies, J. Geophys. Res., 79, 581, 1974.

O'Brien, B.J., High-latitude geophysical studies with satellite
    Injun 3, Precipitation of electrons into the atmosphere,
    J. Geophys. Res., 69, 13, 1964.

Page, D.E. and M.L. Shaw, Some parameters affecting the poleward
    boundary of trapped electrons, in Earth's Magnetospheric
    Processes, (Ed. B.M. McCormac), D. Reidel Publishing Co.,
    Dordrecht, Holland, 1972.

Riedler, W., B. Hultqvist and S. Olsen, A satellite instrument for
    auroral particle measurements using channel multipliers,
    Arkiv för Geofysik 5, 619, 1971.

Smith, P.H. and R.A. Hoffman, Direct observations in the dusk hours
    of the characteristics of the storm time ring current particles
    during the beginning of magnetic storms, J. Geophys. Res., 79,
    966, 1974.

Swift, D.W., A new interpretation of VLF chorus, J. Geophys. Res.,
    73, 7447, 1968.

Swift, D.W., Particle acceleration by electrostatic waves,
    J. Geophys. Res., 75, 6324, 1970.

Søraas, F., ESRO 1 A/B observations at high latitudes of trapped
    and precipitating protons with energies above 100 keV, in
    Earth's Magnetospheric Processes, (Ed. B.M. McCormac),
    D. Reidel Publishing Co., Dordrecht, Holland, 1972.

Søraas, F. and L.E. Berg, Correlated satellite measurements of
    proton precipitation and plasma density, J. Geophys. Res.,
    79, 5171, 1974.

Williams, D.J., J.N. Barfield and T.A. Fritz, Initial Explorer 45
    substorm observations and electric field considerations,
    J. Geophys. Res., 79, 554, 1974.

Williams, D.J. and L. R. Lyons, The proton ring current and its
    interaction with the plasmapause; storm recovery phase,
    J. Geophys. Res., 79, 4195, 1974a.

Williams, D.J. and L.R. Lyons, Further aspects of the proton ring
    current interaction with the plasmapause: main and recovery
    phases, J. Geophys. Res., 79, 4791, 1974b.

EVIDENCE FOR THE LOW ALTITUDE ACCELERATION OF AURORAL PARTICLES

David S. Evans

Space Environment Laboratory, ERL, NOAA

Boulder, Colorado 80302

INTRODUCTION

It has long been surmised that no matter where the origin of auroral particles, there must be particle energization mechanisms local to the magnetosphere in order to account for the height distribution of the auroral light these particles produce. The problem of describing these acceleration mechanisms remains a central one in auroral physics.

The first theories of the energization of auroral particles often involved low altitude acceleration mechanisms. Alfven (1950) suggested that auroral particles were produced by a current driven discharge occurring at the top of the atmosphere. Chamberlain (1956) attempted to put such a discharge theory on a quantatative basis.

The first satellite launch in 1957 was quickly followed by the discovery of the trapped radiation zones and the preliminary mapping of the geomagnetic field geometry at large distances from the earth. Also beginning at about this time, attention became concentrated toward the regions at high altitude in the geomagnetic equatorial plane as the location for auroral particle acceleration mechanisms. For example, Akasofu and Chapman (1961) proposed that auroral particles originated from and were accelerated by an x-type or neutral line magnetic field geometry located in the geomagnetic tail.

Speiser (1967) assumed a reasonable magnetic and electric field geometry in the geomagnetic tail and solved for charged particle trajectories in this region. He showed that particle

acceleration is likely. Dungey, in a recent review (1972),
stresses that particle acceleration in neutral sheets located
in the tail is likely to relate to auroral precipitation.

Those theories which model particle acceleration mechanisms
operating at large distances from the earth have proven very valuable in explaining the source of particles in the outer radiation zone, the appearance of the ring current, the characteristic
behavior in the plasma sheet during substorms, and similar phenomena. Indeed, there has been success in explaining some features of auroral precipitation in terms of scattering into the
atmosphere of those trapped electrons which were energized by
some process in the tail (Lyons, 1974).

However, over the past decade there have been an ever increasing number of observations which strongly suggest that particle acceleration may also occur at low altitude; less than 6000
km above the atmosphere along the high latitude magnetic line of
force. This paper is intended to review these observational
data and to briefly describe the nature of the low altitude
acceleration mechanisms that have been proposed. The theoretical
explanations of such mechanisms will not be discussed in detail.

GENERAL COMMENTS CONCERNING
LOW ALTITUDE PARTICLE ACCELERATION

If an auroral charged particle is to be accelerated significantly at low altitudes along an auroral line of force (perhaps
as low as 2000 km) rather strong electric fields must be responsible. This is because of the short available path length over which
the accelerating force can be applied to the particle (at most
several 1000 km) and the short time that an energetic particle
can spend at low altitudes on the line of force.

It is unlikely that a particle could be energized over a long
period of time by virtue of successive encounters with a low altitude mechanism, each encounter providing only a small energy increment (say a resonant interaction between particles and low amplitude fluctuating electric fields). Each increment in energy
would probably change the particle's pitch angle as well, and,
thus, the particle would be quickly lost to the atmosphere.

In spite of the fact that the electric field strengths, associated with the low altitude acceleration of auroral particles,
are large, virtually all knowledge of these processes has come
from the observations of the accelerated particles rather than
the direct observation of the electric fields. Probably the only
exceptions to this are the reports of ionospheric level, static
electric fields directed parallel to the magnetic field (c.f.
the review by Fahleson, 1972). These observations remain contro-
versial, however.

Many aspects of the auroral particle population have been used to infer the locality of the particle energization. Among these are unusual angular and energy distributions, rapid temporal variations in the auroral flux as observed in the ionosphere; especially steep spacial gradients in the flux, correlations between ion and electron flux variations, and, to some extent, ion species measurements.

The particle observations discussed here were made by a variety of balloon borne, rocket and satellite techniques. Each of these has its own unique advantage either in providing for high resolution measurements (temporal, spacial, energy, or angular) or in allowing long term surveys to be made of particle precipitation over large areas.

The particle observations tend to divide into two groups depending upon which of two general types of low altitude acceleration mechanisms the particular observation supports. Rapid time variations in the auroral particle flux, particularly when rather energetic particles are involved, are thought associated with particles being energized by short lived resonant interactions with large amplitude fluctuating electric fields.

The alternative form of low altitude particle energization is due to the unimpeded acceleration of charged particles by quasi-static electric fields aligned along the local geomagnetic field. There is very strong support for this from observations of unusual auroral particle pitch angle and energy distributions.

There is a third set of particle observations which suggest no particular acceleration process beyond the fact that low altitude mechanisms are involved.

While the association between the type of low altitude acceleration mechanism and the specific nature of the observational evidence is probably not unique, the trend is used to order the following discussion.

LOW ALTITUDE PARTICLE ACCELERATION DUE TO A RESONANT INTERACTION WITH FLUCTUATING ELECTRIC FIELDS

The earliest definite evidence for low altitude particle energization came from the study of rapid temporal variations in the precipitation of energetic auroral electrons. The conclusion that short lived transients in the precipitation fluxes implied low altitude processes was linked to an absence of velocity dispersion in the arrival of electrons of differing energies at the top of the atmosphere.

Much of the knowledge concerning rapid time fluctuations
in auroral electrons came from the study of the bremsstrahlung
x-rays which are generated when the electrons strike the atmosphere. Scintillation detectors which are employed for these studies measure a flux of x-rays having energies above fixed thresholds. Thus the detector response is connected with the intensity
of precipitated electrons as integrated over a broad energy range
extending upwards from some minimum kinetic energy. Hence, velocity dispersion among electrons of various velocities cannot be
directly measured, but fast rise times and short durations of
transient x-ray events will reflect a lack of dispersion amongst
the parent electrons.

Venkatesan et al. (1968) observed auroral x-ray microbursts
which had rise times shorter than .01 sec. Moreover, there was
direct evidence from the x-ray spectrum that electrons having a
wide range of energies participated in these events. These facts
led to the conclusion that the events were initiated by and associated with a near earth process.

More recently Sheldon et al. (1973) reported auroral x-ray
bursts which had rise times of .001 sec and widths of only .02
sec. From additional knowledge of x-ray spectrum, they inferred
that the source location of the transient in the electrons lay
at an altitude of < 500 km.

Lampton (1967) was able to determine directly, by means of
a rocket-borne instrument, that there was usually no observable
velocity dispersion in the arrival of electrons at the top of the
atmosphere during microburst transients in the electron precipitation.   The source of the transient was located within a few
earth radii of the atmosphere, an upper limit given by the instrument time resolution. An interesting aspect of these results was
that those microburst events which did display measurable velocity dispersion appeared to have originated near the conjugate
auroral ionosphere and not at high altitude along the magnetic
line of force.

Evans (1967) reported the observation of 10 cps quasi-peri-
odic variations in the precipitation of energetic auroral electrons and suggested that this reflected the time history of the
process that energized these electrons. Figure 1 displays an
episode of these intensity variations as detected at a number
of electron energies. Periodic variations in electron precipitation, similar in appearance to Figure 1, have also been reported by Winiecki (1967) and Arnoldy (1970). These variations
are almost certainly associated with the "flicker" aurora which
is reported by Beach et al. (1969) as being a common feature in
bright, active auroral displays.

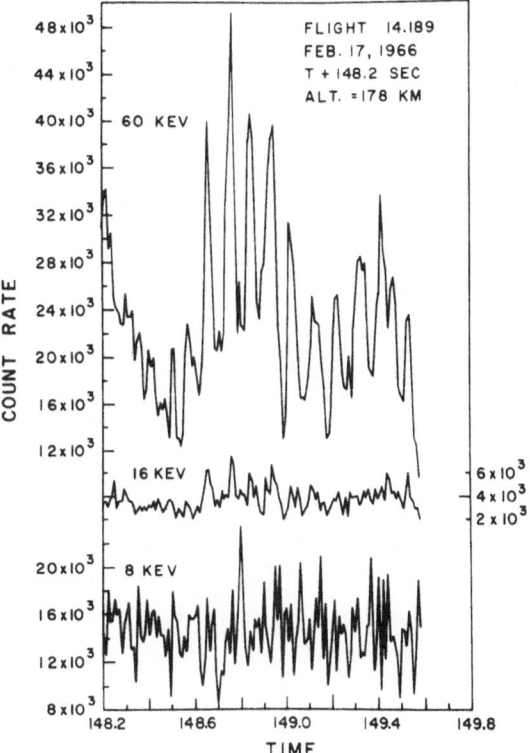

Figure 1. An example of rapid variations in the flux of energetic auroral electrons as observed at several energies. The fluctuations were not prominent in the <8 keV electron precipitation.

Evans performed a cross-correlation analysis between the intensity fluctuations as detected at various electron energies. A typical result is shown in Figure 2. This analysis showed that electrons ranging in energy from 8 keV to 120 keV arrived at the detectors in phase to within 10 msec. This coincidence in arrival times located the source in the modulation to a point at less than 1500 km altitude.

It was further noted that the large amplitude flux variations were a feature of the energetic (> 10 keV) electron population but not of the intense keV elecron beam which was largely responsible for the auroral light. This seemed to speak against the fluctuations simply reflecting variations in the "dumping" of already energized and trapped electrons as particles of all energies might be expected to participate in such a process.

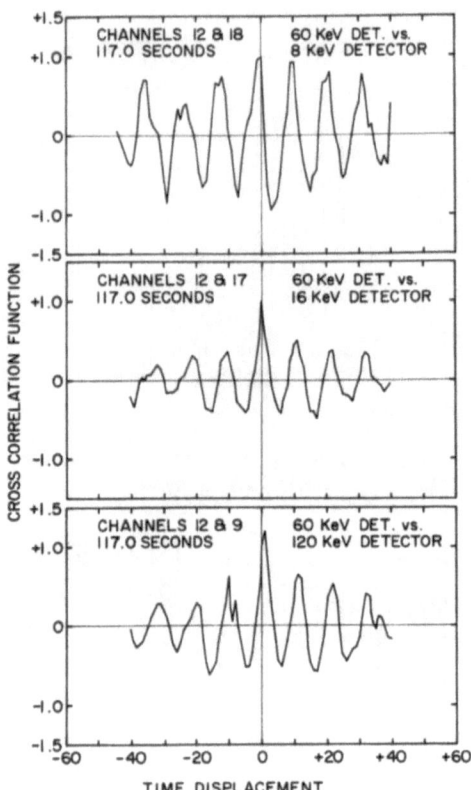

Figure 2. The result of a cross-correlation performed to expose
differences in the arrival of electrons of differing velocities
at the detector. The apparent phase shift in the bottom panel is
in fact introduced by the telemetry system, and thus electrons
from 16 keV to 120 keV arrived at the atmosphere in phase to within .010 sec during these fluctuations.

It was finally proposed that the energetic electrons had
actually been produced by a low altitude acceleration mechanism
and that the pulsations represented a time variation in the
strength of this acceleration. Specifically it was thought that
an intense beam of keV energy electrons had driven a beam-plasma
instability in the upper ionosphere and that energetic "runaway"
electrons were produced as a result of this instability.

Perkins (1968) attempted to put this model on a firm theoretical basis. He concluded that beams of keV energy auroral
electrons would produce instabilities in the auroral ionosphere
providing the fluxes exceeded certain thresholds. Moreover, intense electrostatic waves associated with the beam-plasma

instability could couple to certain favored electrons and accelerate these to very high energies in very short times.

The acceleration of electrons to very high energies by processes operating at low altitude in the auroral ionosphere is sub-
stantiated in both experimental and theoretical results. However,
whether or not such processes contribute significantly to either
the auroral electron or trapped electron populations is not known.
It may be that runaway electrons, energized at low altitudes in
this fashion, are more in the nature of curiosities.

LOW ALTITUDE PARTICLE ACCELERATION
BY PARALLEL ELECTRIC FIELDS

Classically, the electrical conductivity parallel to the
magnetic field in a magnetized plasma is governed by Coulomb collisions alone. In the earth's magnetosphere and ionosphere this
component of the conductivity has been calculated to be many
orders of magnitude greater than the electrical conductivity transverse to the magnetic field. Thus if a potential difference were
imposed across the magnetosphere by some external source, that
portion which would appear along the magnetic field lines would
be a negligible fraction of the total available. This line of
reasoning leads to the conclusion that magnetic field-aligned
potential differences should not exceed a few volts within the
magnetosphere.

In spite of these arguments, much data in the form of high
quality observations of auroral particle pitch angle distributions
strongly suggest that quasi-state electrical potential differences in excess of several thousand Volts exist along the auroral
lines of force at fairly low altitudes. Moreover, these electric
fields do work upon the charged particle population by freely
accelerating them and producing precipitation into the atmosphere.

Low altitude parallel electric felds have been proposed to
account for the "anomalous" backscatter of precipitating particles from the atmosphere (McDiarmid et al., 1961; Mozer and
Bruston, 1966; Reasoner and Chappel, 1973). The nature of each
of these observations was that the flux of upgoing particles
originating from points below the altitude of observations
($\stackrel{<}{\sim}$ 1000km) was larger than could be accounted for by either
the magnetic mirroring or Coulomb backscattering of the observed
primary precipitation flux. The role of the hypothesized parallel
electric field was to accelerate charged particles upwards and
cause "premature" magnetic mirroring of the precipitation. With
few exceptions, such enormous parallel electric field strengths
are required to explain "anomalous" backscatter that these observations represent rather poor evidence for the existence of
such fields.

Albert and Lindstrom (1970) reported observing an auroral electron pitch angle distribution that exhibited a minimum within a pitch angle range around 90^0, the electron flux increasing at both larger (upgoing) and smaller pitch angles. This feature was attributed to the existence of a field-aligned potential drop which accelerated electrons downwards at some altitude not far above the point of observation ($\sim$ 250 km). The field-alignment in the precipitation flux that is imposed by this acceleration removed electrons from pitch angles near 90^0. Neither backscatter from the atmosphere well below the rocket, nor magnetic mirroring of precipitating electrons could populate pitch angles near 90^0 at the point of observation. The very existence of a parallel electric field which freely accelerates particles causes a "forbidden" region of pitch angles near 90^0, thus explaining the observations.

This paper deserves more attention than it has gotten because many of the concepts first discussed there have re-appeared in more recent models (e. g. Evans, 1974).

The most unambiguous evidence that electric fields (or electrical potential differences) exist along magnetic fields comes from the observations of charged particle precipitation which have field-aligned pitch angle distributions (that is to say that the particle intensities maximize at pitch angles contained within a small cone centered at either 0^0 or 180^0 to the local magnetic field line). The first such observation was reported by Hoffman and Evans (1968) and concerned a high degree of magnetic field alignment in the precipitation of 2.3 keV electrons (the only particle species and energy observable by the experiment) as detected by an instrument on the low altitude OGO-4 satellite. Since then a very large number of observations have been reported which have provided knowledge about electron angular distributions as a function of energy (Evans et al., 1972; Whalen and McDiarmid, 1972; Paschmann et al., 1972; Ackerson and Frank, 1972; Berko, 1973; Maehlum and Moestue, 1973; Berko and Hoffman, 1974; Bosqued et al., 1974; Arnoldy et al., 1974) as well as data about field-aligned distributions on the part of positive ion precipitation (Hultqvist et al., 1971; Réme and Bosqued, 1971). There can be no longer any doubt that magnetic field-aligned particle precipitation is a common feature of the auroral latitude particle population.

Several explanations have been proposed to explain field-aligned particle precipitation. Among these are particle acceleration by means of <u>velocity</u> resonant interactions with electrostatic waves (Landau damping) and Fermi acceleration of particles during the substorm associated relaxation of the geomagnetic field.

In general, such mechanisms are unable to provide sufficient increase in the parallel component of a particle's energy to explain
the degree of field alignment that is observed. Currently the
parallel electric field, causing the free acceleration of some
charged particles, appears to be the only viable explanation for
field-aligned particle precipitation.

The fact that magnetic field alignment in particle precipitation is observed at low altitudes is usually taken to mean that
the region of parallel electric field causing the field alignment
is also at fairly low altitudes, not far from the point of observation. The arguments leading to this conclusion have been presented in most of the papers cited above but they are once again
summarized here.

Assume the situation of a particle beam of energy W having
pitch angles collimated to within an angle α of the geomagnetic
field direction as observed at a point where the magnetic field
intensity was B. Typical values of these perameters might be
W = 2000 eV, α = 20^0, and B = .4 G corresponding to the position
of a low altitude satellite or sounding rocket. To account for
the field alignment in the beam, it is envisioned that some appropriate particle population (perhaps a Maxwellian characterized
by a temperature) had been introduced into a region of parallel
electric field and was accelerated downwards. It is presumed
that the scale length of the parallel electric field is small compared to the scale length in the magnetic field so that magnetic
field gradients may be neglected. Moreover it is presumed that
the particle always conserves its magnetic moment which is to say
that scattering in pitch angle and energy is neglected.

In the course of being accelerated by the parallel electric
field each particle acquires an equal increment in energy entirely
in the direction of the magnetic field. The particle beam emerges
from this "accelerator" with pitch angles contained within some
collimation cone whose width is a measure of both the original
energy the particles had perpendicular to $\vec{B}$ and the parallel energy
acquired during the acceleration.

The characteristics of the particle beam at the point of observation may be transformed upwards along the magnetic field
line to any suspected "accelerator exit point" by demanding conservation of the particle's magnetic moment. If a field-aligned
particle beam of the nature specified earlier were transformed
to a point in the equatorial regions along an auroral line of
force (B $\simeq$ .0015 G), the width of the pitch angle collimation
cone would be only 1^0. Moreover, if the particle had a total
energy of 2000 eV, the component of energy perpendicular to B
would be limited to less than 1 eV. Both the degree of collimation and the very low limit put on the perpendicular energy of

the un-accelerated particle are thought much too unreasonable,
and an equatorial parallel electric field is thus precluded from
explaining the observations.

   If a more reasonable value is chosen for the average perpendicular energy of the primeval particles -- say 200 eV based upon
observations of plasma sheet particles -- then particles freely
accelerated to 2000 eV by a parallel electric field would be
contained within a collimation cone of perhaps 18^0 to the magnetic
field. In this case no further broadening in the beam collimation
is required, or allowed, to explain the observations. The region
of parallel electric field must lie near the point of observation,
viz. at low altitude. More sophisticated arguments, for example,
allowing magnetic field gradients within the region of parallel
electric field, do not change the basic conclusion.

   Figures 3 (taken from Evans et al., 1972) and 4 (courtesy
of Arnoldy et al., 1974) show instances of field-aligned electron
events as observed by a sounding rocket instrument. These data
sets are in the form of electron energy spectrums sorted into
various pitch angle ranges. It is seen in both figures that the
field alignment in the electron beam is a feature which is confined to a limited range of electron energies. The electron population approaches isotropy at both higher <u>and</u> lower kinetic energies.

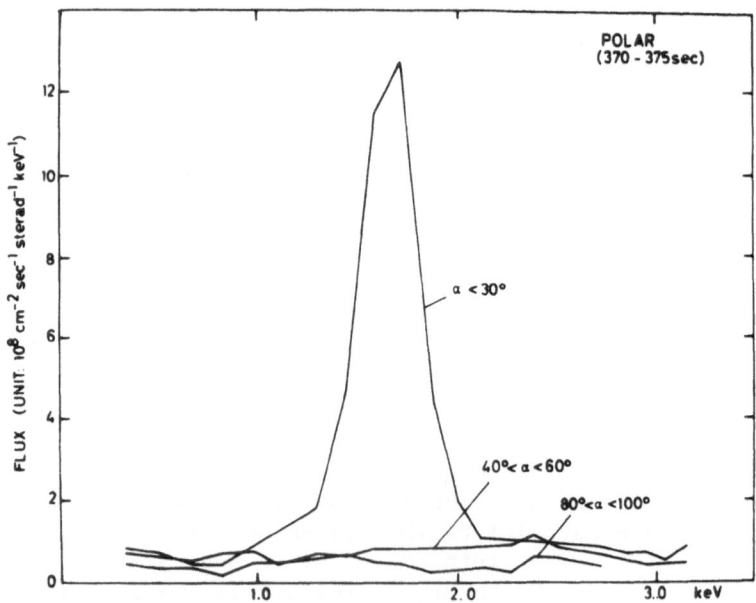

Figure 3. An example of magnetic field aligned electron precipitation displayed in the form of electron energy spectrums sorted
into pitch angle intervals (Evans et al., 1972).

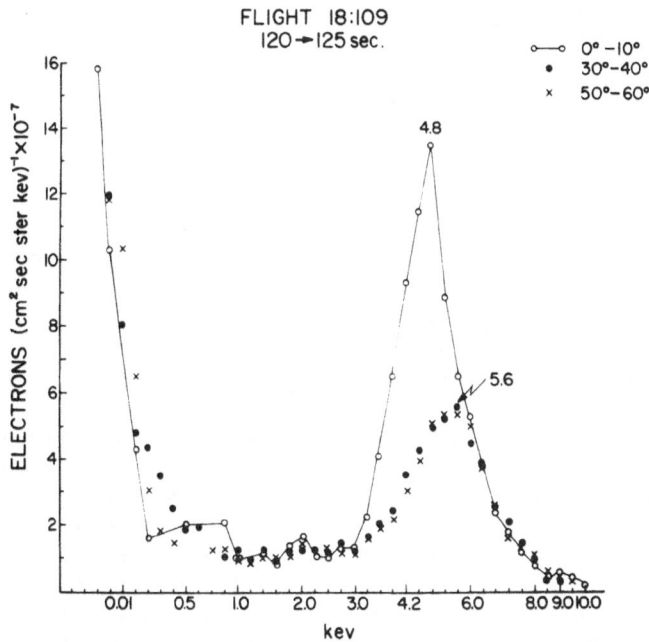

Figure 4. Data similar to that in Figure 3 (courtesy of Arnoldy et al., 1974). Note that the field-aligned precipitation is a feature of a limited range of electron energies.

It is not surprising that the high energy part of the electron beam approaches isotropy. These electrons must have had considerable energy prior to being accelerated by the $E_{\parallel}$. Because of this, the parallel potential drop could not impose a high degree of collimation on these particles. For lower electron energies however, the pitch-angle collimation produced by the $E_{\parallel}$ acceleration would become much more marked. The highest degree of collimation would appear at electron energies corresponding to the field-aligned potential difference that was available (i.e. electrons which had no initial energy at all).

This, most simple, model predicts that there should be no precipitating electrons of energies <u>less</u> than that provided by the available potential drop. Observations of field-aligned electron precipitation, however, invariably show large fluxes of low energy electron precipitation as well. Whalen and McDiarmid (1972) and also Paschman et al., (1972) noted the presence of these low energy electrons in conjunction with field-aligned events. Both papers suggested that these electrons had originated from locations within the region of $E_{\parallel}$ (perhaps by ionization of neutrals) and were then accelerated through only a fraction of the available potential. Even so, such low energy

electrons would also display a high degree of field-aligned collimation and could not directly account for the isotropic low
energy precipitation so clearly displayed in Figure 4.

O'Brien (1970) felt that the existence of large fluxes of
isotropic, low energy electrons was totally inconsistent with
parallel electric fields and the explanation for field-aligned
particle precipitation lay with some other mechanism. Part of the
difficulty was in explaining the complex variations in pitch angle
distribution which were dependent upon the electron energy. The
other part was that there appeared no source of electrons from any
region along the auroral line of force which could supply the low
energy electrons in the numbers that are observed.

Many of these objections were overcome by Evans (1974) who
suggested that the low energy electrons were in fact secondaries
and backscattered primaries which had been produced in the local
atmosphere by the primary field aligned electron beam. These
electrons re-appeared as precipitating particles because they
had been reflected downwards by the parallel electric field which
in effect represented a potential barrier to upgoing low energy
electrons. In this fashion, a large isotropic population of low
energy electrons was created (the atmosphere being a diffusive
source).

There has been some success in explaining the character of
actual auroral electron precipitation using a numerical model
which includes parallel electric fields and the effect of atmospheric backscatter. The data points in Figure 5 reproduce the $0°$
pitch-angle energy spectrum observed by Arnoldy et al. (1974) and
shown above in Figure 4. The solid line is the modeled energy
spectrum. The energetic portion of the model spectrum was generated
by allowing a Maxwellian electron population ($n_e \cong .2$ cm^{-3},
$T \cong 750$ eV) to freely accelerate through a ~ 4500 V parallel potential difference. This portion of the electron beam would have
a field-aligned angular distribution. That portion of the electron spectrum below 4500 V was computed assuming these electrons
were secondaries and backscatter created from the atmosphere by
the primary electron beam. The modeled spectrum closely repro
duces the observed spectral shapes and flux intensities and also
accounts for the pitch angle isotropy at low energy and field-
aligned anisotropy at higher energy.

Implicit in this model is that there should be a symmetry
between the upward and downward going low energy electron population for, basically, they are the same population. Thus, the model
can conveniently account for the anomalous backscatter of low
energy electrons (that is an equality between upward and downward
going electron populations) such as has been reported by Arnoldy
and Choy (1973) and Reasoner and Chappel (1973).

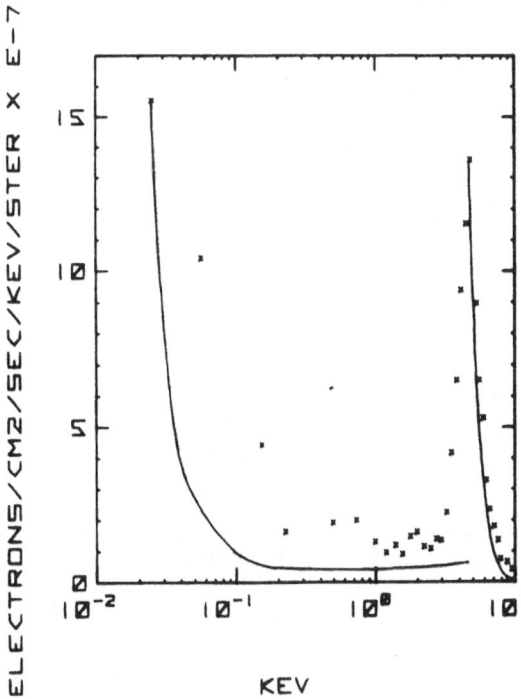

Figure 5. A 0° pitch angle electron energy spectrum measured
by Arnoldy et al. (1974) (data points) compared to a model energy
spectrum (solid line) computed assuming that the energetic electrons were created by free acceleration through a parallel potential drop The lower energy electron fluxes were calculated
from the secondary and backscatter produced from the atmosphere
by the primary beam.

     Some data concerning field-aligned electron fluxes at higher
altitude have been obtained from satellite instruments. Figure 6
(courtesy of Ackerson and Frank, 1972), shows a representative
set of electron energy spectrums obtained at 2000 km altitude during an inverted V event. Each panel displays energy spectrums
taken at 90° (trapped) and 0° (precipitated) pitch angles. The
upper right hand panel shows a pair of spectrums which are in
excellent agreement with what would be expected if the primary
electron flux was caused by a field-aligned potential drop of
about 500 V located above the satellite. At this altitude, field-
alignment is not just a feature confined to the primary, accelerated electrons, but also includes those electrons of atmospheric
origin. This is because these electrons, nearly isotropic just
above the atmosphere, will become collimated to within 45° of B
as they move upwards to 2000 km altitude.

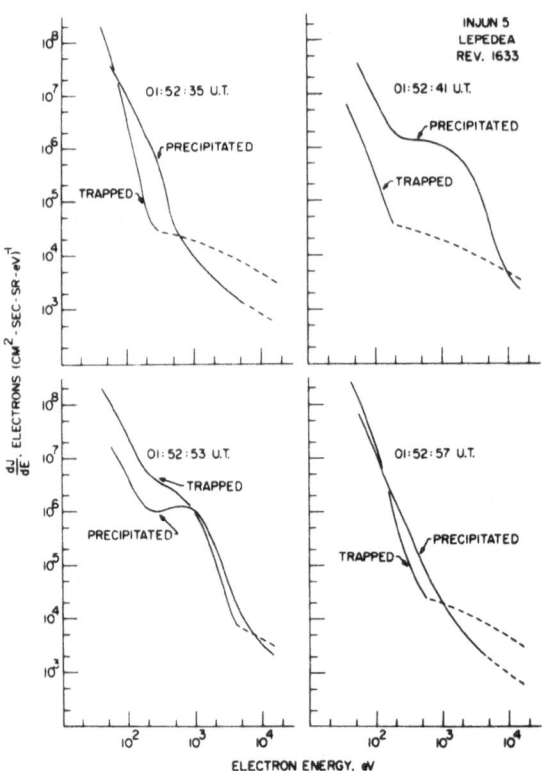

Figure 6. A series of electron energy spectrums obtained during an "inverted V" event by the Injun 5 satellite at 2000 km altitude (courtesy of Ackerson and Frank, 1972). In each panel "precipitated" is 0^0 pitch-angle, "trapped" is 90^0 pitch angle. The pair of spectrums in the upper right panel is consistent with particle acceleration by $E_{\parallel}$. The pair in the lower left panel is consistent with an $E_{\parallel}$ only if pitch angle scattering processes, after free acceleration, are involved.

Of course, the lower left panel in Figure 6 shows data which do not agree with this model. The low energy electrons (of atmospheric origin?) have maximum intensities at 90^0 to the $\vec{B}$. Clearly this data cannot be explained simply in terms of parallel electric fields producing one part of the electron beam while the other part originates from the atmosphere, all electrons conserving their magnetic moment except when within the atmosphere. It may be that this data shows the effect of pitch angle scattering of the electrons at some point between the $E_{\parallel}$ and the atmosphere. Such scattering would tend to populate local pitch angles near 90^0, and these electrons would become quasi-trapped between a high altitude electrostatic mirror and a lower altitude magnetic mirror (Albert and Lindstrom, 1970). Just pitch angle scattering, arising out of beam

plasma instabilities in the high ionosphere, has been discussed by
Papadopoulous and Coffey (1974) and proposed to explain certain
features observed in an auroral electron beam by Reasoner and
Chappel (1973).

It should be pointed out that if a field-aligned beam of electrons can be isotropized by beam plasma instabilities of the sort
suggested by Papadopoulous and Coffey, then the absence of a collimation in auroral electrons need not eliminate parallel electric
fields as the energization mechanism for the electrons. The observation of highly collimated particle beams might then be regarded as sufficient evidence for parallel electric fields but not
necessary evidence.

This point is important because observations of field-aligned
electron beams at ionospheric levels seem to be associated with
either the edges or boundaries of discrete auroral displays (Maehlum and Moestue, 1973; Arnoldy et al., 1974) or with widespread,
low intensity precipitation well displaced from discrete auroral
forms (Evans et al., 1972). In the center of discrete auroral
forms, the electron precipitation is observed to be isotropic
(Maehlum and Moestue, 1973). The isotropy in electron fluxes probably should not be used to eliminate parallel electric fields from
consideration as the accelerating mechanism for such electrons.

Compared to the extensive results concerning field-aligned
electron precipitation, observations of field-aligned ion precipitation are very sparse. Hultqvist et al. (1971) have reported that
strongly field-aligned beams of energetic ions (6 keV) are frequently observed at 1000 km altitude and high geomagnetic latitudes. The only other report in the literature on this subject is
by Réme and Bosqued (1971) who describe field-aligned ion precipitation observed at 400 km over an aurora. Neither of these observations describe the ion beam in any great detail, particularly in
terms of ion energy spectrums as a function of pitch angle.

The reason why field-aligned ion beams have not been observed
to nearly the extent or detail as electron beams may have to do
with the difficulty in making ion observations. Ion fluxes are
generally much lower than electron fluxes, and charge exchange
within the upper atmosphere complicate the measurements even
further. Of course the lack of observations might be due to a
preferential direction in the parallel electric field (i.e. most
often directed upwards so as to accelerate electrons downwards.

McIlwain has presented at this meeting recent observations of
field-aligned charged particle fluxes as measured by ATS-6 at
high altitudes along the high latitude line of force. It seems
that these events may be interpreted in terms of parallel electric
fields located at some distance away from ATS-6 along the line of
force toward the atmosphere.

There has been no accepted explanation for the origin of parallel electric fields within the magnetosphere. Hultqvist (1971)
has presented a theory which considers the interface boundary between the energetic plasma which populates the outer magnetosphere
and the very low energy plasma in the topside ionosphere. Hultqvist imposed the condition that no net current flow across this
boundary and showed that this required that a potential difference develop in the sense to accelerate positive ions downwards.

At present two theories are in vogue to explain parallel
electric fields which accelerate electrons downwards. The first
invokes ionospheric level plasma instabilities driven by field-
aligned currents to greatly decrease the electrical conductivity
parallel to the magnetic field. Kindel and Kennel (1971) outline
a representative theory of this sort. The effect of the decreased
conductivity is to allow a greater fraction of the available magnetosphere potential to appear along the magnetic field line.
Fast electrons introduced into the region of $E_\parallel$ may be freely
accelerated downwards (in contrast to the low energy plasma whose
motion is presumably impeded by the instability).

The alternative theory involves double layers (Block, 1972;
Carlqvist, 1972). In this theory a large fraction of the available magnetospheric potential drop appears at some discrete point
along the magnetic line of force simply because there is insufficient free charge available at that point to complete the current loop.

The experimental evidence that electric potential differences
exist along magnetic lines of force in the magnetosphere seems
uncontestable. Whatever the theory for these electric fields
is, it must account for

1. Parallel potential drops as large as 10 kV.

2. A location of the $E_\parallel$ which is often, if not exclusively
    at low altitude in the auroral line of force.

3. A lifetime for sustaining the $E_\parallel$ which is much longer
    than the time taken by a charged particle in being freely
    accelerated through it.

4. The fact that work is extracted from the electric field in
    order to accelerate charged particles.

5. The latitudinal extent of the regions of $E_\parallel$ which, judg-
    ing from Evans et al. (1972), and Ackerson and Frank
    (1972) may be in excess of 100 km at ionospheric levels.

ADDITIONAL EVIDENCE FOR LOW ALTITUDE
ACCELERATION OF AURORAL PARTICLES

There have been occasional reports in the literature of auroral particle behavior which suggested that some sort of unspecific low altitude acceleration mechanism was involved. Taken individually such reports do not seem significant, but in light of the discussion thus far they do provide additional evidence in favor of the low altitude acceleration of at least some portion of the auroral particle population.

Bernstein and Wax (1970) describe short lived enhancements ($\sim$ 0.1 sec) in the flux of low energy protons observed at 200 km altitude above an active auroral display. Moreover, these enhancements appeared anti-correlated with the intensity of the electron precipitation. The rise times and short durations of the proton flux enhancements restricted their origin to a process located no further than 40 km from the rocket payload; i. e. at an altitude of less than 250 km. The fact that the electron fluxes appeared anti-correlated with the protons during these events led Bernstein and Wax to suggest that a transient parallel electric field was responsible for the local acceleration of ions and deceleration of precipitating electrons.

Johnstone (1971) also observed a correlation between auroral proton and electron variations, in this case flux enhancements occurring simultaneously in both species. Johnstone considered the possibility that simultaneity in the flux variation was due to the rocket moving across a spacial boundary in the auroral particle precipitation. This explanation was discarded because the observed proton intensity variations and pitch angle behavior did not agree with those predicted at the border of proton precipitation based upon charge exchange calculations. The flux variations having been identified as temporal, the time coincidence between proton and electron transients placed an upper limit of 3000 km on the altitude at which the fluctuations were introduced into the beam. Johnstone favored the mechanism described by Swift (1970) which is an instability driven by ring current protons which are mirroring at low altitudes. The net effect of the instability is to accelerate electrons and to enhance the precipitation of protons.

Johnstone and Davis (1974) have also analyzed in detail the behavior of auroral electrons observed during transient flux enhancements unambiguously identified as temporal in nature. On the basis of a general lack of velocity dispersion on the part of the incident electrons, it was concluded that the transient was introduced no further than a few R_E along the auroral line of force. Johnstone and Davis essentially considered every postulated particle acceleration or pitch angle scattering mechanism

in order to explain the observations. Every single one was eliminated except for free acceleration of particles by parallel electric fields. The unique aspect of the parallel electric field that
they proposed was that the field was relatively short lived, per-
sisting for only the 10 seconds or so that would be required to
accelerate all the particles available on an auroral latitude tube
of force. Johnstone (1975) has recently developed a numerical
model describing the effects of a time dependent parallel electric
field upon the pitch angle, energy, and precipitation rates of a
magnetospheric plasma.

Sharp et al. (1971) attempted a correlative study between
the energetic electron population as observed in the equatorial
plane on ATS-5 and as observed by a similar instrument on board a
low altitude polar orbiting satellite. Data for study was restricted to those instances when the low altitude satellite was
near the low altitude location of the ATS-5 line of force. The one
feature which distinguished the two electron populations was the
existence of a peak in the energy spectrum of the low altitude
electron population, a peak which was not a feature in the equatorial plane electron population.

Choy, et al. (1971), also attempted to compare detailed observations of the equatorial electron population made on board
ATS-5 with electron observations performed on board a rocket
launched from Churchill, Man. They found that often the electron
energy spectrums were very similar at these two points. However,
one example was shown where a well-defined peak in the auroral
electron spectrum was observed at low altitude but was not present
in the equatorial plane spectrum.

Intercomparisons of particle populations at two points in
space, distant from one another, is fraught with ambiguities and
a lack of agreement would ordinarily be of little significance.
However, in light of the great mass of evidence for the acceleration of particles by low altitude parallel electric fields, the
two comparisons just cited could be regarded as some additional
favorable evidence.

Finally, the observation of energetic, singly ionized oxygen
within the magnetosphere (Shelley, et al., 1972; Sharp et al.,
1974) should be mentioned. The significance of these observations is not that they suggest a low altitude energization (there
is no direct evidence for this) but that they identify the ionosphere as the origin of some of the particles which are accelerated in situ or by some magnetospheric process.

In this sense this last observation supports the implicit
theme of this review. That is that the ionosphere cannot in any
way be considered as a passive element in magnetospheric processes
but can be seen to play a very active role.

SUMMARY AND CONCLUSIONS

There can be no doubt that mechanisms which accelerate charged particles at low altitude on the auroral latitude lines of force are a significant, if not a dominant, contributor to the production of of auroral particles.

The existence of at least one low altitude process which can energize electrons to greater than 100 keV has been established. It has been suggested that this energization is a result of a beam-plasma instability produced by unusually intense fluxes of keV energy auroral electrons as they pass through the topside ionosphere. Whether such low altitude electron energization can contribute significantly to the population of > 40 keV electrons trapped in the magnetosphere is questionable.

Of far greater importance to both auroral and magnetospheric physics is the virtually certain fact that large, quasi-static potential differences exist parallel to the magnetic field at auroral latitudes. The impact upon auroral physics is obvious. These parallel electric fields appear capable of freely accelerating particles downwards toward the atmosphere (or upwards as well) thus directly producing auroral particle beams. Indeed, the principal evidence for the very existence of such electric fields has been the unique pitch-angle distribution they impose upon the accelerated particles.

To date, observations of auroral electron precipitation which exhibits the field-aligned signature of an $E_\parallel$ have been confined to the edges of discrete auroral forms or to fairly extensive regions of rather weak precipitation detached from the bright discrete aurora. However, the isotropic nature of the electron precipitation over the center of discrete auroral forms need not exclude an $E_\parallel$ from having accelerated these particles as well. It may be that such electrons had been scattered in pitch-angle after acceleration by, perhaps, some plasma instability.

If it can be shown that field-aligned electron fluxes may be rapidly isotropized in passing through the topside ionosphere, then a strong argument can be made that parallel electric fields, often located near the auroral ionosphere, are the prime energization mechanism for those electrons which cause discrete auroral forms.

On a magnetospheric scale, the principal impact of large potential differences along the magnetic field lines would be the negation of the assumptions underlying the "frozen in" magnetic field concept. This in turn would alter some of the concepts and descriptions presently used in magnetospheric physics. Some examples would be the mapping of convection flow patterns along the magnetic field and theories about the magnetospheric current system.

REFERENCES

Ackerson, K. L. and L. A. Frank, Correlated satellite measurements of low-energy electron precipitation and ground based observations of a visable auroral arc, J. Geophys. Res., 77, 1128, 1972.

Akasofu, S. I., and S. Chapman, A neutral line discharge theory of the aurora polaris, Royal Society Phil. Trans. A, 253, 359, 1961.

Albert, R. D. and P. J. Lindstrom, Auroral-particle precipitation, and trapping caused by electrostatic double layers in the ionosphere, Science, 170, 1398, 1970.

Alfven, H., Cosmical Electrodynamics, pa. 204, first edition, Clarendon Press, Oxford, 1950.

Arnoldy, R. L., Rapid fluctuations of energetic auroral particles J. Geophys. Res., 75, 228, 1970.

Arnoldy, R. L., and L. W. Choy, Auroral electrons of energy less than 1 keV observed at Rocket Altitudes, J. Geophys. Res., 78, 2187, 1973.

Arnoldy, R. L., P. B. Lewis, and P. O. Isaacson, Field-aligned auroral electron fluxes, J. Geophys. Res., 79, 4208, 1974.

Beach, R., G. R. Cresswell, T. N. Davis, T. J. Halliman, and L. R. Sweet, Flickering, a 10 cps fluctuation within bright auroras, Planet. Space Sci., 16, 1525, 1969.

Berko, F. W., Distributions and characteristics of high-latitude field-aligned electron precipitation, J. Geophys. Res., 78, 1615, 1973.

Berko, F. W. and R. A. Hoffman, Dependence of field-aligned electron precipitation occurrence on season and altitude, J. Geophys. Res., 79, 3749, 1974.

Bernstein, W., and R. L. Wax, Impulsive precipitation events during an auroral breakup., J. Geophys. Res., 75, 3915, 1970.

Block, L. P., Potential double layers in the ionosphere, Cosmic Electrodyn. 3, 349, 1972.

Bosqued, J. M., G. Cardona, and H. Réme, Auroral electron fluxes parallel to the geomagnetic field lines, J. Geophys. Res., 79, 98, 1974.

Carlqvist, P., On the formation of double layers in plasma, Cosmic Electrodyn., 3, 377, 1972.

Chamberlain, J. W., Discharge theory of auroral rays, the aurora and the airglow, pa. 206, The Universities Press, Belfast, 1956,

Choy, L. W., R. L. Arnoldy, W. Potter, R. Kintner, and L. J. Cahill, Field-aligned particle currents near an auroral arc, J. Geophys Res., 76, 8279, 1971.

Dungey, J. W., The theory of neutral sheets, pa. 210, Earth's Magnetospheric Processes, Ed. by B. M. McCormac, D. Reidel, Dordrecht, 1972.

Evans, D. S., A 10 cps periodicity in the precipitation of auroral zone electrons, J. Geophys. Res., 72, 4281, 1967.

Evans, D. S., B. Maehlum, and T. Wedde, High-latitude observations of field-aligned electron beams (abstract) EOS Trans. AGU, 53, 731, 1972.

Evans, D. S., Precipitating electron fluxes formed by a magnetic field-aligned potential difference, J. Geophys. Res., 79, 2853, 1974.

Fahleson, U. V., Critical review of electric field measurements, pa. 210, Earth's Magnetospheric Processes, Ed. by B. M. Mc Cormac, D. Reidel, Dordrecht, 1972.

Hoffman, R. A., and D. S. Evans, Field-aligned electron bursts at high latitude observed by OGO-4, J. Geophys. Res., 73, 6201, 1968.

Hultqvist, B., On the production of magnetic field-aligned electric fields by the interaction between the hot magnetospheric plasma and the cold ionosphere, Planet. Space Sci., 19, 279, 1971.

Hultqvist, B., H. Borg, W. Riedler, and P. Christophersen, Observations of a magnetic field-aligned anisotropy for 1 and 6 keV positive ions in the upper ionosphere, Planet. Space Sci., 19, 279, 1971.

Johnstone, A. D., Correlation between electron and proton fluxes in postbreakup aurora, J. Geophys. Res., 75, 3915, 1970.

Johnstone, A. D., and T. N. Davis, Low-altitude acceleration of auroral electrons during breakup observed by a Mother-Daughter rocket, J. Geophys. Res., 79, 1416, 1974.

Johnstone, A. D., Time dependent effects of a parallel electric field, preprint, Dept. of Physics and Astronomy, University College, London, 1975.

Kindel, J. M., and C. F. Kennel, Topside current instabilities, J. Geophys. Res., 76, 3055, 1971.

Lampton, M., Daytime observations of energetic auroral-zone electrons, J. Geophys. Res., 72, 5817, 1967.

Lyons, L. R., Electron diffusion driven by magnetospheric electrostatic waves, J. Geophys. Res., 79, 575, 1974.

Maehlum, B. N. and H. Moestue, High temporal and spatial resolution observations of low energy electrons by a Mother-Daughter rocket in the vicinity of two quiescent auroral arcs, Planet. Space Sci., 21, 1957, 1973.

McDiarmid, I. B., D. C. Rose, and E. E. Budzinski, Direct measurement of charged particles associated with auroral-zone radio absorption, Can. J. Phys., 39, 1888, 1961.

Mozer, F. S. and P. Bruston, Observation of the low-altitude acceleration of auroral protons, J. Geophys. Res., 71, 2201, 1966.

O'Brien, B. J., Consideration that the source of auroral energetic particles is not a parallel electrostatic field, Planet. Space Sci., 18, 1821, 1970.

Papadopoulous, K., and T. Coffey, Nonthermal features of the auroral plasma due to precipitating electrons, J. Geophys. Res., 79, 674, 1974.

Paschman, G., R. G. Johnson, R. D. Sharp, and E. G. Shelley, Angular distributions of auroral electrons in the energy range 0.8 - 16 keV, J. Geophys. Res., 77, 6111, 1972.

Perkins, F. W., Plasma-wave instabilities in the ionosphere over the aurora, J. Geophys. Res., 73, 6631, 1968.

Reasoner, D. L. and C. R. Chappel, Twin payload observations of incident and backscattered auroral electrons, J. Geophys. Res., 78, 2176, 1973.

Reme, H. and J. M. Bosqued, Evidence near the auroral ionosphere of a parallel electric field deduced from energy and angular distributions of low-energy particles, J. Geophys. Res., 76, 7683, 1971.

Sharp, R. D., D. L. Carr, R. G. Johnson, and E. G. Shelley, Coordinated auroral electron observations from a synchronous and a polar satellite, J. Geophys. Res., 75, 5401, 1970.

Sharp, R. D., R. G. Johnson, E. G. Shelley, and K. K. Harris, Energetic O^+ ions in the magnetosphere, J. Geophys. Res., 79, 1974.

Sheldon, W. R., J. R. Benbrook, and J. W. Kern, Fast bursts of auroral x-rays (5 keV < E < 15 keV) observed at 40 km altitude (abstract), EOS, 54, 436, 1973.

Shelley, E. G., R. G. Johnson, and R. D. Sharp, Satellite observations of energetic heavy ions during a geomagnetic storm, J. Geophys. Res., 77, 6104, 1972.

Speiser, T. W., Particle acceleration in model current sheets, 2, applications to auroras using a geomagnetic tail model, J. Geophys. Res., 72, 3919, 1967.

Swift, D. W., Particle acceleration by electrostatic waves, J. Geophys. Res., 75, 6324, 1970.

Venkatesan, D., M. N. OLiven, P. J. Edwards, K. G. McCracken, and M. Steinbeck, Microburst phenomena, 1. Auroral-zone x-rays, J. Geophys. Res., 73, 2333, 1968.

Whalen, B. S., and I. B. McDiarmid, Observations of magnetic-field-aligned auroral-electron precipitation, J. Geophys. Res., 77, 191, 1972.

Winiecki, T., Analysis of rapid temporal fluctuations in auroral particle fluxes, M. Sc. Thesis, Space Science Department, Rice University, Houston, Texas, 1967.

MECHANISMS FOR DRIVING BIRKELAND CURRENTS

Rolf Boström

Department of Plasma Physics
Royal Institute of Technology
S-100 44 Stockholm 70, Sweden

INTRODUCTION

The hot magnetospheric plasma interacts electrody-
namically with the cold ionospheric plasma by means of
Birkeland currents flowing along geomagnetic field lines
connecting the two plasmas. The fully ionized, collision-
free plasma of the magnetosphere and the partially ionized, collisionally dominated plasma of the ionosphere behave differently and in general there will be an electric mismatch between the two regions resulting in Birkeland current flow. The set of relations governing the
system is summarized in Figure 1. We have a closed system of equations that in principle can be solved given
appropriate boundary conditions (cf. Vasyliunas 1970).
Most studies thus far, including this one, consider only
some links of the framework. Here we will not use the
approach of a boundary value problem. Rather, we will
assume that some of the physical parameters are known
from measurements and we will ask what information about
the other quantities can be extracted from this knowledge
using the governing equations. In particular we will consider possible conclusions regarding mechanisms for driving the Birkeland currents that can be obtained from
the characteristics of these currents as observed by
Zmuda and Armstrong (1974).

For the magnetospheric plasma we will use MHD theory.
Certainly this has shortcomings which must be borne in
mind; for example it cannot be used to study phenomena

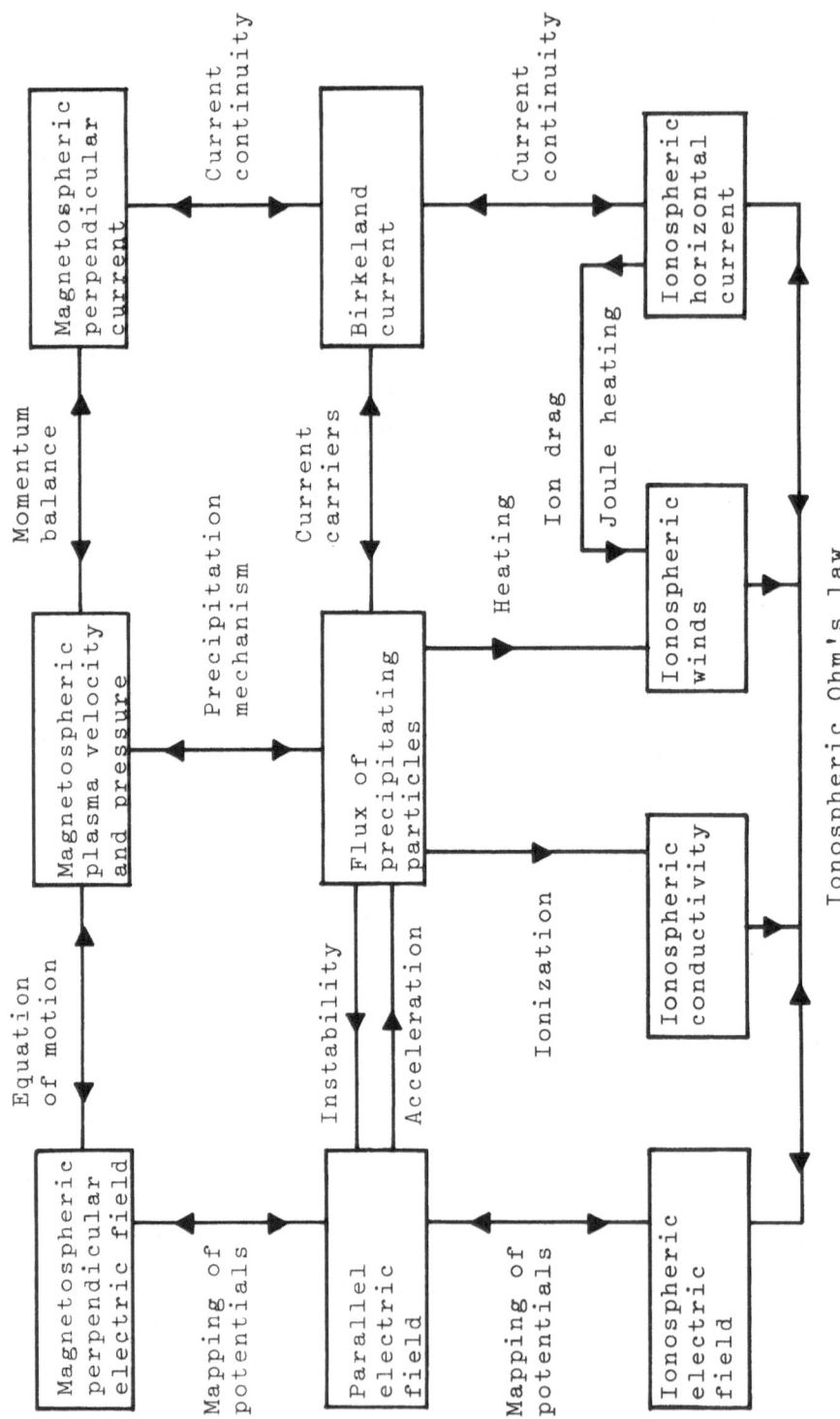

Fig. 1. Interrelationships between magnetospheric and ionospheric parameters.

that depend on the energy distribution of the particles.
It may well be that the formation of forbidden regions
in the particle flow pattern, of a different size for
particles of different energy, is significant for the
Birkeland current system. Nevertheless a discussion in
terms of MHD theory might be useful as we believe that
those phenomena that <u>are</u> predicted using the MHD equations would occur also in a more refined treatment.

   In the magnetosphere there is a balance between pres-
sure and inertial forces and a magnetic $\vec{j} \times \vec{B}$ force due
to the flow of transverse currents. Birkeland currents
originate in the magnetosphere when this transverse current varies spatially so that it has a divergence, caused
by spatial variations of the magnetospheric parameters.
Thus the magnetosphere may act as an electric generator
and the ionosphere is then the load. However, Birkeland
currents may also originate from the ionosphere. Wind
systems in the ionosphere may drive currents there which
could have a divergence giving rise to Birkeland currents.
Here we will not consider this link involving the motion
of the neutral gas. Also secondary Birkeland currents of
ionospheric origin occur when the horizontal ionospheric
currents driven by the magnetospheric generator flow
through regions of varying ionospheric conductivity.
Then polarization electric fields tend to build up, but
they also tend to discharge to the magnetospheric load
by means of Birkeland currents.

                    BASIC EQUATIONS

   For our discussion we will use the very simplest form
of equation for the plasma motion

$$\vec{E} + \vec{v} \times \vec{B} = 0 \tag{1}$$

Although field-aligned electric fields might appear in
regions, and at times, of intense Birkeland currents we
do not believe that they would occur as a regular feature
over large areas affecting the magnetosphere-ionosphere
mapping of the gross-scale electric fields. Thus we neglect the link of Figure 1 involving parallel electric
fields. We will consider relatively quiet (non substorm)
conditions for which we can use an electrostatic potential. If this potential is given over a surface, such as
the ionosphere, cutting all field lines, then the potential, electric field and plasma velocity are determined
everywhere in the magnetosphere, assuming the magnetic
field to be known.

The transverse current in the magnetosphere is related to the plasma pressure and inertial forces by a
momentum balance equation which, if solved for $\vec{j}_\perp$, reads

$$\vec{j}_\perp \left(1 - \mu_o \frac{p_{\parallel}-p_\perp}{B^2}\right) = \frac{\vec{B}}{B^2} \times \left[\rho(\vec{v}\cdot\nabla)\vec{v} + \nabla p_\perp + \right.$$

$$\left. + (p_{\parallel}-p_\perp) \frac{\nabla B}{B}\right] \tag{2}$$

This expression for $\vec{j}_\perp$ may be derived by summing the
contributions from the gradient, curvature and polarization drifts of the individual particles and the magnetization current (cf. Parker, 1957). We will use the ap-
proximation $\varepsilon = \mu_o(p_{\parallel}-p_\perp)/B^2 \ll 1$. Thus the results will
be precise if the pressure is isotropic, while for the
case of anisotropic pressure the results are quantitatively correct only if the kinetic energy density of the
random ("thermal") particle motion is small compared to
the magnetic energy density which implies that the effect of the pressure-driven currents on the magnetic
field should be negligible.

The momentum equation (2) primarily gives information only on the transverse component of the current.
However, we can use this equation also to obtain information on the Birkeland component using the requirement
that the total current is divergence-free, that is,
div $\vec{j}_{\parallel}$ = -div $\vec{j}_\perp$. Integrating along a fluxtube we can
then find $\vec{j}_{\parallel}$.

To specify the anisotropic pressure we need at least
two scalar parameters, one giving the thermal energy
density and the other, or others, characterizing the
form of the pitch-angle distribution of the energetic
particles providing the pressure. If these are known for
example for the equatorial plane the pressure may be
evaluated at any point along the field lines assuming
that all particles in their bounce motion pass this plane.
We will use a simple form of anisotropy corresponding to
a pitch-angle distribution where the number of particles
per sterradian in the equatorial plane for all energies
is proportional to sin$^{2\gamma}$ (Θ) with (Θ) being the pitch
angle. Then the <u>form</u> of the pitch-angle distribution
remains the same at all points along the field line,
although the number density varies. We find

$$p_\perp = (\gamma + 1) p_o \left(\frac{B}{B_N}\right)^{-\gamma}; \quad p_{\parallel} = p_o \left(\frac{B}{B_N}\right)^{-\gamma} \tag{3}$$

where the parameter $p_0 = n_0 m <w^2> (B_0/B_N)^\gamma/(2\gamma+3)$, for a given γ, gives a measure of the kinetic energy density. n_0 is the number density and B_0 the magnetic field at a reference level taken to be the ionosphere, while B_N is a constant magnetic field, taken to be the field at the pole, introduced for the purpose of normalization. The parameters γ and p_0 are constant along each field line but may vary from one field line to another. Obviously the functional form $\sin^{2\gamma} \Theta$ can be used to model a wide range of pitch-angle distributions: $\gamma = 0$ corresponds to an isotropic pressure $p_\perp = p_\parallel = p_0$, which is constant along the field lines, while the limit $\gamma \to \infty$ corresponds to $p_\parallel/p_\perp \to 0$ and the limit $\gamma \to -1$ to $p_\perp/p_\parallel \to 0$.

Our problem of ionosphere-magnetosphere interaction is evidently a two-dimensional one as knowledge of parameters over a surface suffices to determine the field variables at all points. By performing integrations along the field lines of variables deriving from (1) and (2) it is then possible to reduce the study to this surface. We will show in the next section how this reduction appears for the case of currents driven by the pressure terms. It is well known that for the ionosphere a similar reduction can be performed (for phenomena of a spatial scale exceeding a few kilometers) by integrating currents and conductivities along the highly conducting field lines, reducing the influence of the ionosphere to that of a conducting sheet.

BIRKELAND CURRENTS DRIVEN BY THE PLASMA PRESSURE

We will discuss the effects of the pressure and inertial forces separately and we start with the pressure. As the magnetic field lines play a central role for our problem it is convenient to use coordinates that explicitely refer to these. Such coordinates are Euler potentials α and β, labelling the field lines, and the length of arc ℓ measured along the field lines from the ionosphere in the northern hemisphere. With these coordinates

$$\vec{B} = -\nabla\alpha \times \nabla\beta \qquad (4)$$

Let us consider the current continuity in a fluxtube bounded by the surfaces $\alpha = \alpha_0$, $\alpha = \alpha_0 + d\alpha$, $\beta = \beta_0$, $\beta = \beta_0 + d\beta$, the ionosphere ($\ell = 0$), and the equatorial plane. Figure 2 shows a segment of this fluxtube. The total current flowing out from the fluxtube through the

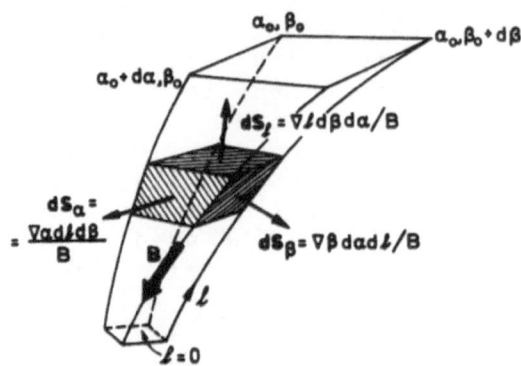

Fig. 2. Segment of fluxtube showing coordinates α, β, and ℓ and surface elements.

surface $\alpha = \alpha_0 + d\alpha$, with the surface element $d\vec{S}_\alpha = \nabla\alpha d\ell d\beta/B$, is

$$J_\alpha d\beta = d\beta \int_0^L \frac{\vec{j}_\perp \cdot \nabla\alpha}{B} d\ell \qquad (5)$$

Using (2), (3), and (4) we find after some reductions, neglecting a term of order ε,

$$J_\alpha d\beta = d\beta \int_0^L \frac{1}{B} \frac{\partial p_\perp}{\partial\alpha} d\ell - d\beta \left[\frac{p_\parallel - p_\perp}{B^2} \frac{\vec{B}\times\nabla\ell \cdot \nabla\alpha}{B^2} \right]_0^L \qquad (6)$$

A corresponding expression applies to the integrated current $J_\beta d\alpha$ flowing through the surface $\beta = \beta_0 + d\beta$. The net inflow through the four sides of the fluxtube due to $\vec{j}_\perp$, which is $-(\partial J_\alpha/\partial\alpha) \, d\beta d\alpha - (\partial J_\beta/\partial\beta) \, d\alpha d\beta$, must be balanced by the flow of vertical current of density j_v through the bottom surface element $dS_\ell = |\nabla_0\ell| d\beta d\alpha/B_0$. We assume that the equatorial plane is a plane of symmetry so that there is no current flow through this plane into the fluxtube considered. With the vertical current counted positive if flowing into the ionosphere we obtain

$$j_v = -\frac{B_0}{|\nabla_0\ell|} \left\{ \frac{\partial}{\partial\alpha} \int_0^L \frac{1}{B} \frac{\partial p_\perp}{\partial\beta} d\ell - \frac{\partial}{\partial\beta} \int_0^L \frac{1}{B} \frac{\partial p_\perp}{\partial\alpha} d\ell \right\} \qquad (7)$$

Introducing the expression (3) for $p_\perp$ we find after some reductions

$$j_v = \nabla_o p_o \times \nabla_o V_1 \cdot \hat{r} - \nabla_o \gamma \times \nabla_o (p_o V_2) \cdot \hat{r} \qquad (8)$$

Here

$$V_1 = \int_0^L \frac{B_N^\gamma \, d\ell}{B^{1+\gamma}} \qquad (9)$$

$$V_2 = \int_0^L \frac{B_N^\gamma \ln(B/B_N) \, d\ell}{B^{1+\gamma}} \qquad (10)$$

and ∇_o is the gradient evaluated for $\ell = 0$ and $\hat{r}$ is a vertical unit vector. In deriving equations (7) and (8) we have neglected terms of order ε and terms which are at most of the same order as the flow of $\vec{j}_\perp$ through the end surfaces of the fluxtube, which we can neglect as we are interested in cases where the contribution to j_v from the Birkeland current $j_\parallel$ dominates over the contribution from $\vec{j}_\perp$. For isotropy ($\gamma=0$) all neglected terms would vanish identically.

As the variables p_o, γ, V_1, and V_2 depend only on α and β but not ℓ, equation (8) demonstrates that it is possible to derive the vertical current density j_v from a study of the variation of these variables in the ionosphere only.

For the particular case of isotropy ($\gamma=0$) Vasyliunas (1970) has given an expression analogous to what (8) and (9) would give for $\gamma=0$ (although off by a factor 2). For the special case of a dipolar field and isotropy we may solve (9) analytically and derive the expression

$$j_v = \frac{4R_e^2}{\mu_o m} (1+4 \cos^2\theta_o -2 \cos^4\theta_o +\frac{4}{5} \cos^6\theta_o -$$

$$-\frac{1}{7} \cos^8\theta_o)\sin^{-10}\theta_o \frac{\partial p}{\partial \phi} \qquad (11)$$

Here m is the magnetic moment of the earth, R_e the radius of the ionosphere, and θ_o the colatitude. A similar expression has been given by Kern (1967) although it differs from (11) by a very small term due to the contribution from the vertical component of $\vec{j}_\perp$ at the ionosphere, neglected in Kern's analysis but included here. This term, although smaller than the contribution from $j_\parallel$ by a factor 5×10^4 for $\theta_o = 20°$, has a certain significance in that it makes the structure of the analytic expression simpler. Only the azimuthal gradient of the pressure is

significant for driving Birkeland currents according to
(11). Evidently j_V integrated along a curve θ_0 = constant
around the globe vanishes. Thus there is as much Birkeland current flowing to the earth as away from the earth
in each strip of constant θ_0. We will show below that a
similar conclusion applies under more general conditions.

   Returning now to the more general case of anisotropic pressure and non-dipolar field geometry, we should
first of all point out that if we have a model of p_0 and
γ and of the magnetic field we may of course use (8),
(9), and (10) to evaluate j_V. However, a more interesting
question is to ask what information about the pressure
parameters p_0 and γ we can obtain from a model, or
measurements, of j_V. Clearly knowledge of one variable,
j_V, is not sufficient to derive the two independent quantities p_0 and γ. Thus, different distributions of p_0 and
γ could give the same j_V. However, if one of the quantities p_0 and γ is given we should be able to derive some
information about the other from measurements of j_V. The
problem is particularly simple if the anisotropy factor
γ is constant. Then equation (8) reduces to

$$j_V = \nabla_0 p_0 \times \nabla_0 V_1 \cdot \hat{r} \tag{12}$$

Equation (12) shows that if j_V is known it is possible
to derive the component of $\nabla_0 p_0$ along contours V_1 =
constant, and by integration, the variation of p_0 along
each such contour. Integrating around a closed curve
V_1 = constant the final p_0 should equal the initial value,
that is, the total variation of p_0 should be zero. However, in general this would not be the case for an arbitrary distribution of j_V. Thus there exists a constraint
on the possible forms of distribution of vertical current
if this is driven by a distribution of pressure with constant anisotropy factor. This can be used as a test if
the observed currents are driven in this way. If we evaluate the total current flowing into the ionosphere within a contour V_1 = constant we find

$$J_V = \iint j_V \, dS_r = \iint -\nabla_0 p_0 \times \nabla_0 V_1 \cdot \vec{dS_r} =$$

$$= \iint \operatorname{curl}(V_1 \nabla_0 p_0) \cdot \vec{dS_r} = -\int V_1 \nabla_0 p_0 \cdot \vec{ds} =$$

$$= -V_1 \int \nabla_0 p_0 \cdot \vec{ds} = 0 \tag{13}$$

Thus the net current to the ionosphere is zero in the
region enclosed by each contour V_1 = constant and thus
also within any strip bounded by contours of constant V_1.
Also in the ionosphere there could be no net flow of the

horizontal currents across these contours. In itself, the existence of contours with this property is not unexpected. However, it is worth noting that we have been able to predict the shape of these contours, so that we can use this as a test on a measured distribution of j_V. Figure 3 shows the contours V_1 = constant evaluated for three different values of γ, namely $\gamma = -0.5$ ($p_\parallel = 2p_\perp$), $\gamma = 0$ ($p_\parallel = p_\perp$), and $\gamma = 1$ ($p_\perp = 2p_\parallel$), using the recent magnetic field model of Olson and Pfitzer (1974). The shape of the contours, although somewhat dependent on γ, in general resembles that of the auroral oval.

Figure 4 shows the average characteristics of the gross-scale Birkeland current flow as observed by Zmuda and Armstrong (1974). There are two separate systems, one in the evening and one in the morning sector of the auroral oval, each consisting of two broad sheets of current. The Birkeland current sheets close by means of horizontal Pedersen currents, flowing northward in the evening sector and southward in the morning sector. In addition there are horizontal, electrojet, Hall currents flowing eastward in the evening and westward in the morning auroral oval, driven by the same electric field as the Pedersen currents. In each one of the two systems there definitely seems to be a net current flow across contours V_1 = constant shown in Figure 3. Thus these current systems could be driven by a magnetospheric pressure distribution of constant γ only if the total current of the evening and morning systems at all times are equal so that there is no net current flow across the contours V_1 = constant. As long as we have observations from one satellite alone this can only be checked on a statistical basis. However, as these sheets of Birkeland currents are the driving agents for the eastward and westward electrojets we should also expect that these jets then should vary in intensity synchronously. Studies of electrojet variations reported in the literature do not always clearly separate substorm variations, which could be of a different nature, from quiet-time variations. If we include substorm events it is obvious that the eastward and westward electrojets do not develop synchronously and, therefore, we would hardly expect that the two systems of Birkeland currents at all times carry the same total current. At quiet times the two systems are separated by the Harang discontinuity and it seems that they are two separate entities. Thus there may be some difficulties explaining the observed current system as being driven by the magnetospheric pressure.

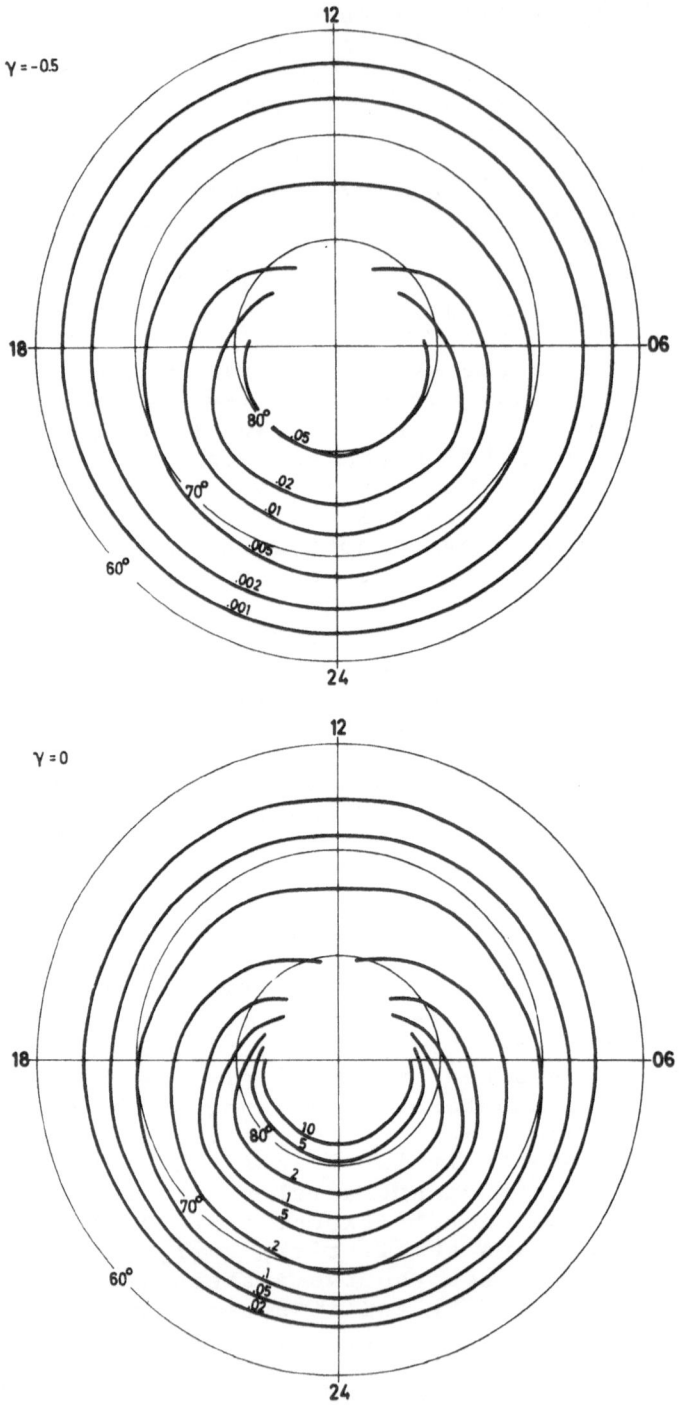

Figure 3.(Caption on page 351)

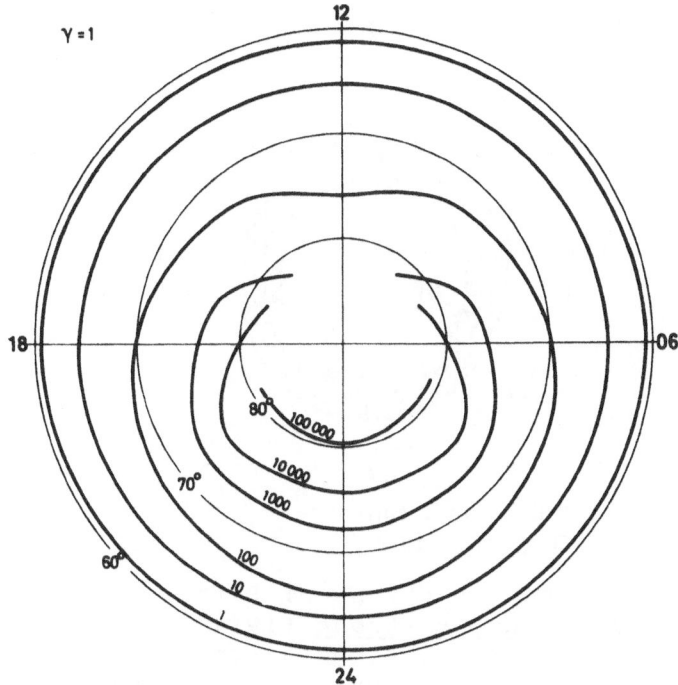

Fig. 3. Contours of constant V_1 (in units of Re/nT) which for constant anisotropy factor γ encircles regions of zero net Birkeland current flow to the ionosphere shown for three different values of γ. Differences in absolute values of V_1 between the three plots have no significance.

We may also make a quantitative estimate of the pressure variations in the magnetosphere required to drive the observed Birkeland currents. According to Zmuda and Armstrong (1974) the Birkeland current densities lie mainly between 3×10^{-7} and 4×10^{-6} A/m^2. Comparing Figure 4 with Figure 3 for $\gamma = 0$ we find that the currents flow in regions where ∇V_1 is of the order $1.5-5 \times 10^9$ T^{-1}. Using equation (12) we then find that the component of $\nabla_0 p_0$ along the V_1-contours, which approximately delineate the Zmuda-Armstrong current sheets, amounts to 6×10^{-17} – 3×10^{-15} Pa/m. The total pressure drop along a sheet extending over about 3000 km is then 1.8×10^{-10} – 9×10^{-9} Pa. For the case of isotropic pressure we would have the same pressure variation in the magnetosphere. A magnetic field of corresponding energy density would be 30-180 nT, thus the magnitude of the required pressure variation is not unrealistic. The form of the pressure variation needed is remarkable. The pressure

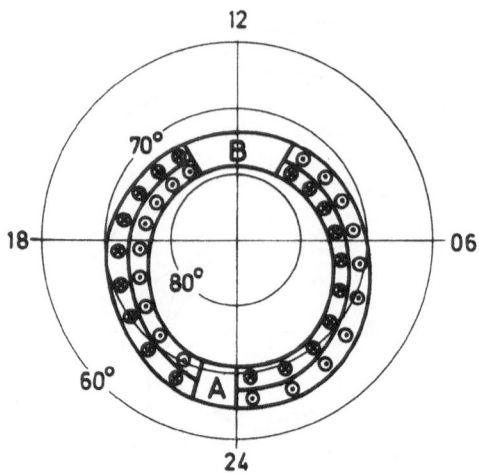

Fig. 4. Regions and directions of Birkeland current flow
observed by Zmuda and Armstrong (1974). In the region of
the Harang discontinuity, A, both types of current patterns are found, while in region B no fixed pattern is
observed.

should <u>increase</u> by the amount evaluated from the nightside to the dayside along the poleward sheets, but <u>de-
crease</u> along the equatorward sheets in both the evening
and morning systems. Thus, somewhere in the systems there
must exist even stronger gradients of pressure perpendicular to the sheets, but these cannot be evaluated from
the Birkeland currents. Figure 5 shows the contours V_1 =
constant of Figure 3 for $\gamma = 0$ mapped onto the equatorial
plane, with the projection of the Birkeland current regions of Figure 4, and the inferred directions of pressure gradients.

BIRKELAND CURRENTS DRIVEN BY THE PLASMA CONVECTION

We start this section with a qualitative discussion
of the observed relationships between the directions of
quiet time current flow, electric fields and plasma velocity. Figure 6 shows schematically, for a dipolar-like
geometry, the current sheets observed by Zmuda and Armstrong (1974). The Birkeland currents close in the ionosphere, presumably by currents transverse to the sheets.

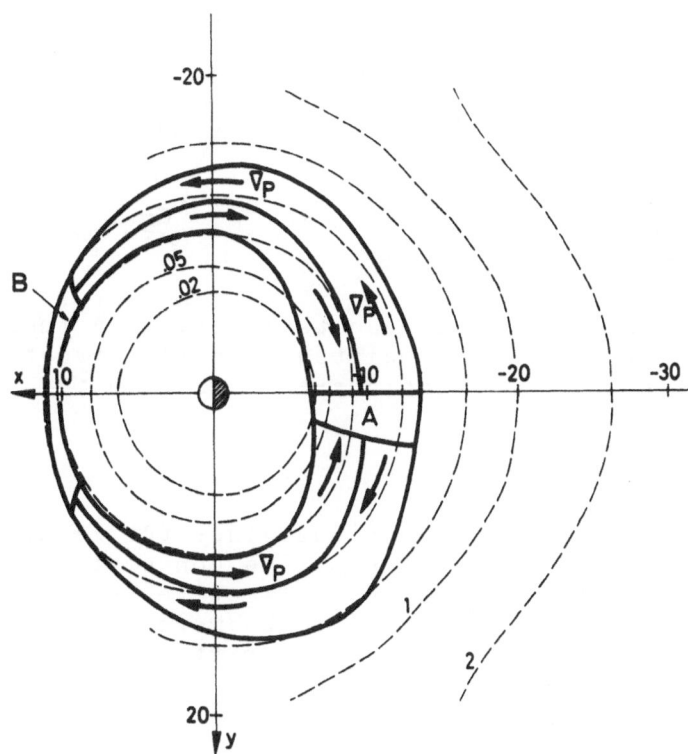

Fig. 5. Projections into equatorial plane of the contours
V_1 = constant of Figure 3 for γ = 0, and projection of
the Birkeland current region of Figure 4 using the field
model of <u>Olson and Pfitzer</u> (1974). Also shown are inferred
directions of pressure gradients along the V_1 contours.

As the electric field in the oval is known to be generally northward in the evening and southward in the morning we conclude that the closure currents dissipate power
and are essentially Pedersen currents. In the magnetosphere the currents must also close by currents transverse to the magnetic field. Although it is not known
that the currents flow along the most natural, shortest
path as shown in the figure, the currents will, whatever
path they take, flow from a region of higher potential
to a region of lower potential through a generator region. This transverse current is then associated with a
$\vec{j} \times \vec{B}$ force directed in such a way that it will brake
the plasma flow associated with the electric field, so
that kinetic energy of the convecting plasma is transferred to electric energy.

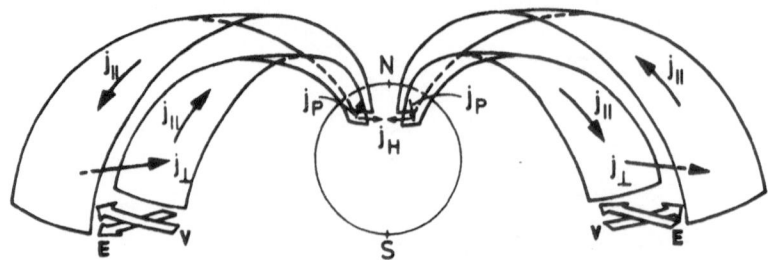

Fig. 6. Birkeland current sheets observed by Zmuda and
Armstrong (1974) with associated closure currents,
electric fields and plasma convection shown for a dipolar
geometry.

     Thus there exists an obvious generator mechanism for
Birkeland currents using the plasma convection. However,
it has been thought to be less effective than pressure
gradients because there seems to be more kinetic energy
available in the random (thermal) particle motion than
in the ordered (convective) motion. Only if the field
lines from the auroral oval extend far back into the
tail the plasma velocities and volumes would be large
enough to support a current system in which the plasma
motion is not stopped too rapidly. Now there is an increasing amount of evidence accumulating showing that
the auroral oval maps into the tail plasma sheet, which
extends at least as far back as the orbit of the moon.
(See Rostoker and Boström (1974) for a discussion and
references. Note, however, that the field model of Olson
and Pfitzer (1974), which applies to the most quiet state
of the magnetosphere, predicts closure of auroral field
lines at rather short distances.) Rostoker and Boström
(1974) have revised the model of Birkeland current generation taking the tail geometry into account and have
demonstrated that with realistic parameters it is quite
feasible to drive Birkeland currents by slowing down convection in the magnetotail. Figure 7 shows the presumed
electric field pattern. In the plasma sheet the $\vec{E} \times \vec{B}/B^2$
velocity must be directed essentially towards the flanks
of the magnetotail. If this velocity is slowed down it
could drive the four systems of Birkeland currents flowing to the morning and evening auroral ovals of the
northern and southern hemispheres. In the tail these currents would flow along the boundaries between the plasma
sheet and tail lobes and, in the opposite direction, in

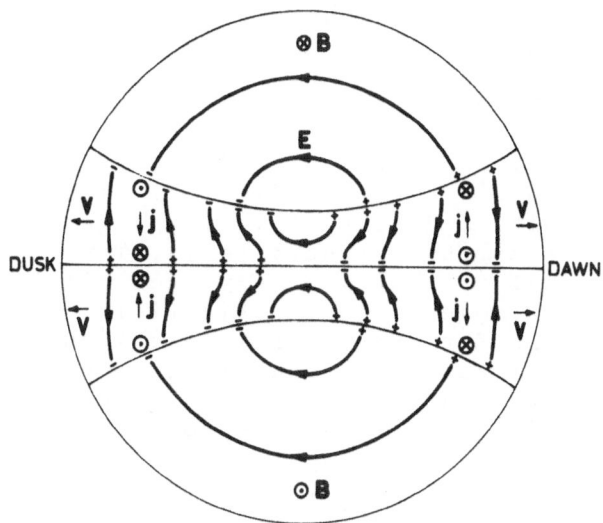

Fig. 7. Cross section of the magnetotail looking towards the earth showing qualitatively the electric field pattern mapped from the ionosphere, and directions of plasma flow and Birkeland closure current (after Rostoker and Boström, 1974).

the central part of the plasma sheet. It is interesting to note that these currents in each system would give the total magnetic field a skew so that the field lines flare outwards in the plasma sheet in agreement with observations.

In the discussion above we have assumed that the poleward and equatorward sheets of Birkeland current balance all along the auroral oval so that the ionospheric closure currents are meridional. Akasofu (1975) reports, however, that the poleward sheets typically, and particularly during substorms, are more intense than the equatorward sheets implying a westward component of the ionospheric closure current flowing across the midnight meridian from the morning to the evening part of the auroral oval. In the tail a corresponding generator current would flow from the dusk to the dawn part of the boundaries between the plasma sheet and the tail lobes braking the flow of plasma from the tail lobes into the plasma sheet, a flow that we expect to be enhanced during substorms.

Having shown that a mechanism similar to that of a magnetohydrodynamic dynamo qualitatively could explain the generation of the gross scale Birkeland currents we will now study the mechanism in some detail. Due to the quadratic nature and complicated structure of the inertial term $(\vec{v} \cdot \nabla)\vec{v}$, of equation (2), where $\vec{v}$ is related to the magnetospheric electric field by (1), it seems somewhat difficult to discuss this in as general terms as we have done for the pressure terms. Thus we will carry out the analysis using a very much simplified geometry with a homogeneous magnetic field. There are two justifications for this. First of all the inertial term could be of importance only if the field lines carrying the Birkeland current are extended far back into the magnetotail and for that region a homogeneous field may not be too bad an approximation. Secondly, the Birkeland currents driven by the inertial term do not depend critically on the existence of gradients in the magnetic field as the pressure-driven currents do.

We introduce orthogonal curvilinear coordinates u_1, u_2 in planes perpendicular to the magnetic field, see Figure 8. Here h_1 and h_2 are scale factors for the coordinates u_1 and u_2 respectively, such that the differential arc lengths along the u_1 and u_2 curves are $h_1 du_1$ and $h_2 du_2$. Unit vectors in directions of increasing u_1 and u_2 are denoted $\hat{u}_1$ and $\hat{u}_2$. We define the coordinate system so that the u_2 coordinate lines are equipotentials, then the electrostatic potential V is a function of u_1 solely. Using equation (1) the plasma velocity is found to be

$$\vec{v} = \frac{1}{Bh_1} \frac{dV}{du_1} \hat{u}_2 \tag{14}$$

and using (2) with $\varepsilon = 0$ the contribution to $\vec{j}_\perp$ from the inertial term is

$$\vec{j}_\perp = \frac{\rho}{B^3} \frac{1}{h_1^3 h_2} (\frac{dV}{du_1})^2 \left[\frac{\partial h_1}{\partial u_2} \hat{u}_1 - \frac{\partial h_2}{\partial u_1} \hat{u}_2 \right] \tag{15}$$

Equation (15) shows that the inertial force gives a current component $j_1 = (\rho v^2 / Bh_1 h_2)(\partial h_1 / \partial u_2) \hat{u}_1$ in the direction of $\hat{u}_1$, which is directed oppositely to the electric field provided $\partial h_1 / \partial u_2 > 0$. Thus the region acts as a generator if the separation of the equipotentials increase in the direction of $\vec{v}$, so that the electric field and plasma velocity decrease. The current component j_1 then gives a magnetic force $\vec{j}_1 \times \vec{B}$ which brakes the plasma

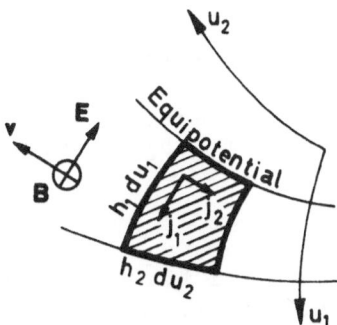

Fig. 8. Cross section of fluxtube, looking along the magnetic field lines towards the earth in the northern hemisphere afternoon sector, showing orthogonal coordinates u_1, u_2 used in study of inertial effect.

convection. In the ionosphere this current component is closed by a Pedersen current dissipating the energy supplied in the MHD dynamo region.

The other current component $\vec{j}_2 = - (\rho v^2 / Bh_1 h_2)$ $(\partial h_2 / \partial u_1)\ \hat{u}_2$, directed parallel or antiparallel to $\vec{v}$, is not involved with any power production or dissipation as it is perpendicular to $\vec{E}$. However, this current component is associated with a magnetic force $\vec{j}_2 \times \vec{B}$ which deflects the plasma flow but does not change its velocity or kinetic energy. This current can be a part of a current circulating in the magnetosphere or it can possibly close in a loop to the ionosphere involving ionospheric Hall currents which are not accompanied by power dissipation. However, as a first order approximation we may neglect the divergence of the Hall currents, compared to that of the Pedersen currents, as the derivatives along the auroral oval (in the direction of the Hall current) generally are smaller than transverse to the oval (in the direction of the Pedersen current).

The power produced by the inertial forces in a fluxtube of length L and cross section $h_1 du_1\ h_2 du_2$ is

$$-\vec{j}_{\perp} \cdot \vec{E}\ h_1 du_1 h_2 du_2 L = \rho v^3 \frac{\partial h_1}{\partial u_2}\ du_1 du_2 L \qquad (16)$$

which also can be seen by considering the difference in
kinetic energy of the plasma flowing in and out of the
fluxtube (the energy and momentum equations are equivalent). Putting this energy equal to the power dissipated
in the ionosphere $\Sigma_P E^2 h_1 du_1 h_2 du_2$ gives

$$\rho v L \frac{1}{h_1 h_2} \frac{\partial h_1}{\partial u_2} = \Sigma_P B^2 \tag{17}$$

Strictly speaking we should here use an effective integrated conductivity mapped from the ionosphere to the
tail, which could differ from the ionospheric one if the
scale factors relating the east-west and north-south
separations of field lines in the ionosphere to corresponding separations in the tail are different. Rostoker
and Boström (1974) estimate that this effect might increase the effective conductivity by a factor of 2. We
will take as typical values $\Sigma_P = 10S$, $\rho = 5 \times 10^{-21}$ kg/m^3
(3 protons cm^{-3}), B = 7 nT and v = 50 km/s (corresponding
to E = 0.35 mV/m). The derivative $(1/h_1 h_2)(\partial h_1/\partial u_2)$ is
interpreted as the relative rate of slowing down of the
plasma $-(1/h_2)(\partial v/\partial u_2)/v$ as can be verified using (14).
We will take this derivative to be 5×10^{-9} m^{-1} corresponding to a characteristic length for stopping the
plasma of about 30 Re which we consider to be a reasonable value as it is somewhat larger than the radius of
the tail. The derivative also describes the divergence
of the equipotential contours. 5×10^{-9} rad/m corresponds
to 1.8 $^\circ$/Re or a tilt of the plasma sheet - tail lobe
boundary of 14° relative to the plane of symmetry (neutral sheet) assuming a plasma sheet half-width of 7.5 Re.
With the parameters given above we may now use equation
(17) to evaluate the length L of the fluxtube, and we
find L $\approx$ 60 Re. This is not an unrealistic value but it
shows that Birkeland currents could be driven by inertial
forces only if the field lines extend far into the magnetotail.

The intensity of the field-aligned current at the
earthward side of the generator region is, neglecting
the possible contributions from $\vec{j}_2$ and ionospheric Hall
currents,

$$j_\parallel = L \, \text{div} \, \vec{j}_\perp = \text{div} \, \Sigma_P \vec{E} \tag{18}$$

Assuming that Σ_P is constant at 10 S and that the characteristic length for variations of $\vec{E}$ across the tail plasma sheet is 5 Re we find, with E = 0.35 mV/m as earlier,
$j_\parallel = 1 \times 10^{-10}$ A/m^2 or mapped to the ionosphere $j_\parallel \cdot$
$\cdot (B_{ionosphere}/B_{tail}) = 9 \times 10^{-7}$ A/m^2 in excellent agree-

ment with the range 3×10^{-7} to 4×10^{-6} A/m^2 quoted by Zmuda and Armstrong (1974).

A more detailed version of the model of the magneto-tail MHD generator has been given by Rostoker and Boström (1974).

Hitherto we have considered Birkeland currents associated with steady state convection. However, substorms are also associated with intense Birkeland current flow. During the reconfiguration of the magnetic field in a sector of the magnetotail from a tail-like to a more dipolar geometry, which is an intrinsic feature of the substorm, magnetic energy is released which accounts for the energy dissipated in the ionosphere. Substantial induction electric fields appear, which set the plasma into a motion, which is braked by the Birkeland current loop to the ionosphere. Assuming that the characteristic time for changing the tail field in a sector is 15 minutes, we find from the Maxwell equation curl $\vec{E} = -\partial \vec{B}/\partial t$ a spatial derivative of $\vec{E}$ of $B \cdot (1.1 \times 10^{-3} \text{ s}^{-1})$ or of the associated plasma velocity of 1.1×10^{-3} s^{-1}. We note that the curl of a transverse electric field is associated with a derivative in the direction of the plasma flow. Assuming the plasma velocity to be 50 km/s as before, the relative rate of slowing down of the plasma is 2.2×10^{-8} m^{-1} or somewhat larger than in our previous analysis. Evidently the induction process can drive significant Birkeland currents.

SECONDARY BIRKELAND CURRENTS OF IONOSPHERIC ORIGIN

When the horizontal ionospheric currents flow through regions of varying conductivity secondary Birkeland currents of ionospheric origin may be generated. Neglecting contributions to the currents from neutral winds, and using the approximation of vertical magnetic field lines, the height-integrated current is

$$\vec{J} = \Sigma_P \vec{E} + \Sigma_H \hat{B} \times \vec{E} \tag{19}$$

The divergence of $\vec{J}$ must match the Birkeland current, thus

$$j_{\shortparallel} = \Sigma_P \text{ div } \vec{E} + \vec{E} \cdot \text{grad } \Sigma_P + \hat{B} \times \vec{E} \cdot \text{grad } \Sigma_H \tag{20}$$

The Birkeland current may close in the magnetosphere by inertial forces, accelerating the plasma if the iono-

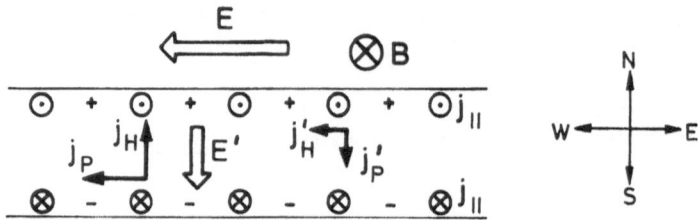

Fig. 9. Secondary Birkeland currents $j_{\shortparallel}$ discharging polarization electric field $\vec{E}'$ developed in region of enhanced ionization in the ionosphere.

sphere acts as a generator. If there is some mechanism
resisting the flow of Birkeland current, such as anomalous resistivity, a polarization electric field will
develop in the ionosphere so that the term Σ_p div $\vec{E}$
partly balances the contributions from the conductivity
gradients. However, we would not expect a complete chok-
ing of $j_{\shortparallel}$. While the opposing term Σ_p div $\vec{E}$ is positive
in regions of positive space charge the net $j_{\shortparallel}$ deriving
from conductivity gradients is negative (out of the
ionosphere). Considering the ionospheric currents connected to these Birkeland currents and the electric
field deriving from the space charges it is clear that
the ionosphere acts as a generator with $\vec{E} \cdot \vec{J} < 0$, although using the total ionospheric current and field
$\vec{E} \cdot \vec{J}$ is always positive.

As an example we may consider the effect of applying
a westward electric field $\vec{E}$ to an ionospheric region
with a slab of enhanced ionization extended in the east-
west direction, see Figure 9. If there is some choking
of Birkeland currents the excess northward Hall current
in the slab builds up a southward polarization electric
field $\vec{E}'$, which drives a Pedersen current $\vec{J}'_p$ opposing
the primary Hall current. However, as the primary northward Hall current dominates, Birkeland currents will flow
out of the ionosphere at the northern edge of the slab.
Evidently the ionosphere acts as a generator for the current loops involving the Birkeland currents and their
closure currents in the ionosphere. Models like this have
been proposed for the large scale current and electric
field structure of the auroral electrojets. However, they
are not consistent with the observations of <u>Zmuda and
Armstrong</u> (1974) which show that the polarity of the

Birkeland current sheets is such that the ionosphere
acts as a load to these currents. Furthermore, to explain an eastward electrojet one would have to impose an
eastward electric field on the ionosphere and this has
not been observed to be a characteristic feature of the
region of the eastward jet.

On the other hand the results of some rocket experiments (e.g. Park and Cloutier, 1971) may be interpreted
as giving evidence that the mechanism discussed here may
operate on a smaller scale in association with individual
auroral arcs. In discussing findings from such experiments, and performing comparisons with satellite observations, it is important to realize their limited latitudinal coverage of the electrojet region, which may
extend over 10° of latitude. Also it is very important
to try to separate Birkeland currents of ionospheric
origin from Birkeland currents of magnetospheric origin,
which could be done if one could decide whether the
closure currents are dissipative or generative. The horizontal ionospheric currents can be inferred from measurements of the electric field and from indirect information
on the conductivity structure provided by measurements
of precipitating particles or observations of visual auroral forms.

SUMMARY AND DISCUSSION

We have considered the basic mechanisms for generat-
ing Birkeland currents in the magnetosphere, that is
mechanisms for redirecting transverse currents to form
Birkeland currents. Using MHD equations we find that
both the plasma convection and pressure can account for
currents of the observed magnitude. For currents driven
by the pressure there exists a constraint on the possible
forms of the current distribution implying that the morning and evening systems of Birkeland current flow must
balance each other at all times so that there is no net
current flow in the ionosphere across certain specified
contours. If the currents are driven by the plasma convection the morning och evening current systems may be
independent. We have suggested geometries for plasma
pressure variations and convective flow that could account for the observed distribution of Birkeland currents
assuming that the currents originate on field lines within the magnetosphere extending into the plasma sheet of
the magnetotail.

We have not discussed how the required plasma convection or pressure is established. The energy dissipated by the Birkeland currents must primarily derive
from the solar wind sweeping by the magnetosphere. Currents flowing through the solar wind plasma will extract
kinetic energy from the solar wind which is transferred
to electric energy. The current driven in this way closes
through the magnetosphere where it can set up the required internal convective flow and pressure distributions which then can drive the internal current system
considered in this paper. If the geometry of the current
flow is different from that envisaged here, and in particular if the Birkeland currents occur on open field
lines extending into the solar wind, some Birkeland currents could be directly linked to the currents in the
solar wind plasma.

We have also considered secondary Birkeland currents
of ionospheric origin which are currents tending to discharge ionospheric polarization electric fields. In interpreting measurements of Birkeland currents it is important to distinguish these secondary currents from the
primary currents of magnetospheric origin which could be
done if one can decide whether the ionospheric closure
currents are dissipative or generative.

REFERENCES

Akasofu, S.-I., this conference, 1975

Kern, J.W., in Physics of Geomagnetic Phenomena, S. Mat-
    sushita and W.H. Campbell (eds.), p. 1037, Academic
    Press, New York, 1967

Olson, W.R., and Pfitzer, K.A., J. Geophys. Res., 79,
    3739, 1974

Park, R.J., and Cloutier, P.A., J. Geophys. Res., 76,
    7714, 1971

Parker, E.N., Phys. Rev., 107, 924, 1957

Rostoker, G., and Boström, R., Report TRITA-EPP-74-25,
    Department of Plasma Physics, Royal Institute of
    Technology, Stockholm, December 1974

Vasyliunas, V.M., in Particles and Fields in the Magneto-
    sphere, B.M. McCormac (ed.), p. 60, D. Reidel Publ.
    Co., Dordrecht, 1970

Zmuda, A.J., and Armstrong J.C., J. Geophys. Res., 79,
    4611, 1974

Akasofu, S-I	Geophysical Institute, Fairbanks, Alaska 99701, USA
Alfvén, H	Inst. för plasmafysik, Kungl. Tekniska Högskolan, S-100 44 Stockholm 70, Sweden
Andersson, L E	Kiruna geofysiska institut, S-981 01 Kiruna 1, Sweden
d'Angelo, N	Danish Space Research Institute, Lundtoftevej 7, DK-2800 Lyngby, Denmark
Ashour-Abdalla, M	Centre National d'Etudes des Telecommunications, F-921 31 Issy-les-Moulineauz, France
Axford, W I	Max-Planck-Institut für Aeronomie, Postfach 60, D-3411 Lindau/Harz, Germany
Block, L	Inst. för mekanik, Kungl. Tekniska Högskolan, S-100 44 Stockholm 70, Sweden
Borg, H	Kiruna geofysiska institut, S-981 01 Kiruna 1, Sweden
Boström, R	Inst. för plasmafysik, Kungl. Tekniska Högskolan, S-100 44 Stockholm 70, Sweden
Brunberg, E-Å	Inst. för plasmafysik, Kungl. Tekniska Högskolan, S-100 44 Stockholm 70, Sweden
Bryant, D A	Appleton Laboratory, Ditton Park, Slough SL3 9 JX, England
Christophersen, P	Kiruna geofysiska institut, S-981 01 Kiruna 1, Sweden
Derblom, H	Uppsala Jonosfärobservatorium, S-755 90 Uppsala, Sweden
Dungey, J W	Imperial College, Department of Physics, Prince Consort Road, London S W 7, England
Dysthe, K	Tromsö Universitet, N-9001 Tromsö, Norway
Egeland, A	Det Norske Institutt for Kosmisk Fysikk, Postboks 1048, Blindern, Oslo 3, Norway
Eliasson, L	Inst. för geokosmofysik, Umeå universitet, Kiruna geofysiska institut, S-901 87 Umeå, Sweden
Evans, D S	Space Environment Lab., NOAA Environment Res. Lab., Boulder, Colorado 80302, USA
Fahleson, D U	Inst. för plasmafysik, Kungl. Tekniska Högskolan, S-100 44 Stockholm 70, Sweden
Fälthammar, C-G	Inst. för plasmafysik, Kungl. Tekniska Högskolan, S-100 44 Stockholm, Sweden
Galeev, A A	Space Research Institute, Academy of Sciences of USSR, Moscow Region, USSR
Gendrin, R	Groupe Recherches Ionospherique, CNET 3 Av de la Republique, 92-Issy-les-Moulineaux, France
Gustafsson, G	Kiruna geofysiska institut, S-981 01 Kiruna 1, Sweden

Haerendel, G	Max-Planck-Institut für Physik und Astrophysik, Institut für Extraterrestrische Physik, D-8046 Garching b München, Germany
Heikkila, W J	University of Texas at Dallas, P O Box 688, Richardson, Texas 75080, USA
Holmgren, L-Å	Kiruna geofysiska institut, S-981 01 Kiruna 1, Sweden
Holt, O	Tromsö Universitet, P O Box 387, N-9001 Tromsö, Norway
Hones Jr, E W	Los Alamos Scientific Lab., Los Alamos, New Mexico 87544, USA
Hultqvist, B	Kiruna geofysiska institut, S-981 01 Kiruna 1, Sweden
Johnson, R G	Lockheed Palo Alto Research Lab., Palo Alto, California 94304, USA
Jurén, C	Kiruna geofysiska institut, S-981 01 Kiruna 1, Sweden
Kennel, C F	Plasma Physics Group, Department of Physics, University of California, Los Angeles, California 90025, USA
Lundin, R	Kiruna geofysiska institut, S-981 01 Kiruna 1, Sweden
Maehlum, B	Försvarets Forskingsinstitutt, Avd för elektronikk, Postboks 25, 2007 Kjeller, Norway
McIlwain, C E	Physics Dept, Univ. of Calif., San Diego, La Jolla, California 92037, USA
Mozer, F S	Physics Dept. and Space Science Lab., Univ. of Calif., Berkley, California 94820, USA
Riedler, W	Institut für Nachrichtentechnik, Technische Hochschule Graz, Inffeldgasse 12, A-8010 Graz, Austria
Rossberg, L	Max-Planck-Institut für Aeronomie, Postfach 60, D-3411 Lindau/Harz, Germany
Rönnmark, K	Avd. för teoretisk plasmafysik, Umeå universitet, Kiruna geofysiska institut, S-901 87 Umeå, Sweden
Sagdeev, R Z	Space Research Institute, Academy of Sciences of USSR, Moscow Region, USSR
Scarf, F L	Space Sciences Department, TRW System Group, Redondo Beach, California 90278, USA
Smith, P	Goddard Space Flight Center, Greenbelt, Maryland 20771, USA
Stenflo, L	Avd. för teoretisk plasmafysik, Umeå universitet, S-901 87 Umeå, Sweden
Søraas, F	Fysisk Institutt avd A, Universitetet i Bergen, Allégaten 53/55, N-5014 Bergen U, Norway

Vasyliunas, V M	Department of Physics and Center for Space Research, Massachusetts Institute of Technology, Cambridge, Massachusetts 02139, USA
Westerlund, S	Kiruna geofysiska institut, S-981 01 Kiruna 1, Sweden
Westin, H	Kiruna geofysiska institut, S-981 01 Kiruna 1, Sweden
Whalen, B A	Division of Physics, National Research Council of Canada, Ottawa, Ontario, Canada
Williams, D	Space Environment Lab. NOAA Environment Research Lab. R43, Boulder, Colorado 80302, USA

SUBJECT INDEX

Acceleration, 187, 229, 251,
  271, 291, 305, 319, 320, 325
Adiabatic processes, 170, 251
Anomalous resistivity, 1, 65,
  76, 187, 196, 202, 233, 240,
  252, 271, 284, 360
ATS-6 satellite, 92, 273
Auroral currents, 13, 14
Auroral electrojets, 2, 113, 129
Auroral particles, 92, 319

Barium ions, 148
Birkeland currents, 2, 229, 240,
  341, 359
Bohm condition, 235
Boundaries, 69, 79, 85, 341

Charge exchange, 168
Collapse, 263
Conductivity, 5, 81, 243, 342,
  361
Convection, 13, 39, 148, 149,
  251, 352
Cosmogonic problem, 16
Coulomb scattering, 176
Current disruption, 1

Diffuse auroras, 127
Diffusion coefficient, 174
Dipole field, 8, 352
Discrete auroras, 127
Dispersion relations, 206, 214,
  265
Dissipation, 76, 251
DMSP-2 satellite, 113
Double layer, 1, 5, 76, 229, 240
  334

Dungey's model, 69, 74, 137, 139

Electric field, 5, 46, 75, 81,
  187, 234, 247, 271, 320, 325,
  334, 342
Electron beams, 91, 100, 241,
  252, 333
Electrostatic waves, 201, 205,
  277
Energy spectrogram, 97, 102-104
Energy transfer, 9, 13, 259
Entry layer, 38
Equivalent circuit, 16
Explorer 45(S^3) satellite, 159,
  252

Fermi acceleration, 326
Field aligned currents, 132,
  238, 284, 334
Field aligned fluxes, 94, 326,
  331, 342
Filaments, 5
Frozen-in fields, 12, 71

Geomagnetic storm, 159, 180, 271
Group velocity, 201, 218
Growth phase, 123, 141
Guiding center, 70

Heliosphere, 17
Helium ions, 45, 48
HEOS 2 experiments, 24

Instabilities, 5, 201, 206, 218,
  229, 237, 271, 277, 306, 310
Interconnection, 75

Interplanetary magnetic field,
   115-116, 137
Ion beams, 333
Ion cyclotron waves, 64, 163,
   202, 252, 273, 304, 310
Ionic components, 45
Ionospheric current, 134
ISIS-2 satellite, 113

Jovian magnetosphere, 19, 287

Kelvin-Helmholtz instability,
   38

Laboratory experiments, 6, 229
Laminar flow, 233
Landau resonance, 174, 256, 326
Langmuir condition, 235
Langmuir waves, 262
Lifetimes, 176
Lorentz transformation, 75
Loss cone, 105, 201, 223, 251,
   291

Magnetic equator, 91
Magnetic field line reconnection,
   5, 9, 75
Magnetic field topology, 74
Magneto-hydrodynamic dynamo, 113,
   243, 354, 359
Magnetopause, 25, 37, 69, 72
Magnetosheet plasma, 69, 71
Magnetospheric cleft, 148
Magnetospheric disturbances, 115
Magnetotail, 14, 137, 355
Maxwell's equations, 7, 359
Mode coupling, 262, 286
Moving plasma, 10

Narrow band emissions, 213, 223,
Neutral line, 74, 137, 145, 156,
   187, 319
Neutral point, 75
Neutral sheet, 12, 17, 187, 190
Non-Maxwellian distribution
   functions, 5, 207, 252, 271-
   272, 311

Ohm's law, 81
Oxygen ions, 51, 336

Parametric interactions, 262,
   286
Phase space density, 35, 166
Pitch angle, 98-99, 101, 105-
   107, 163, 177-179, 203, 271,
   291, 294, 298, 315, 331, 344
Plasma cavity, 263
Plasma flow, 28, 33, 143, 147
Plasma heating, 251, 261, 271
Plasma interactions, 163
Plasma layer, 25
Plasma pause, 252, 302
Plasma sheet, 79, 137, 139, 141,
   145, 152, 187
Plasmon condensate, 262
Polar cusp, 23-24, 39
Polar rain, 84
Polar shower, 84
Polar squall, 85
Poynting vector, 81
Precipitating electrons, 261,
   291, 299, 328
Precipitating ions, 55, 259,
   294, 299, 312
Protons, 26, 46, 47, 143, 160,
   252, 294
Pseudo-plasma, 1, 3, 6

Quasi-linear diffusion, 38, 203,
   251, 291

Radiation belt electrons, 174
Reconnection, 75, 148
Recovery phase, 164
Resonant particles, 253, 273,
   321
Ring current, 2, 16, 165, 252,
   259, 302

SAR arcs, 170, 261, 273
Shock waves, 71, 115, 281
Solar wind, 9, 14, 23, 71, 137
Substorm, 118, 137-139, 146, 187,
   271

Trajectories, 191
Trapped particles, 233
Turbulence spectrum, 266, 271,
   291, 306

Turbulent collision frequency,   Vlasov theory, 195, 236
    233, 267, 284
Turbulent dissipation, 252   Whistler waves, 203-204, 273,
Turbulent flow, 30   286

Upper hybrid frequency, 201, 214   X-line, 74, 118, 319